Financial Enterprise Risk Management

Financial Enterprise Risk Management provides all the tools needed to build and maintain a comprehensive ERM framework. As well as outlining the construction of such frameworks, it discusses the internal and external contexts within which risk management must be carried out. It also covers a range of qualitative and quantitative techniques that can be used to identify, model and measure risks, and describes a range of risk mitigation strategies. Over 100 diagrams are used to help describe the range of approaches available, and risk management issues are further highlighted by various case studies. A number of proprietary, advisory and mandatory risk management frameworks are also discussed, including Solvency II, Basel III and ISO 31000:2009.

This book is an excellent resource for actuarial students studying for examinations, for risk management practitioners and for any academic looking for an up-to-date reference to current techniques.

PAUL SWEETING is a Managing Director at JP Morgan Asset Management. Prior to this, he was a Professor of Actuarial Science at the University of Kent and he still holds a chair at the university. Before moving to academia, Paul held a number of roles in pensions, insurance and investment. Most recently he was responsible for developing the longevity reinsurance strategy for Munich Reinsurance, before which he was Director of Research at Fidelity Investments' Retirement Institute.

In his early career, Paul gained extensive experience as a consulting actuary advising on pensions and investment issues for a range of pension schemes and their corporate sponsors. He is affiliated to a number of professional bodies being a Fellow of the Institute of Actuaries, a Fellow of the Royal Statistical Society, a Fellow of the Securities and Investment Institute and a CFA Charterholder. Paul has written extensively on a range of pensions, investment and risk issues and is a regular contributor to the print and broadcast media.

INTERNATIONAL SERIES ON ACTUARIAL SCIENCE

The *International Series on Actuarial Science*, published by Cambridge University Press in conjunction with the Institute and Faculty of Actuaries, contains textbooks for students taking courses in or related to actuarial science, as well as more advanced works designed for continuing professional development or for describing and synthesizing research. The series is a vehicle for publishing books that reflect changes and developments in the curriculum, that encourage the introduction of courses on actuarial science in universities, and that show how actuarial science can be used in all areas where there is long-term financial risk.

A complete list of books in the series can be found at www.cambridge.org/statistics. Recent titles include the following:

Regression Modeling with Actuarial and Financial Applications
EDWARD W. FREES

Actuarial Mathematics for Life Contingent Risks
DAVID C.M. DICKSON, MARY R. HARDY & HOWARD R. WATERS

Nonlife Actuarial Models
YIU-KUEN TSE

Generalized Linear Models for Insurance Data
PIET DE JONG & GILLIAN Z. HELLER

Market-Valuation Methods in Life and Pension Insurance
THOMAS MØLLER & MOGENS STEFFENSEN

Insurance Risk and Ruin
DAVID C.M. DICKSON

FINANCIAL ENTERPRISE RISK MANAGEMENT

PAUL SWEETING
University of Kent, Canterbury

CAMBRIDGE
UNIVERSITY PRESS

CAMBRIDGE UNIVERSITY PRESS
Cambridge, New York, Melbourne, Madrid, Cape Town,
Singapore, São Paulo, Delhi, Tokyo, Mexico City

Cambridge University Press
The Edinburgh Building, Cambridge CB2 8RU, UK

Published in the United States of America by Cambridge University Press, New York

www.cambridge.org
Information on this title: www.cambridge.org/9780521111645

First published 2011

Printed in the United Kingdom at the University Press, Cambridge

A catalogue record for this publication is available from the British Library

Library of Congress Cataloguing in Publication data
Sweeting, Paul.
Financial enterprise risk management / Paul Sweeting.
p. cm. – (International series on actuarial science)
Includes bibliographical references and index.
ISBN 978-0-521-11164-5 (hardback)
1. Financial institutions–Risk management. 2. Financial services industry–Risk
management. I. Title.
HG173.S94 2011
332.1068′1–dc23   2011025050

ISBN 978-0-521-11164-5 Hardback

Contents

+ Class work

+ Class work

Preface

This book began life as a sessional paper presented to the Institute of Actuaries in Manchester and, some months later, to the Faculty of Actuaries in Edinburgh. Its presentation occurred at around the same time that a new subject on enterprise risk management was being developed for the UK actuarial exams. This made it a good time to expand the paper into something more substantial, with detailed information on many of the techniques that were only mentioned in the initial work. It also means that the book has benefited greatly from the work done by the syllabus development working party, led by Andrew Cairns and managed by Lindsay Smitherman.

I found myself writing this book during a time of crisis for financial institutions around the world. Financial models have been blamed for a large part of this crisis, and this criticism is, to an extent, well-founded. It is certainly tempting to place far too much reliance on very complex models, ignoring the fact that they merely represent rather than replicate the real world. Some senior executives have also been guilty of seeing the output of these models but not understanding the underlying approaches and their limitations. Finally, many models have been designed seemingly ignorant of the fact that the data histories needed to provide parameters for these models are simply not available. However, at least as big an issue is that many non-financial risks were allowed to thrive in the years before the crisis.

Many of the techniques described in this book are quantitative, and such risk modelling and management techniques can be very helpful. However, there are a number of ways in which risk can be quantified. Furthermore, these risk measures do not paint a complete picture. It is important to appreciate the limitations of these types of models, the circumstances in which they might fail and the implications of such failure. It is also crucial to understand that just because a risk is unquantifiable, it does not mean that it should be ignored. Some of the most important – and dangerous – risks cannot be modelled; however, they can frequently be identified and often managed.

All risks should be considered together: this holistic approach is fundamental to enterprise risk management. Whilst identifying the extent – or even the existence – of individual risks is important, looking at the bigger picture is vital. Looking at the interaction between risks can highlight concentrations of risk, but also the potential diversifying or even hedging effect of different risks. It is also important to recognise that risk is not necessarily synonymous with uncertainty. Risk is only bad if the outcome is adverse, and these types of risks can be described as downside risks. Upside risks also occur – these are opportunities – and without them, there would be no point in taking risks at all.

1

An introduction to enterprise risk management

1.1 Definitions and concepts of risk

The word 'risk' has a number of meanings, and it is important to avoid ambiguity when risk is referred to. One concept of risk is uncertainty over the range of possible outcomes. However, in many cases uncertainty is a rather crude measure of risk, and it is important to distinguish between upside and downside risks.

Risk can also mean the quantifiable probability associated with a particular outcome or range of outcomes; conversely, it can refer to the unquantifiable possibility of gains or losses associated with different future events, or even just the possibility of adverse outcomes.

Rather than the probability of a particular outcome, it can also refer to the likely severity of a loss, given that a loss occurs. When multiplied, the probability and the severity give the expected value of a loss.

A similar meaning of risk is exposure to loss, in effect the maximum loss that could be suffered. This could be regarded as the maximum possible severity, although the two are not necessarily equal. For example, in buildings insurance, the exposure is the cost of clearing the site of a destroyed house and building a replacement; however, the severity might be equivalent only to the cost of repairing the roof.

Risk can also refer to the problems and opportunities that arise as a result of an outcome not being as expected. In this case, it is the event itself rather than the likelihood of the event that is the subject of the discussion. Similarly, risk can refer to the negative impact of an adverse event.

Risks can also be divided into whether or not they depend on future uncertain events, on past events that have yet to be assessed or on past events that have already been assessed. There is even the risk that another risk has not yet been identified.

When dealing with risks it is important to consider the time horizon over which they occur, in terms of the period during which an organisation is exposed to a particular risk, or the way in which a risk is likely to change over time. The link between one risk and others is also important. In particular, it is crucial to recognise the extent to which any risk involves a concentration with or can act as a diversifier to other risks.

In the same way that risk can mean different things to different people, so can enterprise risk management (ERM). The key concept here is the management of all risks on a holistic basis, not just the individual management of each risk. Furthermore, this should include both easily quantifiable risks such as those relating to investments and those which are more difficult to assess such as the risk of loss due to reputational damage.

A part of managing risks on a holistic basis is assessing risks consistently across an organisation. This means recognising both diversifications and concentrations of risk. Such effects can be lost if a 'silo' approach to risk management is used, where risk is managed only within each individual department or business unit. Not only might enterprise-wide concentration and diversification be missed, but there is also a risk that different levels of risk appetite might exist in different silos. Furthermore enterprise-wide risks might not be managed adequately with some risks being missed altogether due to a lack of ownership.

The term 'enterprise risk management' also implies some sort of process – not just the management of risk itself, but the broader approach of:

- recognising the context;
- identifying the risks;
- assessing and comparing the risks with the risk appetite;
- deciding on the extent to which risks are managed;
- taking the appropriate action; and
- reporting on and reviewing the action taken.

When formalised into a process, with detail added on how to accomplish each stage, then the result is an ERM framework. However, the above list raises another important issue about ERM: that it is not just a one-off event that is carried out and forgotten, but that it is an ongoing process with constant monitoring and with the results being fed back into the process.

It is important that ERM is integrated into the everyday way in which a firm carries out its business and not carried out as an afterthought. This means that risk management should be incorporated at an early stage into new projects.

Such integration also relates to the way in which risks are treated since it recognises hedging and diversification, and should be applied at an enterprise rather than at a lower level.

ERM also requires the presence of a central risk function (CRF), headed by chief risk officer. This function should cover all things risk related, and in recognition of its importance, the chief risk officer should have access to or, ideally, be a member of board of the organisation.

Putting an ERM framework into place takes time, and requires commitment from the highest level of an organisation. It is also important to note that it is not some sort of 'magic bullet', and even the best risk management frameworks can break down or even be deliberately circumvented. However, an ERM framework can significantly improve the risk and return profile of an organisation.

1.2 Why manage risk?

With this discussion of ERM, it is important to consider why it might be desirable to manage risk in the first place. At the broadest level, risk management can benefit society as a whole. The effect on the economy of risk management failures in banking, as shown by the global liquidity crisis, give a clear illustration of this point.

It could also be argued that risk management is what boards have been appointed to implement, particularly in the case of non-executive directors. This does not mean that they should remove all risk, but they should aim to meet return targets using as little risk as possible. This is a key part of their role as agents of shareholders. It is in fact in the interests of directors to ensure that risks are managed properly, since it reduces the risk of them losing their jobs, although there are remuneration structures that can reward undue levels of risk.

On a practical level, risk management can also reduce the volatility in an organisation's returns. This could help to increase the value of a firm, by reducing the risk of bankruptcy and perhaps the tax liability. This can also have a positive impact on a firm's credit rating, and can reduce the risk of regulatory interference. Reduced volatility also avoids large swings in the number of employees required – thus limiting recruitment and redundancy costs – and reduces the amount of risk capital needed. If less risk capital is needed, then returns to shareholders or other providers of capital can be improved or, for insurance companies and banks, lower profit margins can be added to make products more competitive.

Improved risk management can lead to a better trade-off between risk and return. Firms are more likely to choose the projects with the best risk-adjusted rates of return, and to ensure that the risk taken is consistent with the corporate appetite for risk. Again, this benefits shareholders.

These points apply to all types of risk management, but ERM involves an added dimension. It ensures not only that all risks are covered, but also that they are covered consistently in terms of the way they are identified, reported and treated. ERM also involves the recognition of concentrations and diversifications arising from the interactions between risks. ERM therefore offers a better chance of the overall risk level being consistent with an organisation's risk appetite.

Treating risks in a consistent manner and allowing for these interactions can be particularly important for banks, insurers and even pension schemes, as this means that the amount of capital needed for protection against adverse events can be determined more accurately.

ERM also implies a degree of centralisation, and this is an important aspect of the process that can help firms react more quickly to emerging risks. Centralisation also helps firms to prioritise the various risks arising from various areas of an organisation. Furthermore, it can save significant costs if extended to risk responses. If these are dealt with across the firm as a whole rather than within individual business lines, then not only can this reduce transaction costs, but potentially offsetting transactions need not be executed at all. Going even further, ERM can uncover potential internal hedges arising from different lines of business that reduce or remove the need to hedge either risk.

Having a rigorous ERM process also means that the choices of response are more likely to be consistent across the organisation, as well as more carefully chosen.

Another important advantage of ERM is that it is flexible – an ERM framework can be designed to suit the individual circumstances of each particular organisation

ERM processes are sometimes implemented in response to a previous risk management failure in an organisation. This does mean that there is an element of closing the stable door after the horse has bolted, and perhaps of too great a focus on the risk that was faced rather than potential future risks. It might also lead to excessive risk aversion, although introducing a framework where none has existed previously is generally going to be an improvement.

A risk management failure in one's own organisation is not necessarily the precursor to an ERM framework. A high-profile failure in another firm, particularly a similar one, might prompt other firms to protect themselves against a similar event. An ERM framework might also be required by an industry regulator, or by a firm's auditors or investors.

l   ERM can be used in a variety of contexts. It should be considered when developing a strategy for an organisation as a whole and within individual departments. Once it has been decided what an organisation's objectives are, the organisation must consider what risks might exist to stop them being achieved. The organisation must then consider how to assess and deal with the risks, considering the impact on performance both before and after treating the risks identified. Importantly, the organisation needs to ensure that there is a framework in place for carrying out each of these stages effectively.

z   ERM can also be used when developing new products or undertaking new projects by considering both the objectives and the risks that they will not be met. Here, it is also possible to determine the levels of risk at which it is desirable to undertake a project. This is not just about deciding whether risks are acceptable or not; it is also about achieving an adequate risk-adjusted return on capital, or choosing between two or more projects.

Finally, ERM is also important for pricing insurance and banking products. This involves avoiding pricing differentials being exploited by customers, but also ensuring that premiums include an adequate margin for risk.

1.3 Enterprise risk management frameworks

l   ERM frameworks typically share a number of common features. The first stage is to assess the context in which the framework is operating. This means understanding the internal risk management environment of an organisation, which in turn requires an understanding of the nature of an organisation and the interests of various stakeholders. It is important to do this so that potential risk management issues can be understood. The context also includes the external environment, which consists of the broader cultural and regulatory environment, as well as the views of external stakeholders.

z   Then, a consistent risk taxonomy is needed so that any discussions on risk are carried out with an organisation-wide understanding. This becomes increasingly important as organisations get larger and more diverse, especially if an organisation operates in a number of countries. However, whilst a consistent taxonomy can allow risk discussions to be carried out in shorthand, it is important to avoid excessive use of jargon so that a framework can be externally validated.

3   Once a taxonomy has been defined, the risks to which an organisation is exposed must be identified. The risks can then be divided into those which are quantifiable and those which are not, following which the risks are assessed. These assessments are then compared with target levels of risk – which must also be determined – and a decision must be taken on how to deal with risks beyond those targets. Finally there is implementation, which involves taking agreed measures to manage risk.

However, it is also important to ensure that the effectiveness of the approaches used is monitored. Changes in the characteristics of existing risks need to be highlighted, as do the emergence of new risks. In other words, risk management is a continual process. The process also needs to be documented. This is important for external validation, and for when elements of the process are reviewed. Finally, communication is important. This includes internal communication to ensure good risk management and external communication to demonstrate the quality of risk management to a number of stakeholders.

1.4 Corporate governance

Corporate governance is the name given to the process of running of an organisation. It is important to have good standards of corporate governance if an ERM framework is to be implemented successfully. Corporate governance is important not only for company boards, but also for any group leading an organisation. This includes the trustees of pension schemes, foundations and endowments. Their considerations are different because they have different constitutions and stakeholders, but many of the same issues are important.

The regulatory aspects of corporate governance are discussed in depth with the regulatory environment, whilst board composition is described as part of an organisation's structure. However, regardless of what is required, it is worth commenting briefly on what constitutes good corporate governance.

1.4.1 Board constitution

The way in which the board of an organisation is formed gives the foundation of good corporate governance. Whilst the principles are generally expressed in relation to companies, analogies can be found in other organisations such as pension schemes.

A key principle of good corporate governance is that different people should hold the roles of chairman and chief executives. A chief executive is responsible for the running of the firm, whilst the chairman is responsible for running the board. It can be argued that having an executive chairman ensures consistency between the derivation of a strategy and its implementation. However, since the board is intended to monitor the running of the firm, there is a clear conflict of interest if the roles of chief executive and chairman are combined.

It is also good practice for the majority of directors to be non-executives. This means that the board is firmly focussed on the shareholders' interests. Ideally, the majority of directors should also be independent, with no links to the company beyond their role on the board. Furthermore, independent directors should be the sole members of committees such as remuneration, audit and appointment, where independence is important. The chief risk officer should be a board member.

1.4.2 Board education and performance

Whilst the composition of the board is important, it is also vital that the members of the board perform their roles to a high standard. One way of achieving this is to ensure that directors have sufficient knowledge and experience to carry out their duties effectively. Detailed specialist industry knowledge is needed only by executive members of the board – for non-executive directors it is more important that they have the generic skills necessary to hold executives to account. These skills are not innate, and new directors should receive training to help them perform their roles.

It is also important that all directors receive continuing education so that they remain well equipped, and that their performance is appraised regularly. So that these appraisals are effective, it is important to set out exactly what is expected of the directors. This means that the chairman should agree a series of goals with each director on appointment and at regular intervals. The chairman's performance should be assessed by other members of the board.

1.4.3 Board compensation

An important way of influencing the performance of directors is through compensation. Compensation should be linked to the individual performance of a director and to the performance of the firm as a whole. The latter can be achieved by basing an element of remuneration on the share price. Averaging this element over several periods can reduce the risk of short-termism.

A similar way of incentivising directors is to encourage or even oblige them to buy shares in the firm on whose board they sit.

1.4.4 Board transparency

Good corporate governance implies transparency in dealings with stakeholders who include shareholders, regulators, customers and employees to name but a few. This means sharing information as openly as possible, including the minutes of board meetings, as far as this can be done without the disclosure of commercially sensitive information.

$\mathcal{X}$ 1.5 Models of risk management

In an ERM framework, the way in which the department responsible for risk management – the central risk function (CRF) – interacts with the rest of the organisation can have a big impact on the extent to which risk is managed. The role of the CRF is discussed in more detail later, but it is worth exploring the higher level issue of interaction here first.

| 1.5.1 The 'three lines of defence' model

One common distinction involves classifying the various parts of an organisation into one of three lines of defence, each of which has a role in managing risk. The first line of defence is carried out as part of the day-to-day management of an organisation, for example those pricing and selling investment products. Their work is overseen on an ongoing basis, with a greater or lesser degree of intervention, by an independent second tier of risk management carried out by the CRF. Finally, both of these areas are overseen on a less frequent basis by the third tier, audit.

This model explains the division of responsibilities well. However, it leaves open the degree of interaction between the three different lines, in particular the first and second.

2 1.5.2 The 'offence and defence' model

One view of the interaction of the first-line business units and the CRF is that the former should try and take as much risk as it can get away with to maximise returns, whilst the CRF should reduce risk as much as possible to minimise losses. This is the offence and defence model, where the first and second lines are set up in opposition.

The results of such an approach are rarely optimal. There is no incentive for the first-line units to consider risk since they regard this as the role of the CRF. Conversely, the CRF has an incentive to stifle any risk taking – even though taking risk is what an organisation must often do to gain a return.

It is better for first-line units to consider risk whilst making their decisions. It is also preferable for the CRF to maximise the effectiveness of the risk budget rather than to try to minimise the level of risk taken. This means that, whilst the offence and defence model might reflect the reality in some organisations, it should be avoided.

3 **1.5.3 The policy and policing model**

A different approach involves the CRF setting risk management policies and then monitoring the extent to which those policies are complied with. This avoids the outright confrontation that can arise in the offence and defence model, but is not an ideal solution.

The problem with this approach is that it can be too 'hands-off'. To be effective, it is essential that the CRF is heavily involved in the way in which business is carried out, and this model might lead to a system that leaves the CRF too detached.

4 **1.5.4 The partnership model**

This is supposed to be the way in which a CRF interacts with the first-line business units, with each working together to maximise returns subject to an acceptable level of risk. It can be achieved by embedding risk professionals in the first-line teams and ensuring that there is a constant dialogue between the teams and the CRF.

However, even this approach is not without its problems. In particular, there is the risk that members of the CRF will become so involved in managing risk within the first-line units that they will no longer be in a position to give an independent assessment of the risk management approaches carried out by those units. The degree to which the CRF and the first line units work together is therefore an important issue that must be resolved.

1.6 The risk management time horizon

Risk occurs because situations develop over time. This means that the time horizon chosen for risk measurement is important.

The level of risk over a one-year time horizon might not the same as that faced after ten years – this is clear. However, as well as considering the risk present over a time horizon in terms of the likelihood of a particular outcome at the end of that period, it is also important to consider what might happen in the intervening period. Are there any significant outflows whose timing might cause a solvency or a liquidity problem?

It is also important to consider the length of time it takes to recover from a particular loss event, either in terms of regaining financial ground or in terms of reinstating protection if it has been lost. For example, if a derivatives counterparty fails, how long will it take to put a similar derivative in place – in other words, for how long must a risk remain uncovered?

Finally, the time horizon itself must be interpreted correctly. For example, Solvency II – a mandatory risk framework that is being introduced for insurance companies – requires that firms have a 99.5% probability of solvency over a one-year time horizon. However, this is sometimes interpreted as being able to withstand anything up to a one in two-hundred-year event. Is this an accurate interpretation of the solvency standard? Would one interpretation be modelled differently from the other? All of these questions must be considered carefully.

1.7 Further reading

There are a number of books that discuss approaches to enterprise risk management and the issues that ought to be considered. Lam (2003) and Chapman (2006) give good overviews, whilst McNeil *et al.* (2005) concentrates on some of the more mathematical aspects of enterprise risk management.

It is also important to remember that risk management frameworks can be used to gain an understanding of the broader risk management process. This is particularly true of the advisory risk frameworks such as ISO 31000:2009.

2

Types of financial institution

2.1 Introduction

Whilst ERM can be applied to any organisation, this book concentrates on financial institutions. There is, of course, an enormous range of such institutions; however, detailed analysis is limited to four broad categories of organisation:

- banks;
- insurance companies;
- pension schemes; and
- foundations and endowments.

Before looking at the risks that these organisations face, it is important to understand their nature. By looking at the business that they conduct and the various relationships they have, the ways in which they are affected by risk can be appreciated more fully. This is the first – and broadest – aspect of the context within which the risk management process is carried out.

2.2 Banks

A direct line can be drawn to current commercial banks from the merchant banks that originated in Italy in the twelfth century. These organisations provided a way for businessmen to invest their accumulated wealth: bankers lent their own money to merchants, occasionally supplemented by additional funds that they had themselves borrowed. The provision of funds to commercial enterprises remains a core business of commercial banks today.

By the thirteenth century, bankers from Lombardy in Italy were also operating in London. However, a series of bankruptcies resulted in the Lombard bankers leaving the United Kingdom towards the end of the sixteenth century, at which point they were replaced by Tudor and Stuart goldsmiths. These gold-smiths had moved away from their traditional business of fashioning items from gold, starting instead to take custody of customers' gold for safekeeping. Following on from a practice devised by the Italian bankers, these goldsmith-bankers gave their customers notes by in exchange for the deposited gold, the notes being the basis of the paper currency used today. There also existed a clearing network for settling payments between the goldsmith-bankers. Much of the deposited gold was then invested, with only a proportion retained by the goldsmith-bankers. This forms the basis for what is known as fractional bank-ing, where only a proportion of the currency in issue is supported by reserves held.

Over time, the banking industry grew. In London, goldsmith-bankers were joined by money scriveners who acted as a link between investors and borrowers, and by the early eighteenth century the first cheque accounts appeared.

For much of the history of banks, particularly before the twentieth cen-tury, the industry was characterised by a large number of local banks. This meant that banks did not really need a network of branches. The location of the bank also reflected the clientèle it served. In the United Kingdom, banks based in the City of London were more likely to be merchant banks, whilst banks in the West End of London were more likely to serve the gentry. These West End banks took deposits and made loans (often in the form of residential mortgages), but were mainly involved in settling transactions. Smaller firms, as well as wealthy individuals, often found their needs served by the local (or country) banks of the eighteenth and nineteenth centuries. Following many mergers, these firms developed into the 'high street' banks seen today in the United Kingdom and elsewhere. Today, they raise capital from equity share-holders and bondholders, but also from holders of current and savings accounts with the bank. These funds are then used to fund short-term unsecured loans and longer-term mortgages to individuals and to firms. Many banks also lend funds to each other in order to make use of surplus capital or, as borrowers, to obtain additional finance. This lending is generally done over the short term. A final and important function of many of these institutions is as clearing banks. This is the process by which transactions are settled between as well as within banks, a function that can be traced back to some of the earliest work carried out by the goldsmith-bankers in the seventeenth century.

Although high street banks are now limited liability firms, this structure developed relatively recently. Following legislative changes in the early eigh-teenth century, all banks in England were restricted to partnerships with six or fewer partners. The only exception was the Bank of England, which was a joint-stock bank with limited liability. This restriction remained until legislation allowing the formation of new joint-stock banks was introduced in the nineteenth century. Some banking partnerships do still exist, being more commonly referred to as private banks today, but most banks are now owned by shareholders, being publicly traded companies or corporations. However, another form of bank, predominantly in the retail sector, is the mutual bank. A mutual bank is owned by savers with and borrowers from the bank, rather than by shareholders or partners. In the United Kingdom, the dominant form of mutual bank is the building society, whose main purpose is to raise funds which are then lent out as residential mortgages. The first building societies were set up in the United Kingdom in the late eighteenth century. They were generally small organisations whose customers lived close to each society's headquarters, and whilst there are now building societies operating on a national basis, many of these small, local firms still exist. This is in contrast with the consolidation seen in the rest of the banking sector. "*Traders*"

Compared with building societies, investment banks are a much more recent phenomenon. Their original role was to raise debt and equity funds for customers, and to advise on corporate actions such as mergers and acquisitions. These activities are still undertaken, but today investment banks also buy and sell securities and derivatives. In some cases, this is with the intention of holding a position in a particular market, for example being an investor in equities. However, in other cases the aim is for the bank to hold a 'flat book' – for example, to take on inflation risk from a utility firm and to provide inflation exposure to a pension scheme. The range of investment positions that a bank can hold is huge, and the potential links between the various exposures that a bank holds can lead to large risks. It is important that the impact of each risk on the bank as a whole is well understood. Investment banks are also involved in taking on risk in the form of securities or derivatives and repackaging these risks for sale to other investors. The best-known example of this is the collateralised debt obligation (CDO). This provides a ways of turning a bank's loans into a form of security held at arm's length from the bank. As a result, the risk and reward of the loans is transferred from the bank to a range of investors.

These days investment banks and merchant banks exist together as departments in more general commercial banks. However, this arrangement has only recently become possible internationally. The 1933 Glass–Steagall Act in the United States required the separation of merchant and investment banking activity in that country, the act only effectively being repealed by the 1999 Gramm–Leach–Bliley Act. This latter piece of legislation led to the existence of more broadly based commercial banks, serving all of the needs of commercial customers.

Many of these retail banks also merged with commercial banks, so offering the full range of services to the full range of clients. Furthermore, many banks have merged to form groups catering for both commercial and retail customers, and many have gone further, adding insurance products to their range – the resulting organisations are known as 'bancassurers'. The next section, however, considers the nature of insurance companies as distinct entities.

2.3 Insurance companies

There are two ways in which insurance companies can be classified. First, there are life insurance (or assurance) firms, whose payments are contingent on the death or survival of policyholders; then there are non-life (or general, or property and casualty) firms. Whilst recognising technically that insurance is intended to replace the loss of a policyholder, whilst assurance is intended to compensate for that loss (so a life cannot be insured), and that non-life insurance is not a particularly specific term, only the terms life and non-life insurance are used.

Non-life insurance appears to have started in fourteenth century Sicily, with the insurance of a shipping cargo of wheat, and such policies had made their way to London by the fifteenth century. Life insurance came out of marine insurance, with the cover being extended to people travelling on a voyage. Insurance companies started to appear in the late seventeenth century, initially providing buildings insurance, not least as a response to the Great Fire of London in 1666. At around the same time, a specialist market for marine insurance was forming in what later became Lloyd's of London. Today, Lloyd's and the London Market constitute an international centre not just for marine and aviation insurance, but also for unusual risks such as satellite insurance and, more famously, the body parts of various celebrities (the fingers of Rolling Stones guitarist Keith Richards, for instance).

Lloyd's provides a framework for risks to be covered. The capital for this used to be provided by individuals who had unlimited liability for any losses. More recently, limited liability capital has been used to support risks, this capital coming from insurance companies. Many insurance companies are themselves limited liability organisations. However, not all insurance companies are capitalised solely with shareholder's funds. Many are mutual insurance companies owned by their policyholders. Other proprietary insurance companies with external shareholder capital have classes of (with-profits) policyholders who derive returns, at least partly, from other (non-profit) policyholders due to the fact that the former provide capital to support business written to the latter. The class of mutual insurers also includes friendly societies, which came into existence in the eighteenth century. These institutions offered (and still offer) benefits on sickness and death.

Marine, aviation and satellite insurance have already been discussed. However, the full range of insurance classes is enormous. The three classes above are all forms of non-life insurance and are generally (although not exclusively) written for corporate clients. Car insurance, on the other hand, is predominantly provided to individuals, as are household buildings and contents insurance. A particularly important class is employer liability insurance. This covers, among other things, injury to employees during the course of their work. However, some types of injury may not become apparent until many years after the initial cause. A prime example of this is asbestosis, a lung disease arising from exposure to asbestos dust. Claims on many policies held by firms that used asbestos did not occur until many years after the industrial injuries had occurred. These so-called 'long-tail' liabilities, which resulted in the restructuring of Lloyd's of London, demonstrate another distinction between different classes of insurance. For some classes, such as employer liability insurance, the claims can occur for many years after the policy is written; conversely, the claims for 'short-tail' insurance classes, such as car insurance, are mostly reported very soon after they are incurred. These differences lead to a difference in the importance of the various risks faced by insurers.

Life insurance also has short- and long-tail classes, although most fall into the latter category. An example of a short-tail class would be group life insurance cover, where a lump sum is paid on the death of an employee (often written through a pension scheme for tax reasons). These policies are frequently annual policies, and deaths are generally notified soon after they occur, not least because there is a financial incentive to do so. However, individual life insurance policies can have much longer terms. Term assurance – a life insurance policy often linked to a mortgage – will regularly have an initial term of twenty-five years. Also in existence are whole-life policies, which, as the name suggests, remain in force for the remaining lifetime of the policyholder. On the other side of the equation from these policies that pay out on death are annuities which pay out for as long as an annuitant survives. These too have risk issues linked to their long-term nature.

Life insurance companies also provide a variety of investment policies for individuals and institutions, such as pension schemes. Some of these are unit-linked, where the return for the policyholder is simply the return on the underlying assets (after an allowance for fees). In this sense, the insurance company is acting as an investment or fund manager. However, there are two aspects of life office investment products that can differ from other products. The first is the with-profits policy. As mentioned above, these policies provide a return based not only on the underlying investments held in the with-profits fund, but also from the profits made from writing non-profit business, such as life insurance policies or (non-profit) annuities. However, another important aspect of with-profits policies is that the returns to policyholders are smoothed over time. This is done by paying a low guaranteed rate on funds, and then supplementing this with bonuses. Bonuses are paid each year and at the end of a policy's life. When investment returns are good, not all of these returns are given to policyholders; when they are poor, the bonus may be lower, but a bonus will generally be given. This means that not only is there smoothing, but for most with-profits products, the value cannot fall.

Whilst the typical with-profits products are investment funds, typically in the form of endowment policies which pay out on a fixed date in the future, there are also with-profits annuities which apply a type of bonus structure to annuity payments. Some with-profits policies have also included options allowing investors to buy annuities at a guaranteed price. Since these guarantees were given many years before the options were exercised, the risks taken were significant and, in one case, resulted in the insolvency of the firm writing those policies.

Many insurance companies offer both life and non-life insurance policies. Such providers are known as composite insurers. In the European Union, the creation of new composite insurers is banned further to the First Life Directive of 1979, except when the life component relates only to health insurance.

2.4 Pension schemes

As with banks and insurance companies, pension schemes have a long history. Occupational pension schemes date back to the fourteenth century in the United Kingdom, with schemes providing lifetime pensions on retirement appearing in the seventeenth century in both the United Kingdom and France. The United States eventually followed suit in the nineteenth century. Defined benefit pension schemes, with a format similar to that in place today, also appeared in the nineteenth century in the United Kingdom. These are schemes where the benefit paid is calculated according to some formula, generally relating to the length of an individual's service with a firm and their earnings. The most common form of defined benefit arrangement is a final salary scheme, where the benefits are based on the salary immediately prior to retirement.

These types of arrangements were generally pay-as-you-go (PAYG) arrangements, as were the universal pension systems appearing in Germany in the nineteenth century, and in the United Kingdom and the United States in the twentieth century. This means that no assets were set aside to pay for the pensions – the cost was met as pensions fell due. This model is still the typical method used for state pension schemes, particularly in the United Kingdom. Many of these schemes have grown so large in terms of liabilities that capitalisation is no longer a viable proposition.   × *Norway , Spare*

Funded pensions, where assets were set aside to pay for pension benefits, found popularity in the twentieth century with schemes being set up under trust law in the United Kingdom. This arrangement had a number of tax advantages for firms, contributions having been exempt from tax since the mid-nineteenth century. However, investment returns also received exemption in the early twentieth century. With funded pension schemes, this means that both the benefits due and the assets held in respect of those benefits need to be considered. Virtually all defined benefit pension schemes present today in the United Kingdom were set up under trust law. Although set up by an employer, such schemes are governed by a group of trustees on behalf of the beneficiaries.

From the 1970s onwards, the regulation of defined benefit pension schemes increased, particularly in the United Kingdom. What was previously a largely discretionary benefit structure changed to one that carried a large number of guarantees. This changed fundamentally the degree of risk carried by pension schemes, and the employers (sponsors) that were responsible for ensuring that the pension schemes had sufficient assets.

Although it is not always the case, unfunded, PAYG pension schemes are still generally found in the public sector, and funded pension schemes, where assets are held to cover the benefits due, are found in the private sector. A 'middle ground' between these two types of scheme is the book reserve scheme. Here, the capitalised value of the liabilities is assessed but is held as a liability on the balance sheet rather than being run as a financially separate, funded entity. Such schemes have been popular in Germany, particularly prior to the provision of tax incentives for funded arrangements.

Whilst defined benefit pension schemes are still by far the most important type of retirement arrangement, increasing costs and an increasing appreciation of the risk they pose has led to a large increase in defined contribution pensions. Here, assets are accumulated – usually free of tax – and they are then withdrawn at retirement. In the United Kingdom, there is a requirement that 75% of the proceeds are used ultimately to buy a whole-life annuity. Whereas the majority of the risk in a defined benefit arrangement lies with the sponsor, in a defined contribution scheme it rests with the scheme member. In the

United Kingdom, many defined contribution schemes set up in the past were trust-based schemes. However, an increasing number of defined contribution pension arrangements, whether arranged by an employer or not, are actually held as policies with insurance companies. This became even more common after the introduction of personal pensions in 1988.

2.5 Foundations and endowments

The final types of institution are the broad group that can be classed as foundations and endowments. For the purposes of this analysis, these are institutions that hold assets for any number of reasons. They might be charities or individual trust funds; they might have a specific purpose such as funding research, or a more general function such as providing an income to a dependent; however, the common factor is that they do not have any well-defined predetermined financial liability.

Some of these institutions will be funded by a single payment (endowments), whilst others will be open to future payments and may even have ongoing fund-raising programmes (foundations). These imply very different levels of risk.

In the United Kingdom, the most common type of foundation is the charitable trust, this structure giving beneficial tax treatment. Some such organisations, like the British Heart Foundation, have the term 'foundation' in their name; however, this is an exception. Terms such as 'campaign', 'society' and 'trust' are just as likely to be found, as are names which have no reference to their charitable status.

Endowments are most commonly seen in the context of academic posts, such as the Lucasian Chair in Mathematics at the University of Cambridge. This practice has existed since the start of the sixteenth century in the United Kingdom. In the United States, endowments are also used to finance entire institutions, such as universities or hospitals.

2.6 Further reading

Information on the early history of banking was provided by the Goldsmiths' Company in the City of London. They were helpful in directing me to a number of useful publications, including Gilbart (1834) and Green (1989). There are also a number of popular books dealing with the development of individual banks, such as Chernow (2010) (*The House of Morgan: An American Banking Dynasty and the Rise of Modern Finance*) and Fisher (2010) (*When Money*

Was in Fashion: Henry Goldman, Goldman Sachs, and the Founding of Wall Street).

A good early history of pensions and insurance is given by Lewin (2003). The developments in pensions around the start of the twentieth century are covered in detail by Hannah (1986), with more recent legislative developments being discussed by Blake (2003).

3

Stakeholders

3.1 Introduction

The nature of an organisation gives the basis on which other aspects of the risk management context can be built. One of the more important aspects is the nature of the relationships that various stakeholders have with an institution. There are a number of ways in which these relationships can be described, but a good starting point is to classify them into one of several broad types, these types being:

- principal;
- agency;
- controlling;
- advisory; and
- incidental.

In this chapter, these relationships are considered in more detail, to make it easier to understand where risks can occur.

3.2 Principals

All financial institutions need and use capital (as do all non-financial institutions), and the principal relationships describe those parties who either contribute capital to or receive capital from the institution. Providers can be categorised broadly into those who expect a fixed, or at least predetermined, return on their capital (providers of debt capital, debtholders) and those who expect whatever is left (providers of equity capital, shareholders). The former will generally be creditors of the institution. This means that they have lent money to the institution, and are reliant on the institution being able to repay the debt. Shareholders, on the other hand, are not owed money by the institution; rather, they can be regarded as part owners of the institution. On the other side, institutions have relationships with their customers. The customers provide the *raison d'être* of the institution. Financial institutions also have a number of relationships with governments. Among these are direct financial relationships, justifying the inclusion of governments in this category. Whilst these now include the provision of financial support for some institutions, including privatisation, this is really only the government acting as a provider of capital. The relationships that are exclusively governmental more typically involve taxation. Finally, as well as drawing capital from capital markets, financial institutions are unique in that they are also significant investors in capital markets. Similarly, whilst some financial institutions provide insurance, many also purchase insurance, often due to statutory requirements and generally in order to protect their customers. In the context of relationships, markets are generally insensitive to the actions of an individual investor. This means that those whose relationship with capital is broadly a principal one can be summarised as:

- shareholders;
- debtholders;
- customers;
- the government; and
- insurance and financial markets.

Excluded from this list are those with whom the firm's financial relationship is typically incidental (for the firm). This includes trade debtors and creditors, subcontractors and suppliers, and the general public. Figure 3.1 shows the relationship between the main parties.

In broad terms, the theoretical aim of most institutions should be to maximise the profit stream payable to the shareholders from the customers and investment in financial markets, whilst ensuring that the profit stream is stable enough to meet the fixed payments to debtholders.

This will have an impact on the way in which capital will be used. In particular, shareholders will wish to maximise the return on the capital they supply, whereas debtholders and customers will wish to minimise the risk to capital. The former group is concerned with investing aggressively enough and the model used for pricing; the latter group is concerned with matching assets to liabilities and the model used for reserving.

Whilst this categorisation of principals is true in general terms, the individual parties involved with any industry will differ from type to type. A

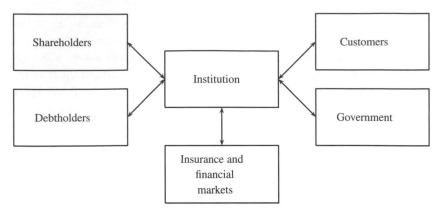

Figure 3.1 Principal relationships of a financial institution

comprehensive list is:

- public shareholders;
- private shareholders;
- public and private debtholders;
- bank customers;
- insurance company policyholders;
- pension scheme sponsors;
- pension scheme members;
- endowment and foundation beneficiaries;
- governments (financial relationships);
- insurance providers; and
- financial markets.

3.2.1 Public shareholders

Many banks and many insurance companies are listed on stock exchanges. This means that they have a large number of public shareholders who can buy and sell the securities that they own. Private shareholders have few direct protections. The key safeguard they have is limited liability – they cannot lose more than their investment in a firm. This gives them an incentive to demand that a firm take more risk since investors have effectively purchased call options on the firm's profits. Some legislative protection available to investors is discussed in Chapter 5, although in many cases litigation will provide the only recourse. Beyond this, a major safeguard for investors is the information used to assess the value of their investments, and to the extent that markets can be said to reflect the true value of investments, the market itself could be said to offer protection to investors through the information it contains; however, markets are very often wrong.

3.2.2 Private shareholders

Private shareholders are subject to the same restrictions as public shareholders, but these restrictions are less likely to be relevant as private shareholders tend to be long-term investors. They are also frequently directors or even managers of the firms that they own, but they still have the same protection afforded by the limited liability nature of being a shareholder.

This is not necessarily the case if the organisation is structured as a partnership. Traditionally, partners are jointly and severally liable for the each others' losses. This means that the private assets of all partners are at risk if a firm becomes insolvent.

The structure of limited liability partnerships can reduce or remove this risk. These types of institution exist largely to allow firms that must exist as partnerships for statutory reasons, or tend to exist as partnerships for tax reasons, to continue with the threat of personal insolvency lessened.

For most of the United Kingdom, the Limited Liability Partnerships Act 2000 allows this type of firm to exist. In effect, this converts a partnership to a private limited company, which remains as a partnership only in tax terms. This is not necessarily the case in the United States, where the liability differs from state to state, but can simply limit the liability of some rather than all of the partners.

3.2.3 Public and private debtholders

The other main suppliers of capital to banks and insurance companies are holders of debt issued by these firms. These suppliers of debt capital are creditors of the institutions, and obligations to these parties must be met before any returns can be given to the shareholders. This means that investors in this type of capital want the firm to take enough risk to meet their interest payments but no more – their concern is security.

The priority of payments between the various issues of bonds and bills will depend on the terms specified in this lending. These terms are included in covenants, and covenants provide an important protection for debtholders, covering not just the seniority of different issues but also the way in which each issue is constructed. Debtholders can also get protection from any collateral to which the debt is linked, the degree of protection depending on the nature of the collateral.

Public debt comprises securities sold in the open market, so ownership is typically spread across a large number of investors. Investors in these securities receive the same protection and are subject to the same obligations as investors in public equity. The most common types of public debt are corporate bonds and commercial paper. The distinction between these two types of security is that the former are long-term debt instruments (often issued with terms of many years, or with no set date for redemption), whereas the latter are issued for the short term (often a year or less). With corporate bonds, the issuer will borrow a fixed amount, will make interest payments on that amount that may be fixed, varying with the prevailing interest rates, or linked to some index, and then, assuming that the bond is redeemable, will repay the amount borrowed at some point in the future. With commercial paper, the issuer will specify the amount to be repaid and will borrow a smaller amount, the interest effectively being reflected in the lower amount borrowed. In other words, commercial paper is sold at a discount.

Private debt typically involves using some type of bank facility. Such borrowing is therefore not tradeable. This facility might be pre-arranged or *ad hoc*, short or long term.

The borrowing by one bank from another constitutes the interbank lending market. This is an important source of liquidity that in normal market conditions helps to ensure the smooth functioning of financial markets. A particularly important type of bank that gets involved in this market is the central bank. These can play an important role in ensuring liquidity in financial systems.

When considering debt finance, it is important to recognise that it should be looked at in the context of financing as a whole. There are a number of theories that explain the extent to which debt and equity may be used to finance a firm. A good starting point is the famous proposition from Modigliani and Miller (1959, 1963). This states that the value of any firm is independent from its capital structure. This works well in the first order, but since interest paid on debt is tax-deductible whereas dividends are paid post-tax, allowing for tax suggests that all firms should be funded completely from debt. One argument for why they are not is that insolvency is costly, and funding a firm entirely from debt raises the risk of insolvency to an unacceptably high level. Controlling the tax liability and of the risk of insolvency are therefore two important risks to be considered.

Another theory considers agency costs, which are discussed in more detail in the next section. This view suggests that the freedom that managers of a firm – the agents – have to act in their own interests will have an impact on the ownership structure used. For example, in industries where it is very difficult to monitor the activities of managers, the dominant form of ownership will be private – owners will also be managers, since it is difficult to persuade a range of small shareholders to delegate management responsibility in such circumstances. In industries where it is easy to alter the risk profile of the business, it will be difficult to attract debt capital, since providers of debt know that the managers will have an incentive to act against them; however, in heavily regulated industries, investors should be more willing to supply debt and equity capital, more so the former since the scope for excess profits is more limited. Agency costs, and how to limit them are discussed later in this book.

The level and term of debt might also be designed by management in order to pass on useful information about the firm, and to reduce the incentives of debtholders to force a firm into insolvency. There is also an argument that firm's choices of sources of finance might be different for existing and future business opportunities. In particular, the 'pecking order' theory suggests that firms will be inclined to finance future opportunities with equity share capital so that profits from the investment are not captured by debtholders.

3.2.4 Bank customers

Banks have a wide variety of customers. Consider, for example, counter-parties to derivative transactions. Many derivative contracts will require each party to pay assets over in advance of settlement if the value of the derivative moves significantly. This offers protection in the event that one of the counter-parties becomes insolvent. However, to the extent that this collateral is insufficient, a price move in favour of the customer means that the customer becomes a creditor of the bank, effectively providing debt capital.

The position of individual and commercial bank account holders is even more ambiguous – are they customers or creditors? The answer is, of course, that they are both. Similarly, those holding bank mortgages are customers, but they are also debt-like investments of the bank. The situation for building societies is complicated still further, because bank account holders are also effectively equity shareholders of a building society, as are customers with mortgages, since both are owners of the firm.

In terms of risk appetite, these factors mean that the interests of bank customers are aligned with debtholders – less risk is better.

3.2.5 Insurance company policyholders

The situation for insurance company policyholders is as complex as that of bank customers. Non-profit and non-life policyholders are unambiguously customers of most insurance companies. However, for a mutual insurance company, the shareholders are also customers, being with-profits policyholders; and even for a proprietary insurance company, part of the equity capital is provided by with-profits policyholders (if they exist) in addition to that provided by more traditional shareholders. The situation is slightly different for friendly societies, where all policyholders are part-owners of the firm as well. This means that with-profits policyholders and the policyholders of friendly societies will tend to have risk preferences that are similar to those of equity shareholders, since they all receive a share of the excess profits earned. For with-profits policyholders, the extent to which they will prefer more risk will depend on the bonus policy of the insurance company. All other things being equal, a greater degree of smoothing of bonus rates over time will lead to a reduction in risk tolerance as the maturity date of the policy approaches.

3.2.6 Pension scheme sponsors

The sponsor of a defined benefit pension scheme can also be regarded as the provider of equity capital to that scheme, being the party that must make up any shortfall and that receives the benefit of any surplus of assets over liabilities (usually through a reduction in contributions payable). Sponsors set the initial levels of benefits that they are willing to fund when the scheme is set up. With trust-based arrangements, these benefits are included in the pension scheme's trust deed and rules, although for many older pension schemes legislation has increased the level of these benefits – what might have originally been offered on a discretionary basis has often subsequently been turned into a guarantee.

An important concept here is the concept of the pensions-augmented balance sheet, where the values of pension assets and liabilities are added to the value of firm assets and liabilities with the value of corporate equity being the balancing item. In this context, a pension deficit can be regarded as a put option and a surplus a call option for the employer and, therefore, the shareholders of the firm. The deficit as a put option is a particularly important concept. It comes about by recognising that a pension scheme deficit is money owed by the company. The firm has the option to default on the deficit in the same way that it has the option to default on debt, and this option has value to the firm. The firm will only default on the deficit when it is insolvent (so the value of its liabilities exceeds that of its assets) and when a deficit exists (so the value of the pension scheme's liabilities exceeds that of its assets). The greater the deficit and the less financially secure the sponsoring employer, the greater the value of this put option. In addition to the economic impact of pension schemes on their sponsors, there are the accounting impacts. For example, increasing pensions costs (in the accounting sense) affects the retained profit of firms. It is possible that losses could be so large as to reduce the free reserves to such a level that the ability to pay dividends is affected; even if the situation does not reach this level, the pension scheme might adversely affect profitability or other key financial indicators.

Regarding the deficit as a put option implies that the riskier a firm is the greater the incentive to increase the value of this put option. This can be done by changing the strike price of the option (by reducing pension scheme contributions and increasing the deficit) and by increasing its volatility (by encouraging the pension scheme to invest in riskier assets). This is the opposite course of actions to those that the members ought to prefer, which is full funding and low risk investments.

At the other end of the scale, a financially sound sponsor has reasons to remove risk from the pension scheme and to put in as much money as possible. To the extent that pension benefits are guaranteed and the sponsor is responsible for meeting these benefits, they constitute a debt owed to the members and, as such, debt financing for the sponsor. The assets in the pension scheme can be regarded as collateral held against the pension scheme liabilities. To the extent that the assets do not match the liabilities, those liabilities represent an increase in debt funding. This reduces the extent to which a firm can use true debt funding in place of equity funding. This is important as the interest payments on debt are tax-deductible, whereas dividend payments are not; there is, though, no corresponding disadvantage or advantage to investing in debt in a pension scheme – all returns are generally free of tax. This means that if a sponsor is financially secure and can therefore borrow cheaply, then there is an incentive for the sponsor to fully match the liabilities in the pension scheme with bonds whilst increasing the level of debt funding relative to equity funding for the firm itself. This strategy is known as Tepper–Black tax arbitrage (Tepper, 1981; Black, 1980).

However, if the members of a pension scheme are entitled to the surplus in a pension scheme, then there is an incentive for them to demand a more aggressive investment strategy. A financially strong sponsor is unlikely to default on its pension promise so the risk of pension benefits not being met is small; however, the potential increase in benefits is significant.

3.2.7 Pension scheme members

Defined contribution pension schemes can be thought of as non-profit or with-profits investments with a life insurance company or investment firm. This means that their members can be thought of as policyholders or investors, except to the extent that a sponsoring employer is late in the payment of contributions. However, when considering members of defined benefit pension schemes, a change in perspective is needed. Pension scheme liabilities can be regarded as collateralised borrowing against scheme member's future benefit payments. This being the case, pension scheme members might be regarded as debtholders as much as customers. This places them in a similar position to the customers of banks, but with arguably less security. Pension schemes are allowed to rely to an extent on the continued existence of their sponsors for solvency, but because banks have no such recourse to their shareholders, the capitalisation requirements are much stricter.

3.2.8 Foundation and endowment beneficiaries

Endowments are a little different in that the customers are also the equity shareholders, the main relationship is with markets through the investments used and there are no obvious debtholders. The benefits of any profit or loss are reflected entirely in the returns to the beneficiary or beneficiaries for whom the endowment is run, who therefore hold the dual role of customer and shareholder. This means that in the absence of any contractual payout requirement, as would be seen with, say, pension scheme benefits, the choice of investment objective needs to be carefully considered. For a foundation, the situation is only slightly more complicated, since contributors might also be regarded as customers – the investment and divestment strategies adopted by the charity will influence the level of contributions made.

3.2.9 Governments (financial relationships)

Governments have a large number of relationships with financial institutions, their customers and those funding them. However, it is the financial relationships alone that are of interest here. A government's incentive here can broadly be summarised as seeking to maximise its income from corporate taxation. However, this is not so straightforward as simply raising the rate of taxation. Too high a figure might lead to insolvencies, reduced incentives to increase profits or incentives to move an organisation to a more favourable tax regime. In relation to pension schemes, it might lead to unpopularity and the risk of electoral losses.

A government's regulatory role is also relevant here, since it will wish to maximise the risks that firms can take (in order to generate higher taxable profits), whilst limiting the risk of insolvency to an acceptable (and hopefully negligible) level.

Taxation affects the choice of defined contribution vehicle in a number of ways. In many cases, people will move from a higher to a lower tax band when retiring. In this case, tax relief on contributions will be most attractive. This attraction is enhanced if there is an additional tax-free lump sum available, as in the United Kingdom. However, some people's tax positions might move in the opposite direction, particularly those of young people who expect large increases in their income over time. Furthermore, vehicles that take post-tax contributions often allow individuals to withdraw their assets at any time. To the extent that this additional liquidity is valuable, it might outweigh any tax advantages.

The effect on risk of tax limits is also interesting. If tax relief is only available on assets up to a particular level, then as the accumulated fund gets closer to this level the incentive to reduce risk increases as the potential upside is reduced by a potential tax liability, thus increasing the asymmetry of returns distribution.

3.2.10 Insurance providers

Many types of insurance taken out by financial institutions will be incidental to the nature of the business. For example, banks in the United Kingdom must, lie all employers, have employer liability insurance.

Financial institutions can also use insurance as a customer. Pension schemes might choose to insure some of the benefits provided, such as death in service pensions or lump sums. Insurance companies are also important purchasers of insurance in the form of reinsurance.

However, an important area of insurance for financial institutions is that of statutory insurance that must be purchased to protect an organisation's members, customers or policyholders in the event that the financial institution becomes unable to meet its obligations.

Statutory insurance schemes are often (though not always) government sponsored, but they are generally set up to be financially separate organisations. As such, they are responsible for ensuring that the premiums received cover the benefits paid. This means that, whilst the various statutory insurance schemes are discussed in terms of the protection, they grant to members, it is worth considering the risks that such institutions face in their own right.

The first risk is that such institutions will not collect enough premiums to cover the benefits. Providing the premiums can be reset to recoup any losses (or reduced if any surplus becomes embarrassingly large), then this ought not to be a problem in most circumstances. However, this is not always the case. For example, the premium rates for the Pension Benefit Guaranty Corporation

(PBGC) in the United States are set out in primary legislation, so changing them is not straightforward. There are also circumstances where being able to change premium rates is not sufficient to stop problems. In particular, if a new insurance arrangement is set up from scratch then there is a risk that exceptionally large claims in early years will be sufficient to bankrupt the fund before it has accumulated sufficient reserves to protect itself from volatility.

An ongoing risk for some schemes is the risk of moral hazard. This occurs when the presence of insurance gives the insured party an incentive to increase the level of risk taken. This is unlikely to be the case for schemes set up for customers of banks and insurance companies, where the premium is ultimately paid by shareholders who would extract no benefit from insolvency. However, in pension insurance schemes this can be an issue.

The initial design of the PBGC caused particular problems. It provided insurance for pension scheme members if the sponsoring employer became insolvent or terminated the pension scheme voluntarily, although termination allowed the PBGC to recover up to 30% of the net worth of the company to fund any deficit. This is a valuable double American option, where either party can force termination at any point in time (as distinct from a European option, which can be exercised only at a single point in time). This ability of both parties to exercise the option at any point in time is at the root of most of the complaints of moral hazard arising from the PBGC, giving an incentive for the pension schemes of riskier firms to invest in riskier assets, knowing that the option of termination would limit any downside risk.

This formulation of the PBGC also allowed a number of ways in which moral hazard could occur, for example by firms spinning off divisions with deficits and collecting the surplus on retained divisions. A firm might also have split a pension scheme into two parts, one for active members and one for pensioners, placing all surplus assets into the pensioner plan, terminating it and capturing the entire surplus. The original structure also used a flat premium structure that took into account neither the riskiness of the employer nor the riskiness of the pension scheme – this too led to moral hazard.

The Single Employer Pension Plan Amendments Act 1986 removed a major flaw from the original legislation by effectively requiring the sponsoring employer to be insolvent, thus removing the ability for solvent employers to pass their deficits on to the PBGC. A risk-based element was introduced to the PBGC premium calculation in 1988, although it ignored the risk of bankruptcy and had an upper limit. This meant that it is more exposure- than risk-related. The increase in premium for schemes with deficits was increased again with the Pension Protection Act 2006.

The Pension Protection Fund (PPF) in the United Kingdom also has a risk-based component to its levy, although the levy is applied to the deficit rather than the liabilities, and ignores the asset allocation of the pension fund. This also means that the risk-based levy could more accurately be described as exposure related. There is only marginally more incentive to invest in bonds or other assets that match the liabilities, since the only risk is that the assets will under-perform and the levy will increase next year – the premium at the outset does not change.

3.2.11 Financial markets

There are two ways in which an institution might have exposure to the financial markets. The first is in directly investing customers' assets, giving customers an indirect relationship to these markets. The key example of this is where an investor has a unit-linked policy. Here, the vast majority of the risk is faced by the customer, the impact of returns on the institution's fee income generally being secondary.

The second way in which institutions have exposure is when they invest assets to make a profit for shareholders. Here, the institution's exposure to investment returns is much more direct.

The nature of an institution's exposure to financial risk, and that of agents, has a major impact on the investment approaches taken. For example, some pension scheme sponsors will prefer a risky investment strategy, as this can increase the opacity of the pension scheme's funding position. This gives the sponsor freedom to increase contributions for tax purposes, or to decrease them to ease cash flow problems or to leave more funds for investment in the sponsor's business.

If investing in the markets on behalf of another party, there is often an incentive to reduce investment risks relative to competitors for fear of under-performing them. This is particularly evident with pension schemes, where assets are ultimately invested for the good of members. In particular, this has been seen in the form of 'peer group benchmarks', where pension schemes measure the performance of their portfolios – and, more importantly, set their asset allocations – relative to other pension schemes, regardless of the extent to which the liabilities of the two schemes are similar. This approach became less common with the advent of liability-driven investment at the start of the twenty-first century.

3.3 Agents

As the name suggests, those parties with an agency relationship act on behalf of a principal. The main risks that occur can therefore be classified as agency risks, and the costs arising from these issues are agency costs. However, many of the interests of principal parties are delegated to agents, and without such arrangements large firms would find it impossible to operate. The agents considered are:

- company directors;
- trustees;
- company managers;
- company employees;
- trade unions;
- central risk functions;
- pricing teams;
- auditors;
- pension scheme administrators; and
- investment managers.

| 3.3.1 Company directors

Company directors are generally appointed by shareholders, or owners in the case of a mutual organisation, to act on their behalf. This means that for banks and insurance companies they are some of the most important agents. As discussed above, the one organisation considered so far where this might not be the case is a private bank, where the directors are likely to include the shareholders, so there is no distinction between principal and agent. However, for all other banks and insurance companies, the shareholders must rely on the directors to determine the strategic direction of the organisation.

This is in fact eminently sensible. For most of these firms, the number of shareholders will be so large that for these parties to make any decisions in relation to firm strategy would be impractical. Furthermore, many of these shareholders, particularly private individuals, will not have the knowledge or skills to make such decisions. For public limited companies, there is also the issue that shares are bought and sold frequently, meaning that the ultimate owners of the firm change far too frequently for there to be any continuity of decision making.

 The approach taken by directors to running an organisation is known as corporate governance. ERM is a fundamental part of good corporate governance and it is important that boards recognise this – it is easy for risk management to be squeezed out by the many other concerns faced by boards. This means disseminating a system and culture of risk management through an organisation, as well as taking more specific actions.

In financial institutions, directors also have an additional responsibility: determining the value of the assets and liabilities held. Advice is taken on these issues, but the responsibility remains with the directors. However, there is a risk that the directors aggregating the assets and liabilities will not understand the products to as great an extent as the groups creating them. The distribution of returns might not be understood, or greater diversification of positions might be assumed than is actually the case. This needs to be recognised in any ERM structure.

Boards of directors will delegate many important functions to committees of board members. This is important in some cases as the correct constitution – independent, non-executive members in the majority or exclusively – ensures that there is sufficient independence within these committees from the executive members of the board.

Whilst many of the considerations of the roles of directors can be applied to trustees, the legislative framework in which they operate is very different, and it is trustees that are discussed next.

3.3.2 Trustees

In pension schemes, where the 'shareholder' is the scheme sponsor, the 'directors' are pension scheme trustees. However, trustees are not necessarily appointed only by shareholders, as discussed below. Pension scheme trustees are responsible, together with the sponsoring employer, for pricing in the context of a defined benefit pension scheme. Whilst the guaranteed benefits are specified by the sponsoring employer, as amended by legislation, trustees may still be involved in the provision of discretionary benefits, but this depends on the terms set out in the pension scheme's trust deed and rules.

Since trustees act on behalf of beneficiaries they should generally want the pension scheme to be as well funded as possible. However, the question of asset allocation is more complex. If the risk of sponsor insolvency is low and if the pension scheme is entitled to spend any surplus on discretionary benefits for members, then the trustees ought to prefer as risky an investment strategy as possible; if, however, the sponsor is weak, then there is an incentive for the trustees to match benefits as closely as possible. It is interesting to note that the opposite is true for the sponsor in each case, as discussed earlier – Tepper–Black tax arbitrage implies that a solvent sponsor should prefer a low risk investment strategy for the pension scheme, and the pension put implies that a risky sponsor should prefer risky investments. The addition of statutory pension insurance, such as that available with the PPF in the United Kingdom, or the PBGC in the United States complicates this decision. If the scheme has insufficient assets to fund even the insured benefits, then no matter how high the risk of sponsor insolvency, more risk is better for the sponsor and the trustees.

However, there are potential conflicts of interest here. Many pension scheme trustees are also potential beneficiaries of pension schemes so might act in their own benefit, in particular benefiting one class of member (their own) over another. Many are also trustees as a result as their roles within the sponsoring employer. This might lead them to act in the firm's best interest if their remuneration is based on firm- but not scheme-related metrics. For example, if a lower level of discretionary pension benefits increases a manager's bonus by more than it reduces his or her pension benefits, then he or she is likely to find it attractive. Even independent trustees are not immune – if they are appointed because a pension scheme is winding up, then they are best served by the continuation of the scheme (and their own remuneration), so have an incentive to make any wind-up proceedings last as long as possible.

The trustees' first concern should always be the buyout valuation. This is a valuation that tells the trustees whether there would be sufficient assets in the pension scheme to secure members' benefits with an insurance company if the scheme were to discontinue immediately. This is a risk faced by all private sector pension funds. The scheme's actuary should ensure that the projected contributions will be sufficient to maintain solvency on a buyout basis with an adequate degree of confidence over the projection period, given the proposed investment strategy. The funding valuation can then be assessed with reference to the minimum contribution rate acceptable for each asset allocation on the buyout basis.

The purpose of the funding valuation is to calculate the level of contributions required to maintain or achieve an acceptable level of funding on an ongoing basis with an adequate degree of confidence over a specified time horizon. As alluded to above, the funding valuation should also be considered together with the asset allocation. Provided the contribution rates arrived at are at least as great as those calculated for the buyout valuation, then there is much more freedom in relation to the appropriate range of assumptions.

Trustees must themselves delegate many of their functions and they may choose to delegate more. The delegated roles are discussed later in this chapter. However, first the roles delegated by company directors are considered.

3.3.3 Company managers and employees

Whilst directors are responsible for much of the strategic work involved in running a firm such as a bank or an insurance company, the day-to-day tasks are generally delegated to managers. Managers are responsible for implementing the strategy set out by the directors, and managers will themselves delegate many tasks to other employees. It should be clear that managers and those who report to them, should be more inclined to act for themselves than for those to whom they report and, ultimately, the shareholders. This is a clear example of agency risks and the resulting financial impacts are agency costs. For example, managers and employees might be inclined to use work-issued mobile phones for personal calls, or to use expense accounts for non-business expenditure. As outlined earlier, the extent to which managers and employees can escape scrutiny can have an impact on a firm's capital structure. However, if acting in the interests of shareholders, directors and managers will also be inclined to structure remuneration and working practices in such as way as to minimise these agency costs.

3.3.4 Trade unions

Trade unions have existed since the eighteenth century as a way of representing groups of workers in a consistent manner. They are agents for employees in a number of ways, the most important of which relate to pay and conditions. Here, trade unions can help by acting on behalf of groups of employees, a process known as collective bargaining. By acting on behalf of groups of employees, trade unions can also apply pressure through lobbying and even strikes. Since their beginnings, the strength of trade unions has grown considerably, even becoming the basis of one of the main political parties in the United Kingdom (the Labour Party). However, whilst most of their practices could be regarded as being on behalf of the members they represent, trade unions are also susceptible to agency risk. In particular, an institution known as the closed shop – where union membership is a condition of employment – could arguably be regarded as of as much use to trade unions as to their members.

3.3.5 Central risk functions

Whilst employees have been considered in general terms, there is a particular class of employee that has a central role in ERM: the CRF. In small organisations, this employee could be a single person, but larger firms could have a full team of specialist risk managers. The CRF does not usually cover risks directly – this task is carried out by employees in all areas of the firm. Instead, the CRF requires involvement in risk at all levels.

One of the most important roles is to advise the board on risk. To do this effectively requires that the board is willing to hear about the risk issues faced in an organisation. However, in order to do this effectively, the CRF needs to assess the level of risk at an organisational level, by aggregating information from around the organisation. This again requires a good level of communication. Communication is also fundamental to the CRF in educating mangers and employees on the identification, quantification and management of risks.

The CRF is also responsible for using the information received and processing it for the use of the Board. In particular, this means comparing the actual level of risk with the risk appetite and monitoring progress on risk management.

The CRF is headed by a chief risk officer (CRO). One aspect of this leadership is to co-ordinate the various risk management divisions that might exist, such as credit, liquidity (treasury), investment, operational, insurance and legal. The heads of these groups would in some cases report to the CRO, but even if not managed by the CRO – for example, the treasurer may well report to the chief financial officer (CFO) – they would still provide the appropriate information to the CRO as well as receiving risk management guidance from them.

The CRO is not a leader just in an administrative sense – he or she is ultimately responsible for determining the risk management policy of an organisation and setting the standards to which all employees must adhere. This includes ongoing development of existing approaches in response to the changing nature of an organisation and developments in the world around it. A key part of this is to establish a coherent risk management 'language' to avoid confusion. The CRO is also responsible for monitoring adherence and overseeing the implementation of risk management policies, and for training in risk management techniques.

It is the responsibility of the CRO to collate information on risks received from around an organisation and to determine appropriate actions if existing policies are not sufficient. The CRO should also be on the lookout for new risks as they develop, as well as new techniques for dealing with these and existing risks. These factors are closely linked to another role of the CRO, which is to allocate economic capital around an organisation. Economic or risk capital is the financial cushion that allows an organisation to write business, so the allocation of this capital around a firm determines the target mix of business.

The CRO also forms a link between the CRF and the board. This involves reporting on risks to the board, and ensuring that decisions of the board in relation to risk management are implemented; however, the CRO should also ideally be a member of that board, and should lead the board's risk committee.

There is also an external reporting aspect to the CRO's role. As well as general comments on risk management for corporate accounts, the CRO will often need to liaise with regulators, investors, rating agencies and other outside parties and provide relevant information on the risks faced and managed by an organisation.

The role of the CRO is a new one for many organisations, and the first CRO for an organisation faces a number of challenges. As well as developing a coherent risk management framework for the organisation, he or she will also need to ensure that the CRF is sufficiently large and skilled to act as required. The CRO will also need to ensure that there is agreement with the board of directors over the scope of the CRO's role, including the authority that the CRO has and the availability of important information.

Given the special nature of the CRF, it is worth considering its relationship with other parts of an organisation. The primary role of the CRF is to control risk. Whilst this role will find it aligned with some parts of an organisation, such as the legal and regulatory compliance teams, the CRF is likely to find itself at odds with parts of an organisation focussed on increasing profit – for the CRF, less risk is better; for profit-focussed teams, more risk is preferable. The extent to which this is a problem depends on the extent to which ERM is integrated into the processes of these teams, and also the extent to which individuals in these teams are rewarded for managing risk. Aligning incentives – whilst recognising that such teams must also be rewarded for taking sensible risks – is the best way to avoid acrimony.

An important role of the CRF is regular contact with all other areas of an organisation, and at all levels. Not only can this help avoid misunderstandings by keeping communication channels open, but it also helps to ensure that all departments are using up-to-date risk management practices. It also makes it less likely that risk management issues will be hidden – deliberately or accidentally – from the CRF. This can be done by embedding the CRF into the various departments of an organisation. There is a risk here that such individuals will become isolated, both from the 'core' CRF by virtue of their location, and from other members of the team due to the potential conflict of interest. One way of alleviating this issue is to ensure that such individuals report to a line manager from the department in which they are based, and as such are at arm's length from the CRF. However, the CRO should also have a say both in the objectives of that individual and in regular performance reviews.

$\wp$ **3.3.6 Pricing teams**

Another group of employees worth discussing is the pricing team. This team is the one to whom pricing of products or services is delegated. It is upon these teams that the profitability and even solvency of the organisation depends. It is important that the reward structure for pricing teams recognises the profitability as well as the volume of business sold, over the long as well as the short term.

Pricing teams within banks are concerned with pricing complex instruments such as CDOs, financial futures and options, and other derivative instruments. The models used to price these instruments are used in a variety of ways. For example, each CDO is made up of several tranches offering different combinations of risk and expected return. A pricing team might be used to determine the levels of exposure in CDO tranches. The models can also feed back into regulatory valuation models.

Pricing for a non-life insurance company – more commonly known as premium rating – covers a wide range of insurance classes from short-tail business, such as household contents and motor insurance, to long-tail business, such as employer liability. The key here is to arrive at a premium which will not only be profitable but which makes the best use of the insurer's capital. This means that the opportunity cost of the business must be modelled – in other words, it is important to determine the profit that would have been made if the available capital had been put to some other use. This modelling involves employing a model office. This is not to say that additional capital cannot be raised. Indeed, capital issuance is desirable if particularly profitable opportunities arise. However, frequent issuance and repayment of capital can be costly.

Pricing for a life insurance company involves similar considerations – although practically all business is long-term – with the additional complication that pricing of with-profits policies must also be carried out. With-profits policies do deserve additional consideration. Such policies provide (generally) low guaranteed rates of return with the potential for higher (but smoothed) returns, subject to investment returns. For policyholders, this means good upside potential with limited downside risk. However, it also means that some investors will receive a return higher than that on the underlying investments, whilst the return for other investors will be lower – there is inter-generational cross-subsidy. For mutual insurance companies, these are the usually the only cross-subsidies (although in extreme circumstances bondholders can suffer if the creditworthiness of the insurer is damaged); however, for shareholders there is limited upside but potential for significant downside. This is because with-profits policyholders would (ultimately) expect to receive the bulk of any strong investment returns, whilst if investment returns were poor, shareholders' funds would be needed to support guaranteed rates of return or previously awarded bonus rates. With-profits pricing, in terms of bonus rates, is so important that it is generally not delegated to pricing teams, with the final decision on bonus rates being made by the board of directors. Having said this, advice will be taken from the firm's with-profits actuary whose role is discussed in more detail later.

The extent to which equityholders obtain value for money is also influenced by the pricing models for both banks and insurance companies, for the former in pricing complex instruments and for the latter in pricing insurance products. If incentives are in place to align the interests of shareholders and those pricing the products, then the pricing teams will also be acting in the interest of equity shareholders.

The roles of the sponsor and trustees in the pricing of pension scheme benefits are discussed above. Responsibility for such a major undertaking generally remains with the directors, but the 'pricing team' to which producing a proposal is often delegated is the human resources team, with the assistance of external actuarial advice.

On the other side of the equation, there are pension contributions that must be paid. The majority of defined benefit pension schemes in the United Kingdom use a 'balance of cost' approach to contributions. This means that the contribution rate for members is fixed in the trust deed and rules and the sponsoring employer must pay the balance. The contributions payable are intended to cover the cost of accruing benefits plus or minus an adjustment for any surplus or deficit. This means that the exact contribution depends not just on the assumptions used in the calculation, but also on the period over which any surplus or deficit is amortised. The decision on the final contribution rate is the trustees, but subject to agreement from the sponsoring employer, making it in effect a joint decision.

In all of these roles, the pension scheme actuary is involved in advising the trustees. The pension scheme sponsor has actuarial advice, often from another advising actuary. These advisory roles are covered later.

3.3.7 Internal auditors

The internal audit function has a key role in the risk management of an organisation. It will normally be focussed on financial risks, ensuring that the possibilities for fraud are minimised, and that fraud is detected if it takes place. It is also responsible for ensuring that payments are paid, received and

accounted for in line with internal procedures. Finally, it might also be responsible for checking other systems in the organisation, or ensuring more general
compliance with internal regulations and statutory requirements. These are
important internal checks on the functioning of an organisation. However, the
most valuable verification is carried out by external parties.

|4 3.3.8 External auditors

One of the ways in which directors can best help shareholders is to ensure
that they receive reliable and timely information. This means that the external
auditing process is of paramount importance. For a bank or an insurance company, the provision of information to shareholders is effectively delegated to
the auditor by the directors since the auditor must approve the accounts even
if they are initially prepared by the firm's directors and employees. The auditor can also be regarded as an agent of the shareholders, acting on their behalf
in ensuring the provision of accurate information. Both the auditor and the
directors might have an incentive to influence the final information. The directors might wish to be portrayed in as good a light as possible and the auditor
will wish to keep his or her appointment. Similarly, the trustees of a pension
scheme, foundation or endowment delegate the provision of information to
an auditor, but the auditor is acting on behalf of the beneficiaries. There is also
the possibility that the auditor might have colleagues trying to sell non-auditing
services to their mutual client, and that these colleagues might also put pressure
on the auditor to sign off accounts in a way that is favourable to the directors.
These potential failings have been addressed in a number of ways, particularly
in relation to companies such as banks and insurance companies.

In relation to pension schemes, it is important to consider what the auditors
audit. Pension scheme accounts ignore a key liability of pension schemes – that
relating to accrued pension benefits. This is rational, as the accounts are more
concerned with the assets and avoiding fraud. Furthermore, there is too much
subjectivity in the valuation of pension scheme liabilities – they are more of an
actuarial than an auditing concern.

|5 3.3.9 Pension scheme administrator

The day-to-day functioning of a pension scheme is managed by pension
scheme administrators. They may be a department within the pension scheme
or an outsourced function. Pension administration involves the payment of
benefits and other outgoings, the collection of contributions and other administrative functions.

There are many aspects of an administrator's function where failures can occur. In many cases, these failures can be costly, not least because of the risk of fines for maladministration.

/6 3.3.10 Investment manager

As discussed above, most financial institutions have relationships with financial markets. However, in the case of many pension schemes, charities and insurance companies investment is outsourced to an external investment manager.

The behaviour of investment managers is dependent on the perceived preferences of their clients, as well as the behavioural biases of the managers themselves. There is a tendency for investment managers and their clients to dislike losses more than they like gains of equal sizes. This can create a tendency for investment managers to track indices, mitigating risk to a greater extent than they seek returns. It is possible to create remuneration structures to avoid this, but it is important to get the balance right – too great a performance-related bonus and there is a risk that the bonus will be regarded as the pay-off from an option with a very low premium, and for too much risk to be taken.

There has also historically been a less-than-clear relationship between investment managers and the brokers with whom they trade. In particular, a system known as 'soft commission' has existed where higher commissions are paid to brokers in exchange for additional goods and services. There is a risk here that investment managers will choose services that do not necessarily benefit their clients, but instead benefit the individual investment manager. The CFA Institute seeks to avoid this behaviour amongst its members with guidance in its Code of Ethics. This limits the uses to which soft commission can be put, in particular specifying that it can be used only to buy goods and services that benefit the client.

The Myners Review in the United Kingdom also discusses commission, stating a belief that broker commissions should be treated as management expenses. This makes it more difficult for fund managers to continue to receive soft commission.

3.4 Controlling

The controlling parties are those with some supervisory role over principals or agents. It is their role to minimise the risks faced by the various parties. The controlling parties considered are grouped as follows:

- professional bodies;
- professional regulators;
- industry bodies;
- industry regulators; and
- governments.

The primary aim of supervisors should be to prevent problems before they occur. However, supervision can include a range of components.

A primary tool is the power of license. This allows a supervisor to limit which individuals or organisations can hold a particular role or operate in a particular industry, and ensures that only those with competence above a certain threshold can operate. It can be done through requiring individuals to take examinations, have particular skills or demonstrate other traits, or by requiring firms to hold particular levels of assets, to have particular systems or processes in place, or adhere to some other minimum standards. Once licensed, those holding the permissions must then continue to follow the rules set by the supervisor. As well as rules prohibiting certain actions, there are also requirements to maintain particular levels of competence. It is also the responsibility of the supervisor to oversee the licensed individuals and firms, and to take action against those who do not comply with the rules set out for them.

This is a broad description of how the controlling function is carried out in practice; however, the exact nature of the relationship will depend on the role of the supervisory bodies, as described below.

3.4.1 Professional bodies

Professional bodies have a key role in managing risk. First, they ensure that their members are trained to a suitable level either through a series of professional examinations and relevant experience or through reciprocal arrangements with professional bodies in other industries or countries. Second, they ensure that members continue to learn through a comprehensive system of continuing professional development (CPD) once they have qualified. In areas where a particular profession has no statutory role, the quality of training and CPD can be used to differentiate a profession's skills; where a statutory role exists, training and CPD can be used to justify the continuation of such a role.

CPD is an important tool for ensuring that skills remain up to date. Ideally, a certain proportion of CPD should be carried out by an organisation other than an individual's employer, and also their profession – in other words it should be external rather than internal. This helps to expose individuals to a wide range of views. CPD can also be active or passive, and active CPD – where an individual contributes to an event rather than simply observing proceedings – is also important to ensure that skills are being developed.

3.4.2 Professional regulators

Whilst these professional bodies administer the qualifications, the standards to which professionals must adhere are frequently determined by outside bodies. These can be regarded as professional regulators. There are three aspects to professional standards:

- setting the standards;
- monitoring adherence to standards; and
- disciplining in cases of non-adherence.

In some cases, all stages of the process are run by the professional organisation. The CFA Institute is one such example – it sets its own Code of Ethics and Standards of Professional Conduct. However, the roles are often performed by an independent body. Broadly speaking, the greater the statutory responsibilities of the profession are, the more likely the regulation of that profession is to be external. This is to ensure that standards are maintained given the privileged position of such a profession. In the context of risk management of financial organisations, two of the most important areas are the actuarial and accounting professions. Within the accounting profession, auditing is itself a special case.

Accounting standards and confidence in the audit process are crucial for the shareholders of banks, insurance companies and all firms, as they ensure that shareholders and debtholders are provided with accurate information on which they can base decisions. They also provide policyholders and account holders with information regarding the security of their investments. All of these interests are also supplemented by additional listing requirements imposed by many stock exchanges. It could be argued that market forces play a role, with the market assessing the information available and arriving at an appropriate price; clearly, the earlier word of warning about market efficiency still holds here. Pension scheme members and those involved with charities are similarly served by the accounts provided to them, these being the main way in which fraud can be avoided.

3.4.3 Industry bodies

Whilst professions play an important part in financial sectors, there are also a number of independent cross-profession organisations that reflect interests in particular industries. One of main purposes of these bodies is to lobby on behalf of their members. This means that there is always a risk that the vested interests served will be those of the members rather than of the industry's customers or shareholders. In this context it is interesting to note that these types of organisations tend to represent firms rather than individuals. This is in contrast to professional organisations, where membership is at an individual level. However, the firms will themselves be represented by individuals, meaning that there is also a risk that the interests of the individual representatives will be high on the agenda.

3.4.4 Industry regulators

Regardless of whether industry bodies exist, industries are themselves often regulated. Again, it is often the firms that are regulated here, but individuals are also subject to codes that must be followed. In the same way that professional regulators control those individuals working in a profession, industry regulators limit what firms and individuals are allowed to do, monitor compliance and take action against those firms and individuals breaking the rules. However, they also act a little like professional bodies in controlling which firms can enter a particular industry and which individuals can hold particular roles in the first place.

Regulation occurs on a national and an international level, with international regulations often being implemented by national regulators. Two broad risk frameworks – Basel II and Solvency II – are discussed later; here the regulators and their remit are considered.

A key difference between many regulatory structures is the division of responsibilities between different regulators. At one extreme, different authorities might oversee the activities of banks, insurance companies, pension schemes and charities, a system known as functional regulation. The system in the United Kingdom used to be similar to this, with the Financial Intermediaries, Managers and Brokers Regulatory Association (FIMBRA), the Investment Management Regulatory Organisation (IMRO), the Life Assurance and Unit Trust Regulatory Organisation (LAUTRO) and the Securities and Futures Authority (SFA) all existing as a result of the Financial Services Act 1986.

At the other extreme, a single regulator may be used for all financial industries, a system known as unified regulation. This is the case in Australia with the Australian Prudential Regulation Authority (APRA), which regulates banks, credit unions, building societies, life and non-life insurance and reinsurance (including friendly societies) and most of the pension industry. This latter approach has clear advantages. For a start it makes the regulation of financial conglomerates, which might otherwise require regulation by a number of parties, much easier. It avoids conflicting approaches being taken in these cases, and ensures consistency between different firms operating in different industries. If properly arranged, it can also limit the incentives for regulatory arbitrage and can provide a good environment for the cross-subsidy of ideas between staff working in different areas. This approach should also improve accountability, since there should be less chance of disagreement over who has authority over a particular issue. However, for this all to be true, it is essential that the different departments within a single regulator do not simply act as independent regulators.

Unified regulation can also be more efficient, but not necessarily so – larger organisations can give rise to additional bureaucracy and dis-economies of scale. This suggests that the most appropriate form of regulation depends on the country in question – in particular, it depends on the extent to which there are large financial conglomerates operating with complex regulatory needs.

Regulators tend to spend more time on institutions where the risk is higher. This might be because a firm has had past risk management failures, or has lower than average resources either in terms of assets or in terms of systems and processes. However, higher risks do not occur only in 'bad' firms. For example, larger, more complex organisations pose a higher risk by their very nature. Also, firms operating in complex areas, or entering areas that are new for those firms, face increased levels of risk so require greater regulatory oversight.

There are a number of aspects of an organisation that a regulator likes to understand. At a strategic level is concern with a firm's overall business plan, taking particular interest if it involves movement into a new area or continued operation in areas where problems appear to be developing across the industry. There is also interest in the nature and standard of corporate governance and the risk management processes in place. Finally, is interest in the financial situation of the firm.

There are, in fact, three ways in which an organisation is likely to interact with a regulator. The first is on a procedural level. This involves regular interactions in relation to any statutory reporting or other ordinary dealings between the regulator and the organisation. There are also non-standard interactions in response to the development of new products, entry into new markets, changes to key employees or in the event of problems arising. Finally, there are less frequent strategic interactions which take place between the regulator and senior members of the organisations, and are designed to give the supervisor an idea of the overall direction of organisation.

7   It is desirable for institutions to work with regulators rather than against them. This can lead to a better understanding by the regulator of the work carried out by the institution, leading to greater trust and less risk of intervention.

To do this, organisations should recognise the regulator's objectives. These generally involve protecting institutional and retail customers by ensuring that they are not sold inappropriate products, and by avoiding individual insolvencies and the failure of the system as a whole.

Financial institutions should ensure that they actively engage with their regulators – waiting for a regulator to intervene will at best give the impression that regulation is not being taken seriously, and at worst will give the impression that an organisation has something to hide. This means that transparency is particularly important. This transparency extends to regulatory breaches, which are bound to occur. The boards of regulated entities should do all they can to ensure that relationships with regulators are entered into in the appropriate spirit. They should also ensure that they are kept fully informed of communications with regulators, and of potential regulatory issues.

The issues of transparency also extends to access to the entity being regulated. The regulatory process should include visits to the firm's sites, so the regulator can gain a clearer idea of how risk is managed. Such site visits can also be more practical if the regulator needs to meet large number of individuals at a firm, which he or she will want to do from time to time. Site visits can also allow for the demonstration of commercially sensitive systems in a secure environment. Regulators should also recognise commercial sensitivities inherent in viewing the inner workings of the firms regulated, and ensure absolute discretion in this matter. Having said this, information may well need to be shared between regulators. This is particularly relevant when multinational organisations are being regulated.

Relationships with regulators can be enhanced if there is a continuity of personnel on both sides. This can help to develop trust, and can ensure that problems are dealt with swiftly, perhaps with informal advice being sought at an early stage.

3.4.5 Governments (controlling relationships)

The financial relationships of governments have already been discussed, but governments also have a number of controlling relationships. Many of these are delegated to regulators for implementation on a day-to-day basis, but governments still intervene directly in some ways.

The clearest form of intervention is through legislation. This includes the legislation that establishes regulators, but also that which deals with policyholder, investor and customer protection, solvency and other issues. The government setting this legislation can be a national or supranational institution. For example, the United Kingdom government has enacted legislation and regulation in relation to financial services, such as the Financial Services and Markets Act 2000 or the Pensions Act 2004, but the European Union has also set in place rules through various directives, such as those comprising the Solvency I framework. Supranational legislation is generally implemented through national legislation in each of the affected countries.

The challenge for any government is to have enough rules to provide adequate protection for investors in firms and customers of them, but not so much that the cost is excessive. Costly legislation cannot only be uneconomical, costing more than the level of protection afforded, but it can also lead firms to base themselves in other countries, thus depriving a country of jobs and a government of taxation revenue.

There are a number of ways in which protection can be implemented, some common ways being:

- requirements to provide information;
- restrictions on insider trading;
- restrictions on the establishment of firms;
- quantitative requirements on the capital adequacy of firms;
- qualitative requirements on the management, systems and processes of firms;
- establishment of industry-wide insurance schemes; and
- intervention in the management or ownership of firms.

These are discussed in more detail below.

The most basic protection that governments or their agencies can provide is to require firms to provide minimum levels of information to customers and policyholders on the one hand, and to investors on the other. This should indicate to these parties how safe their savings, policies or investments are likely to be. There are also frequently restrictions on insider trading to avoid external investors from making investment decisions without the benefit of information available to some internal investors.

Legislation can also limit the establishment of financial institutions in the first place, requiring certain minimum requirements to be met. The same requirements usually exist as the firm continues, and these fall into two categories: quantitative and qualitative. Quantitative requirements relate to the amount and type of capital a firm holds to ensure that it can withstand financial shocks; qualitative requirements relate to the systems and processes that a firm has in place, but also the quality of the directors, management and staff.

This will generally mean that directors need to be of good character according to some definition, and that directors and staff in certain positions might be required to hold particular professional qualifications.

These measures are intended to ensure that financial institutions remain solvent, however measures can be put in place to protect customers, policyholders and investors if they do not. Insurance schemes can be set up to compensate individuals who lose money due to insolvency, or governments can intervene directly through the provision of capital or even full privatisation in order to prevent insolvency occurring in the first place.

3.5 Advisory

As well as those that can directly affect or are directly affected by financial institutions, there are a number of parties acting in an advisory capacity. Although these advisors do not have any statutory right to their roles, they are often subject to statutory requirements controlling the way in which they act. They also have the same incentives to act for themselves as any other party.

There are many different types of advisor, and many ways in which they can be grouped – they are given here by function:

- actuarial;
- investment and finance;
- legal; and
- credit.

3.5.1 Actuarial advisers

Actuaries hold advisory roles in a number of areas. These roles include giving advice on a range of issues, but it is the purely actuarial ones that are discussed here. There are two main groups of institutions that actuaries give advice in relation to, the first of which consists of pension schemes. The obvious clients here are pension scheme trustees, who require advice on scheme valuation, funding and modification. However, whilst pension scheme trustees require actuarial advice, so do pension scheme sponsors – it is, after all, the sponsor who funds pension schemes.

If the actuary advising the trustees also advises the sponsor, then there is likely to be a clear conflict of interests. The risk here is that the scheme's actuary will favour one party to the detriment of the other. Almost as bad is the risk that the advice given will be acceptable to both but suitable for neither. A further issue for actuaries giving advice is that there is an incentive for the actuary to secure his or her position by giving an answer that the client wants – in other words, there is a risk that actuaries could compete on the basis of acquiescence.

The second type of institution is a life insurance company. Actuaries employed by such firms might face a range of conflicts. They typically report to the board of directors and have the aim of maximising shareholder profits. However, they also have statutory responsibilities, as well as responsibilities to policyholders. This is a particular issue where with-profits policyholders are concerned, since their interests conflict directly with those of the shareholders.

3.5.2 Investment and financial advisers

There are two broad categories of advisor in this category: institutional and individual. When considering institutions such as pension schemes, institutional investment consultants advise on a range of investment-related matters, principally the investment strategy and the choice of investment managers. The investment strategy should be determined in relation to the liabilities that the investments are intended to cover, so for pension schemes this aspect of investment consultancy requires actuarial skills. This has tended to mean that investment consulting and actuarial appointments for pension schemes in the United Kingdom are with the same consultant. There is a risk that this can lead to a lack of competition, giving an advantage to large consultancies offering a full range of actuarial and investment services.

The investment strategy decision is generally the most important investment decision taken, as the choice between asset classes has the greatest impact on the returns achieved. The other important aspect of the investment consultant's role is the choice of investment manager. This involves first deciding whether to use active or passive management. If active management is to be used, then the level of risk taken by the investment manager must be considered, either in absolute terms or relative to some benchmark. The incentives of the investment manager, discussed earlier, should be borne in mind.

3.5.3 Legal advisers

Legal advisers form an important category, since they help to mitigate the risk that an institution will find itself on the wrong side of the law. This can be at a very high level, where advice is received on issues relating to a merger of two insurance companies or on the change to the benefit structure of a pension scheme, or at a much lower level such as the discretionary payment of a partic-ular benefit from a pension scheme. If there is any doubt at all as to whether a proposed cause of action will lead to any legal issues, it is important to obtain legal advice.

4 3.5.4 Credit rating agencies

Credit rating agencies provide ratings on debt issues and issuers that are intended to give a broad view on creditworthiness. For most companies there are two types of credit rating: issuer and issue. The issuer rating gives a view of the overall credit risk in relation to an entity as a whole, whilst the issue rating takes into account any particular factors associated with a specific tranche of borrowing. However, banks also receive ratings on the security of their deposits through bank deposit ratings, and insurance companies on the security of the products they sell through insurance financial strength ratings. There are also bank financial strength ratings that consider the likelihood that they will require external support. Rating agencies also rate credit derivatives, hedge funds, supranational organisations and even countries.

Issue ratings differ from issuer ratings by taking into account the terms of each debt issue and its location within a corporate structure. This means allowing for features such as collateralisation (the funds or assets notionally supporting the issue), subordination (where holders of this bond issue would be in the list of creditors) and the presence of any options. The agencies generally use a combination of 'hard' accounting data and 'soft' assessments of factors such as management quality and market position to arrive at forward-looking assessments of creditworthiness, although some use methods based on leverage and the volatility of quoted equity.

Credit ratings are long-term assessments, considering the position of an entity over an economic cycle. This means that, whilst the risk for each firm will change over the economic cycle, the credit rating may well not. An issuer may in fact have a number of different credit ratings. Short- and long-term ratings may differ, and varying levels and type of collateralisation invite different credit ratings.

A conflict of interest exists with credit rating agencies to the extent that such agencies are hired and paid by firms in order to allow those firms to borrow more cheaply. One would hope that competition between rating agencies would be for credibility rather than for favourable ratings.

It could also be argued that credit rating agencies who – for large issues of traded debt at least – monitor the creditworthiness of the issuer, are a source of advice for investors. The purpose of a credit rating is to allow a firm to borrow funds at a more competitive rate of interest, and it is the firms themselves who pay for the credit ratings; however, in order to maintain credibility with debt investors – so that a credit rating is seen as reflective of the creditworthiness of the borrower and therefore worth paying for – a degree of accuracy in the rating process is required, and thus rating agencies are also acting on behalf of debtholders (whether they want to or not).

The interests of bank depositors and insurance company policyholders are also partly served by rating agencies. The assessments of rating agencies are also a key source of information for institutions choosing between banks as counter-parties for derivative transactions. However, in the United Kindgom individuals with bank deposits are more obviously served by the Financial Services Authority (FSA). Insurance company debtholders again use credit rating agencies, but both non- and with-profits policyholders also rely on the FSA, regardless of the extent to which they may be regarded as providers of equity capital. Pension scheme members are also limited users of credit rating agencies, despite the fact that to a greater or lesser extent they are often subject to the creditworthiness of the sponsor; instead, pension scheme members rely on the Pension Schemes Regulator, but on a more practical level, their scheme's actuary for security.

It must be recognised, though, that any benefit that investors and customers receive from credit ratings is purely a by-product from their main purpose, which is to facilitate the sale of debt. Investors in most rated firms will also carry out their own analysis rather than relying on the credit rating. Also, holders of unquoted debt, or smaller quoted issues do not have the benefit of rating agency analysis and so must rely on their own calculations. Credit rating methodologies are discussed later.

3.6 Incidental

Finally, there are those parties that are affected incidentally by the behaviour of financial institutions. These can be categorised as:

- trade creditors;
- subcontractors and suppliers;
- general public; and
- the media.

3.6.1 Trade creditors

Trade creditors are at risk of failure of a financial institution to the same extent that financial creditors such as debtholders are. They therefore have similar desires regarding the risk taking of a financial institution, but generally with less power.

Trade debtors might also exist if a financial institution is owed money from a customer, but such instances are rare except in the case of insurance companies who often provide cover for business taken on through brokers before the premiums are received.

3.6.2 Subcontractors and suppliers

These parties exist as trade creditors, but are also subject to the risk that future income will fall if a financial institution fails. For this reason, trade creditors might choose to withhold goods or services if the risk of failure of the institution increases significantly.

However, subcontractors and suppliers themselves pose a risk to financial institutions if they fail. Replacement can be costly and time consuming, and many risks might be presence in the period of time it takes to put a replacement in place.

3.6.3 General public

As well as having an interest in financial institutions as customers, policyholders and members, members of the general public are also involved in more subtle ways. They are potential future customers, policyholders and members, either through explicit purchase or by virtue of being related to someone currently associated with a financial institution. This means that financial institutions should be aware of potential as well as current stakeholders.

Members of the general public are also usually taxpayers, and so are aligned in this way with governments' roles as recipients of tax from financial institutions. Furthermore, if people do not agree with a government's approach, then they can act through their roles as voters to change the government (assuming that the government is democratically elected).

3.6.4 The media

The media are responsible for communicating information to the general public and also to people in their roles within financial institutions. The media operate through newspapers, television and online. Some information is available to the general public, either freely or for a fee, whereas some is available only to certain groups, for example members of a profession. The cost of some media services can also restrict its availability. This is particularly true for some financial data available from some data providers.

The media are important as they can ensure the prompt and wide dissemination of factual information, helping to ensure the efficient functioning of markets. However, the tone of reporting can affect the impact that a story has.

This is particularly important when news on financial markets or individual firms is being transmitted; however, there is an incentive for journalists to make news as newsworthy as possible, which can lead to volatility, particularly in the short term.

3.7 Further reading

Stakeholders are discussed in other ERM books such as Lam (2003) and Chapman (2006), but there are few books that concentrate exclusively on these issues. Some of the best sources of information are papers written on situations where stakeholder actions are important. For example Jensen and Meckling (1976) wrote a pivotal paper on the role that company ownership structure had on the incentives of various stakeholders, in particular when an owner-manager sells shares in his firm. Jensen (1986) also wrote on the way in which debt issuance can be used to limit the extent to which managers used funds for their own purposes rather than for the benefit of shareholders.

4

The internal environment

4.1 Introduction

The nature of an organisation is important to the risk management context. However, there is no such thing as a simple, featureless institution, nor do any operate in a vacuum. The nature of each organisation and what surrounds it influences its operation fundamentally.

Understanding the internal environment is crucial for understanding the way in which risk management should be approached. An analysis of the various aspects of an organisation's internal risk environment helps risk managers within an organisation to appreciate what they need to do to carry out their roles effectively. It also helps external analysts to determine the risks that an organisation is taking – even if the organisation itself does not appreciate these risks.

4.2 Internal stakeholders

The only internal stakeholders with a principal relationship with an organisation are owner-managers – all other internal stakeholders are agents, acting on behalf of the an organisation's shareholders, customers, clients and so on. Their views of risk form an important aspect of the risk management environment, and they are discussed together with external stakeholders in the next chapter. However, as well as their individual views of risk, the ways in which they interact are an important determinant of the ways in which organisations behave. At the head of a firm, this means the board of directors. This group includes executive directors who have a day-to-day role in managing the firm and who are led by the chief executive, and non-executives who are are responsible for representing the interests of shareholders. The board of directors is led by the chairman.

54

The executive directors delegate much of the running of the firm to managers, and ultimately to employees. Depending on the industry, the employees may be represented to a greater or lesser extent by trade unions. This, too, will affect the internal environment of the firm.

There are also issues for pension schemes through the structure of trustee bodies. The inclusion of member-nominated trustees can lead to a better reflection of the interests of members, whilst trustee boards dominated by employer-nominated trustees can at times give too great an emphasis to the interests of the sponsor. Using independent trustees can add valuable expertise to the trustee group. The trustees of endowments and foundations are similarly affected.

✓ 4.3 Culture

Culture is something that is present in all organisations; however, its impact is felt differently by different types of organisation. For banks and insurance companies, culture is likely to be something felt from board level all the way down through the firm; in a pension scheme, foundation or endowment, it is likely to affect only the board of trustees.

For banks and insurers, the board of directors influences the culture of the firm both directly and indirectly. It is important that this culture puts risk management at its core. At its most fundamental level, it includes the willingness of an organisation to embrace ERM, and it is determined by the board of the organisation. This is partly reflected in the structures that the board puts in place, as discussed later; however, culture is also reflected in more subtle ways.

A board should make sure that risk is considered in all stages and at all levels of the organisation; however, it is should also consider the way in which the members of an organisation relate to one another. An overbearing chairman, or a culture in which the views of non-executives are not given as much weight as those of executive directors can lead to a form of blindness in relation to developing risks. There should be a culture of openness, encouraging dialogues not only between all members of the board, but also between all levels of the firm. This requires good internal communications, and can be characterised by the involvement of all levels when decisions on risk are made, and a willingness of board members and managers to encourage input from those that report to them. Good communication also means that the CRF becomes aware of the emergence of new risks promptly, as well as ideas for mitigating these risks and updating existing systems. In addition, it means the prompt transfer of knowledge from the CRF to the rest of the organisation.

Openness also means an openness to new ideas and a commitment to learning and integrity. Boards should recognise the importance of relevant professional qualifications and the investment in CPD. This is important as the standards set by professional organisations and the requirements they place on their members can ensure that risk management is taken seriously. Both should be encouraged, and the lessons learned should be shared throughout the organisation.

It is also important that the culture is one that allows people to learn from their mistakes – there should be accountability for actions, but not blame. This too is important, as a culture of blame can encourage mistakes to be hidden and, possibly repeated, when instead lessons could be learned.

These ideas reflect the features of a good risk management culture, but also possibly actively affect the culture of an organisation. In relation to CPD and education, time and money can be made available to employees to maintain and develop skills. It is possible to go even further and to require employees to take these opportunities, or to engage in other risk management-related training. Statements on risk management can also be incorporated into job descriptions and performance management indicators, so that employees' remuneration and promotion prospects depend on working in the context of a sound risk management framework.

For this to happen, it is important that specific risk management responsibilities are well defined. It is equally important that individuals know who to turn to with risks that are outside their area of expertise – and that they are commended for passing on information on such risks.

One way of fostering a good risk management culture is to praise people who manage risk well. It is often the case that risk management is only heard about when there are failures, but it is important to recognise the importance of low-key actions that prevent the development of serious risks within an organisation.

Changing a firm's culture is difficult – if people with radically different outlooks are recruited, then they might become frustrated as existing employees grow resentful. However, recruiting people just because they fit in with the existing culture is not necessarily a good thing if the culture should change. Culture can usually change only incrementally, with the views of existing staff changing as the profile of new recruits also changes. As mentioned at the start of this section, it can also change only from the top of an organisation

When changes are made to the management of risk in an organisation, it is important to assess the extent to which the culture is being changed. This can be done through surveys or as part of employees' appraisals on an ongoing basis.

These are all aspects of risk culture that are essentially part of the fabric of an organisation. A related area is the level of risk that the organisation decides to take – in other words, its risk tolerance. This is distinct from its risk capacity, which is how much risk it can actually take, although the combination of risk tolerance and capacity determine the overall risk appetite. The risk tolerance is determined by the level of risk-adjusted return available and also the access to additional capital if it is required.

Many of the aspects described above relate equally to pension schemes, foundations and endowments, particularly the larger ones. However, for many other organisations it is important to recognise that important functions such as fund management and administration are likely to be outsourced. This means that it is important for the trustees to ensure that the cultures of the organisations to which work is being outsourced have a risk management culture that is of a sufficiently high standard.

4.4 Structure

The issue of structure covers a number of aspects of organisations. It relates to the components of the organisation, the way in which they are constituted and the way that they interact. Many aspects relating to structure are, in fact, reflections of the culture of an organisation. However, because they are so important to how an organisation performs, these factors are worth discussing separately.

Many structural aspects of the internal risk management environment relate to the structure and activities of the board of directors. At the highest level, these are about merits of the division of responsibilities between the chief executive and the chairman. A commonly held view is that there is an agency risk in having the same individual running the firm and looking after the shareholders. Furthermore, separating the roles of chairman and chief executive should ensure that the latter's effectiveness is subject to greater scrutiny. However, it could also be argued that there is merit in combining the roles of those responsible for a firm's strategic direction and the implementation of that strategy. In any event, the final decision will have a major impact on the way in which the company is run.

The executive roles meriting appointment to board level vary from industry to industry and from firm to firm. The presence or absence of a particular role at board level can be used to infer the importance that a firm places on that role. One key role which is finding its way onto more boards is that of chief risk officer, the individual responsible for a firm's CRF and the overall risk management of the organisation. The presence of this role at board level should mean that a firm has a strong commitment to risk management.

5   The degree of representation by non-executive directors is also important, and there should be sufficient to ensure that there is an adequate critique of executive directors by individuals acting on behalf of shareholders. However, employing too many might make a board too cumbersome. There is also the risk that non-executives will be as subject to agency risks as their executive counterparts. It is worth noting that non-executive directors are not necessarily independent of the firm, particularly if they have moved into their non-executive roles from previous executive positions with the firm. It is important to recognise such potential conflicts of interest, and to ensure a sufficiently high level of independence on the board.

The non-executive directors effectively form a committee of the board, and within this committee there is a further sub-committee made up of independent directors. It is important to recognise this, as these groups should meet in addition to the full board meetings, in particular to discuss any concerns that they might have. However, there should also be more formal committees to oversee important board- and company-related issues, in particular:

- audit, looking at the provision of accurate information to internal and external stakeholders;
- risk, looking at the level of risk the firm is taking and setting desired levels of risk, considering large- and small-scale issues;
- appointments, looking at the appointment of board members and senior executives, as well as the terms of appointment; and
- remuneration, looking at the remuneration of board members and senior executives.

As with the board as a whole, there should be sufficient representation of non-executive directors, including independents, on these committees. In some cases – in particular, audit, appointments and remuneration – this might mean a complete absence of executive directors. This is because the work of these committees concerns the performance of executive directors, so it is important that such committees can adequately assess the issues before them without the interference of executive directors.

6   The risk committee is different from the other three committees in that independence is less important than a good knowledge of the organisation, although it is important to have non-executive membership to ensure that performance is measured objectively. This committee, which should be chaired by the chief risk officer (CRO) – who should, therefore, be a member of the board of directors – is responsible for the strategic oversight of the firm's risk management. This includes setting policy, but also considering information

received on the risks faced and assessing the treatment of risks by the CRF. The CRO is also responsible, on the board's behalf, for implementing an ERM framework throughout the firm, reporting on compliance with the objectives set out in this framework – including regulatory requirements – and for preparing reports on this subject for the board as a whole.

In order to be effective, these committees must meet regularly. This is particularly true for the audit and risk committees, which have an ongoing role. The frequency of meetings and the constitution of all committees should be included in their terms of reference. It is also important that all committees have clear guidance on how their performance will be assessed, and on the resources to which they will have access.

These areas are dealt with in different countries by legislation, and by a number of reports into corporate governance together with subsequent codes of practice. As these form part of the external risk management environment within which a firm operates, they are dealt with in that chapter.

The structure of the firm itself is also crucial. Having too many departments can lead to a lack of clarity over responsibilities for various functions; too few can mean it is difficult to find the party responsible for particular issues within a department. The structures in place for obtaining approvals for everything from expenses to initiating new projects are also important. These should be rigorous enough to satisfy risk management needs, but also smooth enough to avoid paralysis from excessive bureaucracy.

The interaction between departments and the CRF is also key. For ERM to be effectively implemented in a firm, it is essential that it is used at all levels, and that information can be conveyed quickly and easily from the board to the 'shop floor' – and in the opposite direction. This is a matter of the culture of the firm, but also of the structure – this should be such that communication can take place without messages being lost in the ether.

Whilst the CRF has a key function as the second line of defence, the third line of defence – audit – also merits special discussion. The responsibilities of the audit function will differ from organisation to organisation. As discussed above, they will normally be focussed on financial risks, ensuring that the possibilities for fraud are minimised, and that fraud is detected if it takes place. This means that systems need to be developed to ensure that this is possible. The audit function is also responsible for ensuring that payments are paid, received and accounted for in line with internal procedures. It might also be responsible for checking other systems in the organisation, or ensuring more general compliance with internal regulations and statutory requirements. There is therefore a possibility of an overlap with the CRF. This means that it is important not only to ensure that there is no duplication, but also to guard against duties being missed by both functions.

As mentioned whilst discussing the culture of pension schemes, foundations and endowments, many functions are typically outsourced. This means that the structures of these organisations are not under the control of trustees; however, it also means that the structure of a firm to which business is outsourced should be investigated thoroughly since this will affect the ability of that firm to deliver positive results.

Having said this, there is no reason that trustee bodies should not have many of the same committees as company boards in relation to risk, audit and appointments. Trustees are not typically paid, so there is no need for a remuneration committee; however one additional committee that it is helpful to have is an investment committee. This should consider both the long-term investment strategy of the organisation and the selection of fund managers to implement the chosen policy. It should also monitor both the appropriateness of the strategy and the performance of the fund managers.

4.5 Capabilities

Having risk-aware cultures and structures are high aims – but they will remain only aims if the organisation does not have the capabilities to implement them.

There are many different dimensions to the capabilities of an organisation, but the most crucial are the people. These people should be sufficiently well qualified to fulfil their roles, with opportunities to develop and to change roles as they grow in skills and experience. Conversely, there is little point in implementing a structure that is the last word in risk management but cannot be implemented by the staff currently employed.

Even if the staff are capable, they will be unable to perform to the best of their abilities if the infrastructure – in particular relating to information technology – is inadequate. Furthermore, all of this must sit within processes designed to provide a good risk management environment.

This all means that sufficient monetary resources must be devoted to allow risk management to be properly implemented. However, money alone is not the answer, and good planning together with clear insight can be even more valuable.

4.6 Further reading

Whilst Chapman (2006) and other risk management books include useful content on the internal environment, the advisory risk management frameworks discussed in Chapter 19 offer some of the best insights.

5

The external environment

5.1 Introduction

The external risk management environment refers to everything that can affect
the risks faced by an institution and the way those risks are managed. These
factors are not uniform, and vary by industry and geographical location. Even
within a particular industry in a particular country, different types of firms
might find themselves in different environments. Small firms might be treated
differently from large ones, and privately held ones will certainly be treated
differently from publicly quoted ones. The list of potential firm-specific factors
is extensive – but the important point here is that it is not sufficient simply to
look at the industry and location and decide that all firms will be treated the
same; rather, it is important every time to consider the nature of the firm and
how this affects the external context.

5.2 External stakeholders

Since it was established in the previous chapter that the number of internal
stakeholders was small, it follows that the number of external stakeholders
that might exist is large. All principals except the owner-managers are external to the institutions. This means that the other holders of bank and insurance
company debt and equity are external, as are pension scheme sponsors; all
customers, policyholders, pensioners and other beneficiaries are external; and
clearly the government, the markets and any statutory insurance arrangements
are external.

By contrast, the agents are generally the insiders. This is particularly true for
banks and insurance companies, where only trade unions and external auditors
can be considered external; however, for pension schemes, foundations and
endowments, where more facilities are likely to be outsourced, then functions

such as investment management and benefit administration are also frequently external.

Professional and industry bodies and regulators are also external to the organisations considered here, and both have an important impact on the environment in which they operate. In particular, professional bodies and regulators have an impact on the way in which individuals within organisations must behave, whereas industry bodies and regulators influence the way in which the organisations themselves act.

Advisers to financial organisations also contribute to the environment in which those organisations operate. To a large extent, this is through the context of the regulatory and professional regime in place; however, it can also be more broadly about the way in which various types of advisers have developed in a particular region or industry, or in relation to particular types of firm.

Those with incidental relationships generally have little effect on the external environment, except in times of crisis. Then, the general public and the media can strongly influence the way firms behave, both directly through widespread negative reporting, and indirectly through the perceived effect on votes, translated into an effect on regulation and legislation.

5.3 Political environment

This leads neatly into the discussion of the political environment in which firms operate. There are two aspects to this type area. The first is the broad underlying environment. For example, to what extent is a firm operating in a free market environment, and to what extent is there government control and regulation? Is there a culture of redistribution of wealth, as seen through systematically high taxes and government spending? How great is the requirement for disclosure, in relation both to the organisation and to its customers, policyholders or members? These factors can affect the very attractiveness of operating in an industry in a particular country; at the very least, they can affect the target market for customers.

The second aspect of the political environment that is of interest is the political climate, which can change over time. As discussed above in relation to stakeholders, public and media sentiment can turn against particular institutions. This frequently affects the political climate and can lead to stricter regulation, higher taxes or other restrictions on organisations.

5.4 Economic environment

In this context, the economic environment refers to the point in the economic or business cycle rather than any capitalist/socialist comparison – this is discussed

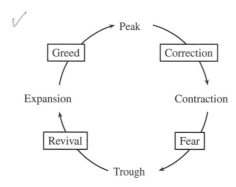

Figure 5.1 The business cycle

under the political environment. There are a number of depictions of business cycles, as well as time scales, with the longest being over fifty years; however, the business cycle of interest here has a span of around a decade and is characterised by periods of expansion and contraction in gross domestic product (GDP), with associated peaks and troughs. As shown in Figure 5.1, expansion can include both a recovery and a 'greed' phase, the former being a return from a trough to some measure of equilibrium and the latter being a continued expansion beyond that point. Similarly, contraction can include both a correction and a 'fear' phase, the former being a return from the peak to some measure of equilibrium and the latter being a continued contraction beyond that point. There are a number of events that can trigger a move from one phase to another – low interest rates and easy credit can cause expansion, whilst the opposite can cause contraction: e.g. catastrophes, stock market shocks. Over-reaction in both directions is a key feature of these cycles, a factor that is particularly clear in financial markets. However, these cycles do not necessarily follow any regular pattern, nor do the phases necessarily follow each other sequentially – a partial recovery might be followed by a further episode of fear rather than full recovery and greed.

The economic environment affects all firms, including non-financial organisations. When the economy is in recession, sales are likely to suffer, and raising capital is likely to be harder. Financial institutions are also affected in a number of ways.

The state of the economy has an impact on the returns achieved on investments. Consider, for example, the effect of an economic downturn on a bank. In a recession, equities are likely to perform poorly whereas bonds will perform well as long-term interest rates fall. Rates of default on loans will increase, and the level of savings may fall – increasing unemployment might mean account holders need to access their savings, and pensioners might need to use their savings to offset falls in returns from other asset classes. Counter-party risk on over-the-counter (OTC) derivatives will increase, as will the requirement for collateral from those counter-parties.

Insurance companies' assets will be similarly affected. However, in addition their long-term liabilities will rise as discount rates fall. There will also be an increase in the level of claims in many insurance classes, partly as a result of redundancies, but also as fraudulent claims increase. This can lead to stricter claim-handling procedures, since the effort put into reducing fraud should be consistent with the amount of money that is likely to be saved. Rising unemployment might also lead to higher lapse rates on policies, which can lead to individual policies losing money if the lapse occurs before the initial costs in setting up the policy have been recouped. The increase in claims and lapses can lead to a fall in profits for insurance companies, although this can be mitigated slightly by a fall in claim inflation (on a per claim basis).

For pension schemes, assets and liabilities will be broadly affected in a similar way to insurance companies. However, rising unemployment might have one of two effects on the liabilities – redundancies might cause a fall in liabilities as individuals move from being active members to being deferred pensioners, but they might cause a rise if people are instead offered early retirement on beneficial terms. A key issue for pension schemes is also the financial health of the sponsoring employer, since sponsor insolvency is more likely in a recession – at the same time any deficit in the pension scheme increases.

The impacts of differing economic climates on the health of a financial institution are clearly important. It should also be clear that it is important to consider the effects in a consistent manner as firms are affected in many different ways. This consistency is an important part of ERM. Considering various economic scenarios can also provide a good basis for arriving at stress-testing scenarios when analysing potential future outcomes for an organisation.

5.5 Social and cultural environment

The social and cultural environment of a country or industry determine a huge range of softer issues, such as the extent to which business is carried out on trust rather than through contract, the importance of inter-personal relationships and the degree to which social hierarchies exist. This final point can be particularly important in a risk management context since strongly hierarchical systems, where there is a deeply ingrained culture of respect for superiors, can mean that bad decisions go unchallenged.

5.6 Competitive environment

The level of competition can, like the political environment, be considered in two distinct ways. The first is the underlying level of competition in the industry and country. For example, occupational pension schemes do not face a great degree of competition; insurance companies and banks generally do. The underlying level of competition can be affected by factors such as the size and power of market participants, particularly in banking and insurance where there can be significant economies of scale. Do dominant companies limit the ability for smaller firms to enter a particular market? Furthermore, to what extent is regulation in place to avoid such dominance, in the form of competition authorities?

A second aspect of competition is the extent to which competition changes through the economic cycle. For banks, this change – as seen through the availability of loans and mortgages to clients – follows the economic cycle closely. This is because changes in credit risk through the cycle, as well as the financial strength of banks, lead banks to compete more in growth years and less in recessions. For insurance companies, there is a separate cycle, known as the underwriting cycle. In fact, different cycles are typically seen for different classes. These too can follow the movement of the economy as a whole, with rates affected by issues such as recession-driven fraud. However, changes in the cycle (usually in terms of a fall in profits) can also be triggered by falls in other markets, such as the housing market or the stock market, or class-specific catastrophes.

The starting phase of the underwriting cycle (although since it is a cycle, it could just as easily be an intermediate phase) is the situation where premium rates are high, profits are high and competition is limited. This situation is unsustainable as additional capital is attracted to the prospect of high profits. This leads to premium rates falling, until rates fall below profitable levels. Eventually, losses become so high that some participants retreat from the market, perhaps as a result of a catastrophe or some other spike in claims. This leaves a small number of competitors who are prepared to face these continued losses. However, with the number of competitors reduced, rates are able to rise again, ultimately to the level where good profits are being made again – resulting in the cycle being completed. This is shown below in Figure 5.2.

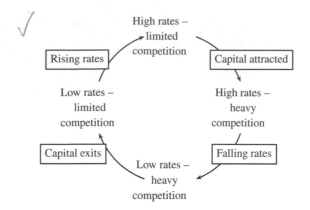

Figure 5.2 The underwriting cycle

The change in capital available can occur so easily because barriers to entry are so low, particularly for participants such as Lloyd's syndicates. This suggests that all insurers would want to exit the market before significant losses were incurred – but they do not. There are several reasons why this might be the case. Some insurers prefer to maintain long-term relationships with clients, so will write business even when it might be temporarily loss-making. Other insurers – particularly larger ones – might find it difficult to deal with the changes in staff numbers required. Cutting costs as profits fell could mean making people redundant. This can be costly and damaging to the morale of those not made redundant. Expanding again when profitability returns means recruiting skilled staff. It might be difficult to find enough candidates in time to allow for the planned increase in business, particularly if a firm has a reputation for making people redundant at the first sign of trouble. If systems also need to be scaled up, then it might be dangerous to expand in advance of this as a loss of goodwill due to poor administration could damage the prospects for new business for years to come. A more mundane reason that firms might not exit the market in a downturn is that they might not realise that the premiums they are charging are too low.

5.7 Regulatory environment

The range of regulatory restrictions on financial firms is extensive. Some of these restrictions are in the form of coherent risk frameworks, in particular Basel II and Solvency II. These are discussed in Chapter 19; however, in this section a number of specific regulatory and legislative issues are considered.

5.7.1 Public shareholders

Public shareholders are affected by legislation to a significant degree. On the one hand, they are offered a degree of protection. For example, in the United Kingdom the Financial Services Act 1986 introduced the Investors Compensation Scheme (ICS), aimed at individual shareholders, which paid the first £30,000 and 90% of the next £20,000 of any loss arising from negligence, theft or fraud. This scheme was taken over by the Financial Services Compensation Scheme (FSCS) following the introduction of the Financial Services and Markets Act 2000, which guarantees all of the first £50,000 lost. However, shareholders are also subject to a number of restrictions, generally to protect other shareholders through promoting the efficiency of markets. The two most important relate to insider trading and to market manipulation, both of which are risks faced by innocent market participants.

Insider trading is the act of buying or selling securities on the basis of knowledge that is not publicly available, whereas market manipulation is the act of generating a false or misleading market in a security or derivative, or otherwise influencing its price. This has been a criminal offence in the United Kingdom only since 1980, with the Companies Act of that year. The provisions in this act are later consolidated in the Company Securities (Insider Dealing) Act 1985, and this itself is strengthened by the Financial Services Act 1986. This act also introduces provisions regarding market manipulation. The Financial Securities and Markets Act 2000 strengthens the provisions of 1986 act, where both offences are classed as market abuse. The European Union has also issued a number of directives in this area, the most recent being the Market Abuse Directive 2003. Legislation in the United Kingdom is stronger than that required by the directive.

In the United States, insider trading and market manipulation have been illegal for much longer, where both have been classed as fraud. This means that primary legislation covering insider trading has existed since the 1933 Securities Act, the reference to both insider trading and market manipulation being made more explicit in the Securities Exchange Act 1934. However, it was not until the introduction of the Insider Trading Sanctions Act 1984 that insider trading was well defined. The provisions of this act are then strengthened by the Insider Trading and Securities Fraud Enforcement Act 1988.

Many other countries have similar laws. Australia and Canada both introduced legislation in 1970, and by the end of the last century 85% of markets had insider trading laws; however, there was evidence of enforcement of these laws in fewer than half of the markets (Bhattacharyya and Daouk, 2002).

Investors in the European Union have also received additional protection through the Market in Financial Instruments Directive (MiFID) (European Commission, 2004), which came into force in 2004, being implemented in 2007. This increase pre-and post-trade transparency, and codifies 'best execution' for trades, which allows for not just the price but also the speed and other relevant factors in the execution of the trade.

5.7.2 Bank customers

Customers of banks in most countries have protection from bank insolvency. The United States was the first country to develop a depositor protection scheme, the Federal Deposit Insurance Corporation (FDIC) which was created with the Glass–Steagall Act 1933. The level of protection has increased steadily since the scheme's inception, the last permanent rise being to $100,000 with the Depository Institutions Deregulation and Monetary Control Act 1980. In October 2008, the level of insurance was temporarily raised to $250,000, but returned to $100,000 at the end of 2009.

Depositors in India have had coverage since 1961, and the Deposit Insurance and Credit Guarantee Corporation (DICGC), formed in 1978, covers deposits of up to Rs. 100,000. Canada followed with the Canada Deposit Insurance Corporation (CDIC) in 1967 which now offers maximum compensation of C$100,000, an increase from C$60,000 in 2005.

The customers of banks and building societies in the United Kingdom received protection at around the same time as investors. In this case, it came about with the Building Societies Act 1987 and the Banking Act 1987. The former set up the Building Societies Investor Protection Scheme (BSIPS) and the latter the Deposit Protection Scheme (DPS). In both cases, 90% of an investor's deposits were secure, up to a limit of £18,000. As with the ICS, these two schemes were absorbed into the FSCS following the introduction of Financial Services and Markets Act 2000, which covers individual investors and small firms. Initially, the level of compensation was lower than that available to investors, being the first £2,000 lost and 90% of the next £33,000. However, following the run on Northern Rock Bank in 2007 and the increasing lack of confidence in financial institutions at that time, the compensation available was changed first to 100% of the first £35,000, then in October 2008 to 100% of the first £50,000 lost. Having said this, the United Kingdom government has offered even higher protection in practice. After the collapse of the Icelandic internet bank Icesave in 2008, it guaranteed all deposits of United Kingdom retail depositors.

Similar schemes exist in other European Union countries, although it is notable that schemes in the Republic of Ireland and Portugal both have formal unlimited protection rather than an undertaking to safeguard all depositors, also in response to the liquidity crisis in 2008. Australia has also offered unlimited protection for the period from 2008 to 2011 with the Financial Compensation Scheme (FCS) set up by the Financial System Legislation Amendment (Financial Claims Scheme and Other Measures) Act 2008. It is important to note, though, that the higher the level of depositor protection, the less incentive a depositor has to ensure that his or her bank is creditworthy.

5.7.3 Insurance company policyholders

The protection available to customers of failed insurance companies started earlier than for customers of other institutions in the United Kingdom with the introduction of Policyholders Protection Scheme (PPS). This was set up by the Policyholders Protection Act 1975. The protection available from this act varied from policy to policy. For compulsory insurance, such as third party motor, 100% of all claims were paid; however, for other policies, 90% of the value of a claim was covered. Protection for friendly society members, mirroring that for insurance company policyholders, was introduced in the Financial Services Act 1986, which created the Friendly Societies Protection Scheme (FSPS). Both of these schemes were subsequently absorbed into the FSCS. The provisions for compulsory insurance are mirrored in this scheme, but the cover for other insurance policies increased slightly with the first £2,000 of all claims being covered and 90% of the excess over that amount.

In Canada, protection has existed since 1998 in the form of the Property and Casualty Insurance Compensation Corporation. This is an industry-run organisation to which all non-life insurers authorised in Canada must contribute. It aims to meet all claims from non-life insurers that have become insolvent.

There is no federal policyholder protection in the United States, but some states where there is greater exposure to non-life insurer insolvency (such as Florida, from hurricane damage) have their own schemes. In Australia, the FCS also gives unlimited coverage for non-life insurance policies until 2011.

5.7.4 Pension schemes

There are three ways in which regulation typically impacts pension schemes: through the requirement to provide certain benefits, through the requirement to hold a certain level of assets in respect of those benefits and through restrictions on the assets that can be held. Pension scheme sponsors and members are considered together here, since regulations that impose restrictions on the former provide protection for the latter.

In the United Kingdom, one of the first movements from discretion to compulsion in relation to the benefits payable was with the 1973 Social Security Act which required the provision of deferred pensions to pension scheme members leaving employment, providing they had at least five years of service in the pension scheme. This limit was reduced to two years with the 1986 Social Security Act. There were also requirements to increase various portions of deferred pensions over the period between leaving service and drawing a pension in the Health and Social Security Act 1984, and the Social Security Acts of 1985 and 1990. This is a high level of protection for deferred pensions when compared with other pensions systems. For example, in the United States the Employee Retirement Income Security Act 1974 (ERISA) provides protection for early leavers, but the vesting period remains at five years, or seven if the pension is guaranteed in stages.

Not only did legislation in the United Kingdom create guaranteed benefits for deferred pensioners, but it also added guaranteed increases to pensions in payment, sometimes known as cost of living adjustments (COLAs). This effectively started with Guaranteed Minimum Pensions (GMPs) with the Social Security Act 1986, moving on to the excess over the GMP with the Pensions Act 1995, with only some respite being given in the shape of a rate reduction in the Pensions Act 2004. Guarantees of such benefits are still unusual in most pension schemes outside the United Kingdom.

The funding requirements for pension schemes differ significantly from country to country, and have developed over time. In the United States, ERISA defines a notional plan balance, the Funding Standard Account, to which contributions are added and interest accrued. However, this does not represent reality, a fact addressed by the Pension Protection Act 2006. This introduced new Minimum Funding Standards, which require deficits to be amortised, and for more immediate action to be taken with severely under-funded pension schemes.

Defined benefit pension scheme members in the United Kingdom had very little in the way of direct protection until the Maxwell Affair beyond the various guaranteed increases described in the section above. After the death in 1991 of Robert Maxwell, it was discovered that assets were missing from the pension schemes in the Mirror Group of companies. This led to the regulations being introduced in 1992 requiring pension schemes to have sufficient funds to secure members' benefits with annuities, and to pay in any difference required – the debt on the employer – if the scheme is wound up. The basis was weakened with the introduction of the Pensions Act 1995 to the MFR (minimum funding requirement) basis, but strengthened again to the buyout basis with the Pensions Act 2004. The debt on the employer is essentially a requirement to hold a particular level of assets. It is worth noting that this means that unlike most equity capital, that provided by the pension scheme sponsor is unlimited in the United Kingdom, since the debt on the employer in the event of a pension scheme wind-up has no explicit limits.

The MFR was the first funding standard for the United Kingdom, requiring benefits to be fully funded according to a defined basis. The defined basis was replaced by a scheme-specific funding requirement in the Pensions Act 2004.

Another important part of the Pensions Act 1995 was the introduction of compensation for pension scheme members whose benefits had been reduced due to fraud. Compensation was in the form of a payment to scheme from the newly formed Pension Compensation Board of up to 90% of the shortfall calculated using the MFR basis.

However, many pension scheme members who lose benefits did so for reasons other than fraud, predominantly due to insolvency of the sponsoring employer. Struggling employers are also less likely to be able to keep their pension schemes adequately funded. Furthermore, there are likely to be more insolvencies when equity market values are depressed, thus increasing the number of pension schemes with deficits. The problem for some members can be compounded by the order with which pension scheme assets are used to provide benefits to members when a scheme is wound up. Following the Pensions Act 1995 when a statutory order of priorities was first introduced, active members and deferred pensioners were not entitled to anything until the guaranteed benefits of pensioners had been secured. In some cases, such as with the Allied Steel and Wire Pension Scheme, this meant that, whilst pensioners benefits were secured, individuals that had worked for the firm for many years but had not yet retired were entitled to only severely reduced pensions. The Pensions Act 2004 altered the statutory order largely to reflect the benefits payable from a new institution created by this Act, the Pension Protection Fund (PPF). This fund, administered by the Board of the Pension Protection Fund, takes on the assets of any pension scheme with an insolvent sponsor and insufficient assets to meet liabilities, and pays benefits up to a maximum statutory level.

Whilst the PPF was a new phenomenon in the United Kingdom, a similar scheme had existed in the United States for thirty years – the Pension Benefit Guaranty Corporation (PBGC). This was launched as part of ERISA. Under this act, each employer is required to fund benefits as they accrue and to amortise any deficit, albeit over a long period. In return for this, and for statutory contributions, pension scheme members are protected against non-payment of their benefits up to a statutory limit.

Switzerland also has a relatively comprehensive and generous insolvency protection scheme, the Law on Occupational Benefits (LOB) Guarantee Fund set up by the Law on Occupational Benefits 1982. However, pension protection in other countries is patchier. In Canada, Ontario has the Pension Benefit Guaranty Fund (PBGF), but nothing elsewhere and the PBGF has only moderate benefit coverage. Germany's Pensions-Sicherungs-Verein Versicherungsverein auf Gegenseitigkeit (PSVaG) has coverage that is good in terms of numbers but modest in terms of benefits, and Japan's Pension Guarantee Programme (PGP) has similar issues. Finally, there is Sweden's Försäkringsbolaget Pensionsgaranti (FPG), an arrangement by which individual members can obtain protection. Here, only a small number of members are covered.

There are two ways in which the restrictions in relation to investments can be implemented. First, there can be limits on investment of the pension scheme in shares of the sponsor, in order to limit the risk of an employee losing both job and pension. In this regard, the OECD recommends a limit on self-investment of 5%, a level adopted in many countries. Interestingly, this restriction does not exist in the United States in relation to 401(k) defined contribution plans, and a number of employees of Enron and Worldcom suffered with the effect of such leverage when these two firms collapsed. The second restriction, which aims to reduce the level of mismatch risk, is on the extent of domestic bond or other matching investment. An interesting implementation of this principle occurs in the Netherlands, where the solvency requirements differ depending on the level of matching.

5.7.5 Government (financial relationships)

Most financial institutions in most countries are taxed on the profits that they make, although the exact definitions of those profits and the deductions that can be made vary hugely. There are several exceptions to the profits-based taxation basis that are of interest. Within United Kingdom, insurance companies, basic life assurance and general (rather than pension) annuity business, or 'BLAGAB' as is it often known, is taxed on the excess of income over expenditure. This means that if the BLAGAB business of a life insurance company has an excess of expenses over income – so is in an 'XSE' position – then it can write policies that take account of this fact as they do not need to allow for the insurer's taxation liability. The products sold in this way are generally short-term insurance bonds, and since the risk here is that the income will exceed the expense, the volumes sold are strictly controlled.

Contributions to most occupational pension schemes get relief from taxation, and investments within a pension scheme are generally allowed to accumulate free of taxation on income and capital gains, the payments to members being taxed as income. This means that the government is at risk of losing out on revenue as tax is deferred. For this reason, there are often restrictions on the volume of assets that can be accumulated in a pension scheme and the time over which they must be extracted.

In the United Kingdom, restrictions on the maximum amount that could be accumulated in a defined benefit pension scheme came with the Finance Act 1986, the rules of which were consolidated in the Income and Corporation Taxes Act 1988. This said that pension schemes with assets worth more than 105% of liabilities (calculated using a prescribed actuarial basis) had to reduce this ratio, the funding level, to 105% or below by increasing benefits, reducing or suspending contributions or refunding assets to the employer. The basis for the calculation was such that any excess, known as a statutory surplus, rarely arose.

With the reduced funding levels experienced by many schemes due to falls in interest rates, increases in longevity and increased levels of benefits, the concept of a statutory surplus became less relevant. Furthermore, as an increasing number of pension schemes closed to new entrants or the accrual of future benefits, company sponsors became less inclined to fully fund pension schemes for fear of creating irrecoverable surpluses. The statutory surplus provisions where therefore repealed in the Finance Act 2004.

In the United Kingdom, defined contribution plans take contributions from pre-tax income, and these accumulate investment returns free of taxation. At retirement, 25% of the accumulated fund may be taken free of tax (as is also the case with United Kingdom defined benefit pension schemes). Income may be taken from the balance, within specified limits, up until age 75. At this point, an annuity must be purchased (if it has not been bought already). Both income withdrawn from the plan and received from the annuity are taxed at the recipient's marginal rate. Following pensions simplification further to the Finance Act 2004, the maximum value of the fund is effectively the only limit. For the 2007–8 tax year, this stood at £1.6m, a figure which was due to rise to £1.8m in the tax year 2010–11.

In the United States, pre-tax income can be contributed to a defined contribution plan, known as a 401(k). As with United Kingdom plans, the contributions accumulate investment returns free of tax, but the accumulated amount can then be drawn down in full at retirement or drawn down over time. Either way, the payments are taxed at the individual's marginal rate as they are received. Most international defined contribution plans follow the

United States model of allowing individuals to take as much income as they like rather than requiring annuitisation at some point in time as in the United Kingdom.

There is also a similar vehicle known as a Roth 401(k), where the contributions in are post-tax, and the accumulation and payouts are tax-free. A similar vehicle in the United Kingdom is an Individual Savings Account (ISA), although this is not specifically designed for retirement.

Foundations and Endowments set up as charitable organisations are generally exempt from tax on their investment returns. In many countries, contributions are partially or entirely tax-deductible. In the United States, foundations and endowments must distribute a minimum of 5% of their assets each year to remain tax exempt.

5.7.6 Financial markets

The various relationships with financial markets are covered in the European Union by the Market in Financial Instruments Directive 2004 (MiFID). This classifies clients as either eligible counter-parties, professional clients or retail clients. Retail clients are helpfully defined as being not professional clients. The definition of professional clients is more informative, and includes a number of types of institution that are by definition in this category together with individuals or firms that might otherwise be classed as retail clients but have the desire and sufficient are experience to be treated as professional clients. Eligible counter-parties are a category of professional clients that deal directly with each other and with other organisations, such as central banks, issuers of government debt or supranational organisations.

In the United States, the Sarbanes–Oxley Act of 2002 – or, more formally, the Public Company Accounting Reform and Investor Protection Act 2002 – implemented a number measures designed to protect shareholders. This was put in place as a response to the Enron and WorldCom scandals, and has a number of far-reaching implications discussed later. However, in the context of financial markets, its main purpose was to require an increase in the level of disclosure required from firms, which, together with the provisions related to auditing discussed in Section 5.7.11, are designed to improve the quality and quantity of information available to shareholders in order to help them reach decisions.

Sarbanes–Oxley also seeks to improve the quality of analysts' stock recommendations by strengthening the separation of analysts and investment bankers. This is important because there might be strong incentives for analysts to give 'buy' ratings to those firms who are investment banking clients.

(notes) ### 5.7.7 Company Directors

As part of good corporate governance, directors must ensure that their firms comply with a wide range of rules, including stock exchange regulations (if the firms are listed), accounting standards and legislation relating to employment, pensions, health and safety and possibly other areas depending on the area in which the firm operates. Directors must also comply with a range of rules themselves. Many of these rules start as reviews, which result in codes, which to a greater or lesser extent must be obeyed.

The general standards of practice in boardrooms have been addressed in a number of these reports and codes, with many important changes starting in the late 1980s and 1990s. The United Kingdom faced a number of corporate scandals in this period. In response, a committee chaired by Sir Adrian Cadbury was set up in 1991 by the Financial Reporting Council (FRC), the London Stock Exchange (LSE) and the UK accountancy profession. The aim of this committee was to recommend a code of best practice for boards of directors, and in 1992 the committee released its report on 'the financial aspects of corporate governance' (Cadbury, 1992). The report highlights the value of regular board meetings and good oversight by the board of the executive management. It also recognises the importance of having checks on power at the top of a company. In particular, the report recommends a strong and independent presence on the board in the absence of separate appointments for the roles of chairman and chief executive. The emphasis on independence is strengthened by a recommendation that the majority of non-executive directors be independent of the firm, so free of any business or other relationship with the company. In addition, it recommends limited-term appointments for both executive and non-executive directors, without automatic reappointment at the end of each term. The UK Corporate Governance Code issued by the Financial Reporting Council (2010) goes further in both of these areas. In respect of independence, it sets out the conditions under which independence could reasonably be questioned, namely:

- if a director has been an employee of the company or group within the last five years;
- has, or has had within the last three years, a material business relationship with the company;
- has received or receives additional remuneration from the company apart from a director's fee, participates in the company's share option or a performance-related pay scheme, or is a member of the company's pension scheme;

- has close family ties with any of the company's advisers, directors or senior employees;
- holds cross-directorships or has significant links with other directors through involvement in other companies or bodies;
- represents a significant shareholder; or
- has served on the board for more than nine years from the date of their first election.

In relation to the term of appointment, this code also recommends that directors of all FTSE 350 companies be put forward for re-election every year. Furthermore, it emphasises the need to appoint members on merit against objective criteria, taking into account the benefits of diversity. Gender diversity is singled out as a particularly important example. The importance of regular development reviews for all board members is also emphasised.

In Canada, the Toronto Stock Exchange commissioned a report at around the same time as the Cadbury Report from a committee chaired by Peter Dey. The report by Dey (1994) – known as the Dey Report – came to similar conclusions to the Cadbury Report, emphasising the role of non-executive ('outside') and independent ('unrelated') directors. The advantages of a non-executive chairman are also recognised, and the report recommends that most committees are composed mainly of non-executive directors with some, such as the nominations and audit committees, consisting only of non-executive directors. In the same year, King (1994) in South Africa was addressing the issue of corporate governance with the first King Report (King I), this time in the context of the social and political changes that were occurring there at that time. The report emphasises disclosure and transparency and, given the unique situation of South Africa at that time, requires firms to have an affirmative action programme. This report was updated in 2002 (King, 2002) with the second King Report (King II) which expands on many of the principles discussed in the first report, and defines what the committee believes to be the characteristics of good corporate governance:

- discipline;
- transparency;
- independence;
- accountability;
- responsibility;
- fairness; and
- social responsibility.

King (2009) gave further guidance in the form of the third King Report (King III), which further strengthens the independence and accountability of boards.

In India, 2002 also saw the publication by Birla (2002) of the Kumar Mangalam Birla (KMB) Report, which is unambiguous in its aim to help shareholders. Its recommendations, which again focus on disclosure and non-executive directors, require that they comprise at least half the board, and also require that at least one-third of the board's directors be independent.

A particular concern in relation to directors and agency risk is directors' pay. This is an area where directors might be particularly tempted to act in their own interests rather than on behalf of the shareholders. In response, the Confederation of British Industry (CBI) set up a committee chaired by Sir Richard Greenbury to look into this area and to propose a code for director remuneration. Greenbury (1995) subsequently issued his report on directors' remuneration – the Greenbury Report – which was also initiated in 1995. As well as recommending the introduction of remuneration committees, consisting solely of non-executive directors, the code suggests much greater disclosure. Disclosure of all benefits is required, including share options and pension benefits calculated on an actuarially sound basis. The code also addresses the level of remuneration. Whilst recognising that pay needed to be sufficient to attract, retain and motivate good directors, the code recommends that regard be given to wider issues, including the pay of other employees. It also builds on the Cadbury recommendations relating to limited terms of appointment. The UK Corporate Governance Code also comments on pay, recommending that performance-related pay be aligned with the long-term interests of the company.

King I in South Africa also considers the remuneration committee, recommending that at least two non-executive directors sit on the remuneration committee, one of whom should be the committee's chairman; King II updates this to recommend that the remuneration committee should consist mainly of independent non-executive directors, and King III goes even further requiring that all members be non-executive directors, the majority of whom being independent, with an independent non-executive chairman. Similarly, the KMB Report recommends that the remuneration committee consist solely of non-executive directors with the chairman being an independent director. King III also rules out the payment of share options to non-executive directors in order to increase independence.

Not long after the publication of the Greenbury Report, a number of parties commissioned a further report into corporate governance in the United

Kingdom. These parties were the LSE, the CBI, the accountancy profession, the National Association of Pension Funds (NAPF) and the Association of British Insurers (ABI). All but the last of these instigated either the Cadbury or Greenbury Report. The committee for this new report was chaired by Sir Ronald Hampel, and it gave its final report in 1998. The Hampel Report (Hampel, 1998) confirms many of the recommendations made in the Cadbury and Greenbury Reports, but also addresses the roles of institutional shareholders, emphasising the role they ought to play given the voting rights that they held.

The Hampel Report was effectively the first iteration of what later became the Combined Code on Corporate Governance issued by the Financial Reporting Council (2008) and is now the UK Corporate Governance Code. Turnbull (1999, 2005) gives guidance to directors on how to comply with the Combined Code, and the London Stock Exchange's Listing Rules require disclosure of the extent of compliance with the Combined Code. However, Pensions and Investment Research Consultants Ltd (2007) found that only around one in three firms complied fully with the code, although the level of compliance was climbing.

As mentioned earlier, both the Cadbury and Greenbury Reports discuss the role of non-executive directors. Both reports recognise that their dual role, encompassing both working with the directors and acting as an independent check, creates a clear conflict. Non-executive directors are considered in Higgs (2003) *Review of the Role and Effectiveness of Non-Executive Directors* (the Higgs Report). According to the report, their role should cover:

- development of corporate strategy with executives;
- monitoring the performance of executives;
- financial reporting and controls; and
- appointment, removal and remuneration of executive directors.

In order to manage the conflict these duties create, the Higgs Report suggests that non-executive directors meet independently of the executives at least annually, and that they have a senior member who can report any concerns to the chairman. The report goes on to recommend amendments to the Combined Code, mainly to reflect its work on non-executive directors. The UK Corporate Governance Code further sets out the responsibility that non-executive directors have to provide constructive challenges to the executives. In respect of all directors, this code also emphasises the time commitment that directorship implies.

In the United Kingdom, Cadbury recommends the presence of an audit committee, giving auditors direct access to non-executive directors, and quarantining audit from other business services provided. Cadbury recommends that the members of the audit committee are only non-executive directors, as does Dey in Canada. Unlike Cadbury, Dey does not suggest that the majority of members be independent. However, the Saucier Report, which in 2001 updates the Dey Report in Canada, recommends that all members of the audit committee be independent, describing independence (or lack of it) in some detail. The KMB Report in India recommends that all members should be non-executive directors and most, including the chairman, should be independent. The King II Report in South Africa recommends that the majority of the audit committee be independent non-executive directors, and King III strengthens this by requiring that there be at least three members meeting at least semi-annually, all of whom hold this status at the holding company level. King III also requires that the chairman of this committee be an independent non-executive director.

The Auditing Practices Board (APB) in the United Kingdom considers the issue of audit. In order to limit the reliance of an auditing firm on any one listed client, which might use such a relationship to influence reported results, ethical standards issued by the APB prohibit auditors from continuing appointments where the annual fee income exceeds or is expected to exceed 10% of total fee income. Hampel suggests strengthening guidance even further, perhaps reducing the 10% limit.

The issue of auditing came to the fore again with the scandals involving Enron, Worldcom and a number of other firms. In the United Kingdom, a committee chaired by Sir Robert Smith was set up by the FRC to look again at the function of audit committees. Smith (2003, 2005) sets out the functions of audit committees, and recommends that these be included in terms of reference of the committee. These functions, now included in the UK Corporate Governance Code, can be summarised as:

- monitoring the integrity of financial statements;
- reviewing the internal financial control and risk management system;
- monitoring and reviewing the effectiveness of the internal audit function;
- recommending to the board the appointment, remuneration and terms of engagement of the external auditor;
- monitoring and reviewing the external auditor's independence, objectivity and effectiveness; and
- developing and implementing policy on the supply of non-audit services by the external auditor.

Many of the principles in these and other reports were encapsulated in a report on the principles of corporate governance by the OECD (1999, 2004). However, since this document is intended to cover a wide range of different countries, it is of a much higher level than the reports discussed above, and some of the principles would be taken for granted in many developed financial markets.

There is also some references to the behaviour of directors in primary legislation. The Sarbanes–Oxley Act of 2002 makes it clear that the chief executive officer (CEO) and chief financial officer (CFO) of a public company are each personally responsible for the disclosures in financial reports, and they must certify that the reports contain no untrue statements of material fact. The CEO and CFO are also legally responsible for setting up, maintaining and evaluating internal controls, and reporting any issues to the external auditors. Directors are also prohibited from interfering in the audit process, and all employees are prohibited from altering, concealing, destroying or falsifying records or documents.

In the United Kingdom, the requirements under primary legislation are at a higher level. According to the 2006 Companies Act, directors are constrained to act within their powers as set out in the articles of association, and are required to act in the best long-term interests of the company, having regard to a wide range of parties such as employees, suppliers and the wider community, and should avoid (or at least declare) conflicts of interest. What the best interests of the shareholders are is something that is open to interpretation. It implies maximising long-term returns subject to some sort of measure of risk. This implies that risk should be measured and mitigated, but the exact measures are not set out in this act; they are, though, explored later in this book.

5.7.8 Trustees

Trustees are the agents responsible for looking after the interests of the trust's beneficiaries in the same way as directors are responsible for looking after the interests of a firm's shareholders. In the United Kingdom their actions are governed by primary legislation, such as the 2000 Trustee Act, but also by a large body of case law.

Compared with 'general' trustees, pension scheme trustees face additional rules and regulations to reflect the fact that the benefits for which they are investing are more complex. Pension scheme trustees have a duty towards scheme members and to fulfil their specific legal obligations. The way in which they are expected to do this varies from country to country. For example, in the United Kingdom, trustees are governed by the 'prudent man' rule. Following the Pensions Act 1995, they are, though, expected to appoint specialists from whom they take advice, in particular relating to actuarial, auditing, investment and legal matters. In the United States, the obligations on the trustees are much greater, with the 'prudent expert' requirements of ERISA.

The way in which pension scheme trustees behave in the United Kingdom was brought to the fore with the Maxwell affair, discussed above in relation to pension scheme members. The outrage that followed led to the creation of the Pension Law Review Committee, chaired by Professor Roy Goode, which reported in September 1993. Among other things, Goode (1993) remarked that pension scheme trustees should be thought of as analogous to company directors, and that legislation should reflect this.

Many of the recommendations in this review were taken up in the 1995 Pensions Act, which had a direct impact on trustees in a number of ways. First, in order to increase the accountability of the trustees to members, the act required one-third of the trustee body to be nominated by members. Some of the opt-outs to this requirement were subsequently removed with the 2004 Pensions Act, which was introduced in response to the 2003 European Union Pensions Directive (European Commission, 2003b). Furthermore, there was a requirement in the 1995 act that at least one trustee of any pension scheme in wind-up should be independent – in other words, external. The act clarified that the only power of investment that could be delegated was the management of assets, which could be given to one or more investment managers. However, it required that other powers be delegated. Trustees were no longer allowed to act as either auditor or actuary to the scheme, and two new statutory roles were created: scheme actuary and scheme auditor.

The act allowed trustees guaranteed time and resources for their duties and for training, but also imposed additional requirements on them. It obliged them to provide greater and more timely disclosure to scheme members, and required them to obtain a valuation from the scheme actuary according to a MFR basis. It also required them to provide a schedule of the future contributions due to the pension scheme and a statement of investment principles. However, the act also gave the trustees additional powers, including allowing them to impose a minimum level of contributions on the scheme sponsor based on the MFR. As discussed elsewhere, the MFR was soon regarded as ineffective and the 2004 Pensions Act replaced it with a scheme-specific funding requirement. The methodology for this requirement, in the 2005 Occupational Pension Schemes (Scheme Funding) Regulations, was general enough that it was up to the pension scheme trustees to ensure that pension scheme members were truly protected. In practice, the trustees rely on the advice of the scheme actuary when considering funding.

Not long after this, in 1998, the Kirby Report was published as a report by the Canadian Senate Committee on Banking, Trade and Commerce. Kirby (1998) came about as part of a broader investigation of issues relating to the Canada Business Corporations Act, in particular concerning corporate governance. However, many of those giving evidence to the committee raised concerns about the behaviour of institutional investors. The committee therefore held a series of meetings considering pension funds alone. Among the key conclusions that the committee come to is that pension scheme trustees (or boards, as they are described in Canada) should have sufficient knowledge to monitor the pension schemes' investment managers. Otherwise, the report concentrates on disclosure and the broader areas of corporate governance as they apply to those responsible for pension schemes.

The Myners Report was commissioned by the United Kingdom Treasury, following comments in the 2000 budget speech. The impetus for the report was a perceived lack of investment in private equity by institutional investors; however, this formed only a small part of the final report, delivered by Paul Myners in 2001. Myners (2001) was aimed at pension schemes and insurance companies, but the bulk of the recommendations applied to occupational pension schemes.

In terms of impact, Myners recognised that legislation can introduce unintentional distortions into financial markets. In particular, he cited the MFR, which was later replaced by scheme-specific measures. He was also concerned by the low level of shareholder activism from institutional investors, and he proposed the adoption of the United States Department of Labor Interpretative Bulletin on ERISA. This required a higher level of intervention by fund managers in corporate decisions in order to maximise shareholder value.

Myners also found that the extent of trustee expertise was limited in key areas relating to investment and thought that more training was needed. In particular, he preferred the 'prudent expert' rule for trustees described in ERISA over the 'prudent man' approach. The latter was first described in the Massachusetts case of *Harvard College* v. *Amory* in 1830, where trustees are expected to have regard to how 'men of prudence, discretion, and intelligence manage their affairs'. In the United Kingdom, similar sentiments were first expressed in the House of Lords decision on *Speight* v. *Gaunt* in 1882. The prudent expert rule, as described in ERISA, requires a trustee to act 'with the care, skill, prudence, and diligence, under the circumstances then prevailing, that a prudent man acting in a like capacity and familiar with such matters would use in the conduct of an enterprise of a like character and with like aims'.

In order to encourage more skilled trustees, Myners not only suggested more in-depth training, but also that trustees should be paid. Some independent trustees (typically those appointed when a pension scheme is being wound up) are paid, but the majority are not. In its response to the Myners review, the National Association of Pension Funds (NAPF) criticised this proposal, suggesting that the implied additional responsibilities would discourage individuals from acting as trustees. The suggestion of routine payment of trustees was not adopted.

5.7.9 Company managers and employees

The agency issues surrounding employees are substantial. However, employers do not have unfettered rights in relation to how they can act with their employees. In the United Kingdom, workers are protected by legislation such as the Employment Rights Act 1996, which covers unfair dismissal, discrimination, employment tribunals and redundancy payments among other areas. There is also considerable case law that has built up over decades. Similar legislation exists in other countries. Finally, of course, employees are represented by trade unions.

5.7.10 Trade unions

As discussed earlier, trade unions can also pose agency risks for firms. However, there are legislative issues here too, since closed shops were made illegal in the United Kingdom in the Employment Rights Act of 1996. Having said this, despite trade unions being an important part of the industrial landscape of the United Kingdom, they have never been a major factor in the financial services industry.

5.7.11 External auditors

Internal auditing is discussed in many of the codes discussed below. However, in the United States where the Enron and Worldcom scandals originated, an even tougher line was taken with the introduction of primary legislation in place of a voluntary code. This was in the form of the Sarbanes–Oxley Act of 2002. This legislation had a number of purposes. The first was to strengthen the power of the audit function. One way in which this was attempted was by limiting the length of appointment of an audit partner within a firm to five years. Auditor rotation was considered by the Smith Report in the United Kingdom, but rejected on the grounds that the resulting loss of trust and continuity would outweigh any benefits from increased independence; legislators in the United States took a different view. Sarbanes–Oxley also bans the provision of audit and non-audit services by the same firm, in order to avoid pressure on the audit partner from other departments of his or her firm. Ironically, this restriction was introduced only three years after the provisions of the 1933 Glass–Steagall Act, separating commercial and retail banking, were repealed for banks. A third provision reflected the codes discussed below in requiring the presence of non-executive directors on audit committees. However, in order to ensure that all of these measures were having the desired effect, the act also established the Public Company Accounting Oversight Board to oversee audit of public companies.

5.7.12 Actuarial advisers

The statutory role for actuaries in relation to pension schemes is long-standing: the role of the scheme actuary has a place in United Kingdom legislation from the Pensions Act 1995, but equivalent roles existed before this here, and continue to exist elsewhere, most notably in the case of the United States' enrolled actuary, as defined in ERISA. Under the Pensions Act 2004, the scheme actuary is responsible for advising pension scheme trustees on the method and assumptions used in calculating technical provisions, funding benefits and modifying the pension scheme. The scheme actuary is appointed by the pension scheme trustees and acts for them.

In relation to life insurance companies, there have been important changes in the United Kingdom in recent years. Until the end of 2003, the role of reserving was delegated to the appointed actuary; however, with the adoption of the FSA's Integrated Prudential Sourcebook on 1 January 2004, two new roles were created: with-profits actuary and actuarial function holder. An actuarial function holder advises the management of an insurance company on the risks affecting ability to meet liabilities to policyholders, and on the valuation methods and assumptions, as well as performing calculations on this basis. If the life insurance company has with-profits business, then a with-profits actuary is required to comment on issues relating to this business, not least in relation to bonus declarations.

5.7.13 Investment and financial advisers

In the United Kingdom, advice relating to the choice of investment manager, whether for an institution or an individual, requires the advisor to be authorised under the Financial Services and Markets Act 2000, and is subject to regulation by the FSA. Advisers must obtain a great deal of information on their clients in order to ensure that the advice that they are giving is appropriate.

5.8 Professional environment

The regulatory environment has a major impact on firms operating in the financial services industry; however, individuals working for these firms are often members of professional bodies. In fact, there are a number of roles that can be held only by individuals with particular professional qualifications. Professionals must fulfil certain requirements – and can be subject to harsh sanctions if they do not. These are the subject of this section.

5.8.1 Professional bodies

The range of professional bodies is very large. Some are worldwide, such as the CFA Institute which (among other things) administers the qualification for financial analysts and provides CPD opportunities for its members; many more organisations are regional. For example, as well as the CFA Institute's United Kingdom branch, there is also the Chartered Institute for Securities and Investment (CISI).

Another profession organised on a regional basis is accountancy and, within this, auditing. In the United Kingdom, external auditors are catered for by the Institutes of Chartered Accountants in England and Wales (ICAEW), Scotland (ICAS) and Ireland (ICAI), or the Association of Chartered Certified Accountants (ACCA). The equivalent organisations in Australia, Canada and South Africa are the Institute of Chartered Accountants in Australia (ICAA), the Canadian Institute of Chartered Accountants (CICA) and the South African Institute of Chartered Accountants (SAICA) respectively. In the United States, external auditors must be Certified Public Accountants. They are regulated on a state-by-state basis, but examined by a national body, the American Institute of Certified Public Accountants (AICPA).

Actuarial bodies are also regionally based, although several umbrella organisations exist. Most actuarial bodies belong to the International Actuarial Association (IAA). In the United Kingdom, the Institute and Faculty of Actuaries is responsible for training actuaries. In the United States, the American Society of Pension Professionals and Actuaries (ASPPA) covers those working in pensions, as does the Society of Actuaries (SoA) which has a broader remit whilst remaining focussed on life contingencies. The Casualty Actuarial Society (CAS) covers those working in non-life insurance. All actuaries can also belong to an umbrella organisation, the American Academy of Actuaries

(AAA). The Canadian Institute of Actuaries (CIA), the Actuarial Society of South Africa (ASSA) and Institute of Actuaries of Australia (another IAA) look after actuaries in their respective countries.

As its name suggests, the remit of the ASPPA is wider than just actuaries, also providing qualifications for pension consultants and administrators. In the United Kingdom, pension administrators can also work towards qualifications with the Pensions Management Institute (PMI). There are also a number of further affiliations that qualified professionals can have reflecting particular specialisms, such as the Association of Consulting Actuaries (ACA) in the United Kingdom, of which many fellows of the Institute or Faculty of Actuaries working in consultancy are members.

All of these bodies either administer professional qualifications, require membership of another body, or both. The also frequently require a minimum level of CPD, although the level and type of CPD vary widely.

Some of these organisations also place restrictions on what their members can do. In some cases, such as the Institute and Faculty of Actuaries, the restrictions are quite general and principle-based, as set out in The Actuaries' Code; however, others, such as the CFA Institute, impose much more specific restrictions on their members. For example, in its Code of Ethics the CFA Institute comments that 'members and candidates who possess material non-public information that could affect the value of an investment must not act or cause others to act on the information', and that 'members and candidates must not engage in practices that distort prices or artificially inflate trading volume with the intent to mislead market participants'. These rules apply even if there are less stringent laws in the country in which the member or candidate is working.

5.8.2 Professional regulators

In the United Kingdom, much of the work of professional regulation is carried out by the Financial Reporting Council (FRC), which effectively regulates the accounting and actuarial professions. It does this by setting standards for these professions, ensuring standards are upheld and running a disciplinary scheme.

On the accounting side, the FRC is responsible for the production of Financial Reporting Standards (FRSs) and their predecessor, Statements of Standard Accounting Practice (SSAPs) through the Accounting Standards Board (ASB). These specify the way in which accounts should be drawn up. These standards are supplemented by Statements of Recommended Practice (SORPs) which, although not issued by the ASB, are under supervision of the ASB. For new issues, abstracts are produced by the Urgent Issues Task Force (UITF). The standard of auditing is controlled by the Auditing Practices Board (APB).

The FRC also encompasses the Board for Actuarial Standards (BAS). This produces technical standards in the form of Guidance Notes (GNs), but not ethical standards, which remain with the Institute and Faculty of Actuaries.

Monitoring of both the actuarial and accountancy professions is carried out by the Professional Oversight Board (POB), with disciplinary proceedings being run by the Accountancy and Actuarial Discipline Board (AADB), both of which are also part of the FRC.

In the United States, the equivalent of the ASB is the Financial Accounting Standards Board (FASB). This produces Financial Accounting Standards (FASs) and, for urgent issues, abstracts drawn up by the Emerging Issues Task Force (EITF). However, in the United States there is no federal body that considers discipline, this being dealt with at a state level.

Actuarial regulation in the United States is only semi-independent. It is carried out by the Actuarial Standards Board (another ASB), appointed by the Council of US Presidents (CUSP), which consists of the presidents and presidents-elect of the AAA, the ASPPA, the CAS, the Conference of Consulting Actuaries (CCA) and the SoA. The CUSP also appoints the Actuarial Board for Counselling and Discipline. The ASB in the United States guides actuaries through the issuance of Actuarial Standards of Practice (ASOPs).

In Australia, accounting standards are set by the Australian Accounting Standards Board (AASB), whilst auditing quality is maintained by the Auditing and Assurance Standards Board (AUASB). These both fall within the remit of the Australian Financial Reporting Council (Australian FRC), a government agency.

The disciplinary process in Australia differentiates between situations where the law has been breached and situations where an action has been legal but nonetheless misconduct is alleged. In the first case, there are a number of external regulators who might be involved, depending on the area of accountancy where the alleged breach occurred. If the allegation relates to pension schemes, insurance companies or banks, then it is within the remit of the Australian Prudential Regulation Authority (APRA); however, if it relates to any other company, then it is within the remit of the Australian Securities and Investments Commission (ASIC) through the Companies Auditors and Liquidators Disciplinary Board (CALDB). These are all independent government bodies. For other professional misconduct, investigations are carried out by the Professional Conduct Section, the disciplinary arm of the Institute of Chartered Accountants in Australia (ICAA).

In contrast to accountants, the Institute of Actuaries of Australia produces its own guidance notes and runs its own disciplinary scheme, in the same way as the Institute and Faculty of Actuaries did until 2006.

In Canada, accounting standards are set by the Accounting Standards Board (AcSB), whose members are appointed by the Accounting Standards Oversights Council (AcSOC). This itself was set up and continues to be funded by the Canadian Institute of Chartered Accountants (CICA), so is only semi-independent, although its membership is drawn from a wide range of disciplines, including various professions other than accountancy. Auditing quality is maintained by the Auditing and Assurance Standards Board (AASB), which itself is overseen by the Auditing and Assurance Standards Oversight Council (AASOC). This too was set up by CICA.

Canadian actuarial standards are set by the Canadian Actuarial Standards Board (yet another ASB). This is independent of the CIA, but is overseen by a body set up by the CIA, the Actuarial Standards Oversight Council (ASOC). The CIA has its own disciplinary scheme.

In South Africa, accounting standards are set by the South African Accounting Standards Board (one more ASB), which is itself responsible to the South African Minster of Finance. However, disciplinary matters are dealt with by the South African Institute of Chartered Accountants (SAICA), which operates a Professional Conduct Committee and a Disciplinary Committee.

For actuaries the Actuarial Society of South Africa (ASSA) produces professional guidance notes, but there is external oversight provided by the Actuarial Governance Board (AGB). Whilst established by ASSA, members of the AGB are also nominated by non-actuarial financial bodies in South Africa in order to increase the level of external scrutiny. Disciplinary investigations are dealt with by the Professional Conduct Committee and Tribunal of ASSA. Again, non-actuaries serve on both bodies.

As well as the country-based accounting standards, including those above, there are also International Financial Reporting Standards (IFRSs) and their predecessors, International Accounting Standards (IASs) drawn up by the International Accounting Standards Board (IASB). These are intended to provide an alternative to country-specific standards, and in most jurisdictions firms can choose to use either their national or the international standards.

5.9 Industry environment

In the same way that members belong to professional bodies, firms often belong to industry bodies. Similarly, they are subject to controls imposed by industry regulators. The contribution of these bodies to the industry environment is discussed below.

5.9.1 Industry bodies

In banking, most countries have a banking association (such as the British Bankers' Association) representing the interests of financial services firms. In the United Kingdom, there is also the National Association of Pension Funds (NAPF) which represents all parties involved in employer-sponsored (rather than individual) retirement benefits, including not only the pension schemes and sponsors, but also other interested parties, such as fund managers.

International bodies also exist such as the International Swaps and Derivatives Association (ISDA), which represents all parties involved with those types of financial instruments.

The purpose of these bodies is lobbying and member assistance rather than the maintenance of a particular level of skill. As such, their main role is to apply pressure on behalf of member institutions on governments, leveraging the power that individual organisations would have. Little of the lobbying is done in the public eye, so it is difficult to judge the success these bodies have; however, they have been a long-standing feature of the industry landscape.

5.9.2 Industry regulators

The United Kingdom has a unified system of regulation for all industries except occupational pensions, which are regulated by The Pensions Regulator. All other aspects of financial services are regulated by the FSA. The FSA started life as the Securities and Investments Board (SIB), which was established by the Financial Services Act 1986. However, the SIB effectively delegated most of its powers to other organisations set up by the same legislation, namely FIM-BRA, IMRO, LAUTRO and the SFA. These four bodies were classed as Self Regulatory Organisations (SROs). SIB also allowed accountants, actuaries and lawyers to carry out a limited amount of investment business without registering with any of these bodies: they could instead be regulated by their relevant Recognised Professional Body (RPB). In 1995, FIMBRA and LAUTRO were merged into a single organisation, the Personal Investment Authority (PIA), and in 1997 SIB became the FSA and took direct control of the areas previously looked after by SROs.

The FSA has two broad aims: to protect customers and to limit the risk of systemic failure. Two of the most important ways in which this is done is through the regulation of banks and of insurance companies. The first of these is through the implementation of the Second Basel Accord (Basel II) via the EU Capital Requirements Directive of 2006, and the second is through the provisions of Solvency II. The implementation of Basel II is reasonably straightforward in that few additional decisions need to be taken relative to the accord itself; however, Solvency II requires more discretion.

Whilst the FSA's remit covers private pensions, occupational pensions are supervised by the Pensions Regulator. The first body established to oversee all occupational pensions was the Occupational Pensions Board (OPB) created in the Social Security Act 1973. The OPB was replaced by the Occupational Pensions Regulatory Authority (OPRA) following the Pensions Act 1995, and this was itself replaced by the Pensions Regulator further to the Pensions Act 2004. The Pensions Regulator has the power to appoint trustees and to freeze or wind-up a pension scheme. It can also influence the actions not only of pension scheme trustees, but also of other parties. In particular, it has the power to intervene where it is thought that employers, directors or majority shareholders are failing to uphold their responsibilities to pension schemes. It can also step in if it believes that an employer no longer has sufficient resources to continue to support a pension scheme.

The FSA also has a role in relation to non-financial firms that are listed in the UK. In this regard, it has the power to ensure that all firms comply with stock exchange listing rules in relation to disclosure and corporate governance, and the power to cancel their listings if they do not.

It is worth recognising that regulators are, in effect, acting on behalf of governments. This means that there is a risk that they will act for their own benefit first, in particular by extending their influence. In theory, this could lead to excessive regulation; in practice, this rarely seems to have been the case.

5.10 Further reading

As with the internal environment, the advisory risk management frameworks discussed in Chapter 19 offer some of the best insights into the considerations surrounding the external environment.

6

Process overview

Once the context has been defined, the ERM process can be implemented. However, this is not to say that the context cannot change. Both internal and external factors will develop over time, so it is important to constantly be aware of the context and its impact on the process

ERM is implemented as a control cycle. This means that it is a continual process rather than one with a defined start and end. The broad process is given in Figure 6.1.

The first stage in a risk management process is identification, but it is important to ensure that this is done using a consistent risk language and taxonomy. This involves not only defining all of the risks, but also grouping them in a coherent fashion. This is important because it ensures that risks have consistent meanings throughout the organisation.

Risk identification itself involves not only working out which risks an organisation faces, but also a description of the broad nature of those risks. It also means recording them in a consistent and complete way to make reviewing them in future a much easier process.

Having identified the risks, it is then time to assess them in the context of the risk appetite of an organisation. In practice, the risk appetite should be agreed and given in clear terms before risks are actually measured. This includes specifying the risk measures to be used, as well as the values of those measures that are thought to be acceptable. However, because it is helpful to understand the way in which risks can be modelled in order to define a risk appetite, the quantification of risk is dealt with before the question of risk appetite in this book.

Risk assessment includes the question of whether a risk can be quantified, as well of the question of how to sensibly aggregate risks. Having assessed all of the risks, it is then time to compare them with the risk appetites defined earlier and, when needed, to manage them somehow. The management of risks

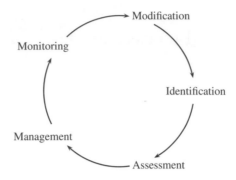

Figure 6.1 The ERM control cycle

is itself not final – the way in which risks are treated should also be kept under constant review. More importantly, if the treatments are not behaving as they should, further action should be taken in respect of that risk.

Constantly reviewing the context has already been discussed. However, there are other ongoing features of the ERM process. The first of these is monitoring. All inputs to and outputs from an ERM process should be reviewed frequently, and if necessary action should be in response to the results. Monitoring includes actively investigating aspects of the process, but can also involve setting trigger points for review, such as significant changes in market indices, or the introduction of new legislation. As part of this process, any losses arising from risks – anticipated or not – should be carefully recorded in order to improve the assessment and treatment of future risks.

A related process is reporting. Monitoring will require information to be produced on a regular basis to the CRF, but broader reporting is also needed. This includes reports for internal stakeholders, such as the board of directors, and external ones, such as regulators and shareholders.

Finally, the ERM process, and all its components should be subject to frequent external audit. This can help validate the system itself as well as the inputs to and outputs from the system.

7

Definitions of risk

7.1 Introduction

When managing risks, it is important to be aware of the range of risks that an institution might face. The particular risks faced will differ from firm to firm, and new risks will develop over time. This means that no list of risks can be exhaustive. However, it is possible to describe the main categories of risk, and the ways in which these risks affect different types of organisation.

7.2 Market and economic risk

Market risk is the risk inherent from exposure to capital markets. This can relate directly to the financial instruments held on the assets side (equities, bonds and so on) and also to the effect of these changes on the valuation of liabilities (long-term interest rates and their effect on life insurance and pensions liabilities being an obvious example). Closely related to market risks are economic risks, such as price and salary inflation. Whilst these risks often affect different aspects of financial institutions – market risk tends to affect the assets and financial risk the liabilities – there is some overlap and both can be modelled in a similar way.

Banks face market risk in particular in two main areas. The first is in relation to the marketable securities held by a bank, where a relatively straightforward asset model will suffice; however, this risk must be assessed in conjunction with market risk relating to positions in various complex instruments to which many banks are counter-parties. It is important both to include all of the positions but also to ensure that any offsetting positions between different risks (for example, long and short positions in similar instruments) is allowed for. *netting*

Market risk for non-life insurance companies again relates to the portfolios of marketable assets held, but is also closely related to assumptions used for claims inflation. Similarly, for life insurance companies and pension schemes, the market risk in the asset portfolios is linked to the various economic assumptions used to value the liabilities, in particular the rate at which those liabilities are discounted. For these two types of institution, market risk is arguably the most significant risk faced.

7.3 Interest rate risk

Interest rate risk is a type of market risk that merits particular consideration. It is the risk arising from unanticipated changes in interest rates of various terms. This can be changes in the overall level of interest rates, or in the shape of the yield curve – that is, in interest rates at different terms by different amounts. As mentioned above, it affects both the value of long-term financial liabilities and the value of fixed interest investments. It is also interesting because expected returns at different points in the future are closely linked through the term structure of interest rates. This means that modelling interest rates brings particular considerations that have resulted in a number of models designed to deal with the issues arising from this term structure.

The term structure of interest rates is an important aspect of interest rate risk. In particular, holding interest-bearing assets to hedge interest sensitive liabilities is only effective if both are affected by various changes in interest rates in a similar way.

7.4 Foreign exchange risk

Foreign exchange risk is another special type of market risk or economic risk. It reflects the risk present when cash flows received are in a currency different from the cash flows due.

Foreign exchange risk is sometimes cited as a component of equity market risk when comparing domestic and overseas equities. However, the underlying cash flows of many domestic equities are from unhedged overseas sources, and in many cases a stock listed on an exchange in one country will have a similar pattern of underlying cash flows to one listed elsewhere. This is particularly true for multinational firms whose main differences are the locations of their head offices. This suggests that unless there are regular, significant arbitrage opportunities, the prices of such stocks should follow each other rather than the currencies of the exchanges on which they are listed.

4 7.5 Credit risk

Credit risk here refers only to default risk. The other main aspect of credit risk – that is, spread risk or the risk of a change in value due to a change in the spread – is covered by market risk. It is also worth noting that there is an element of default risk inherent in traded securities, and this too can be covered by market risk.

For banks, credit risk is often the largest risk in the form of a large number of loans to individuals and small businesses. Another major source of credit risk for many banks is counter-party risk for derivative trades. This is the risk that the opposite side to a derivative transaction will be unable to make a payment if it suffers a loss on that transaction.

Banks also model credit risk for many of the credit-based structured products that they offer, such as CDOs. Complex credit models are needed to accurately model the risk in these products and to correctly divide the tranches.

Whilst credit risk in this context is separate from market risk, it is clear that these risks will be linked, together with economic risk. An economic downturn is likely to increase the risk of default, and particularly for quoted credits an increased risk of default will be higher when the value of the equity stock is lower. It is important to consider these interactions together.

For life and non-life insurance companies, the main credit risk faced is the risk of reinsurer failure. This credit risk is clearly linked to longevity or, more likely, mortality risk for firms writing life insurance business, and to non-life insurance risk for those writing non-life business – when claims experience is worse, then claims from reinsurers are more likely to be made.

Banks and insurance companies also expose their bondholders to credit risk, since they themselves are at risk of insolvency.

The greatest credit risk for most pension schemes is the risk of sponsor insolvency. This is potentially a significant risk, given that the sponsor's covenant can often be in respect of a significant portion of the pension scheme liabilities, and that the creditworthiness of many sponsors leaves much to be desired. An additional credit risk that many pension schemes now face relates to the financial strength of buyout firms. This is an important issue and should be borne in mind by any scheme actuaries considering the buyout firm route.

An important point to note is that credit risk is very similar to non-life insurance risk in that there in both incidence (the probability of default) and intensity (the recovery rate). This is important to bear in mind when considering the techniques used in each to model and manage risk.

ς 7.6 Liquidity risk

Liquidity risk is a risk faced by all financial institutions. Illiquidity can manifest itself through high trading costs, a necessity to accept a substantially reduced price for a quick sale, or the inability to sell at all in a short time scale. This risk, that a firm cannot easily trade due to a lack of market depth or to market disruption is known as market liquidity risk.

However, another aspect of liquidity is the ability of organisations to raise additional finance when required. The risk that a firm cannot meet expected and unexpected current and future cash flows and collateral needs is known as funding liquidity risk.

When assessing the level of liquidity needed from the asset point of view, the timing and the amount of payments together with the uncertainty relating to these factors are key. However, some illiquidity can actually be desirable – if an institution can cope with a lack of marketability in a proportion of its assets, then it might be able to benefit from any premium payable for that illiquidity. However, it must be borne in mind that some illiquid assets also have other issues such as higher transaction costs or greater heterogeneity (real estate and private equity being key examples). Illiquid assets are also less likely to be eligible to count (or at least count fully) towards the regulatory capital of a bank or insurance company.

Assets can provide liquidity in three ways: through sale for cash, through use as collateral and through maturity or periodic payments (such as dividends or coupons).

From the funding point of view, it might sometimes seem attractive to lend over the long term whilst using short-term funding – for example, selling mortgages whilst raising capital in the money markets. This will appeal particularly when long-term interest rates are higher than short-term rates. However, there is a risk here that if the short-term money markets close, then an organisation following such a policy will find itself with insufficient reserves.

This leads us to a discussion of individual institutions. Banks are generally short-term institutions, but, whilst the direction of net cash flow is not clear, it is only in exceptional circumstances that the excess of outflows over inflows will amount to a large proportion of the bank's assets (a 'run on the bank'). This suggests that a degree of asset illiquidity is acceptable. However, reliance on short-term funding can be – and has been – a problem for banks, as it leaves them with insufficient statutory reserves to carry on business.

Life insurance firms generally have long-term liabilities and greater cash flow predictability than banks, so a higher degree of illiquidity is appropriate. Non-life insurance liabilities fall somewhere between bank and life insurance liabilities in terms of both term and predictability, depending on the class of business, so the appropriate level of liability is similarly variable. In both cases, insurance companies are generally less reliant on short-term finance, so financing liquidity is also less of an issue for them than for banks.

Pension schemes are generally long-term institutions; however, a pension scheme which is cash flow positive (where benefits are still being accrued at a higher rate than they are being paid out) can afford to invest a higher proportion of its assets in illiquid investments than can a cash flow negative scheme (a closed or even just a very mature scheme). Having said this, even mature pension schemes or those in wind-up can afford illiquidity in some of their assets: the extent depends on whether those assets match the liability cash flows and, in the case of a wind-up, the extent to which the insurance company is willing to take on illiquid assets.

7.7 Systemic risk

This is the risk of failure of a financial system. Systemic risk occurs when many firms are similarly affected by a particular external risk either directly or through relationships with each other and more broadly. The risk of systemic failure is particularly great if all firms follow similar strategies. In this instance even if all firms are well managed individually, an external event resulting in the insolvency of an individual firm could result in failure of the entire financial system if all firms following the same strategy were similarly affected.

To the extent that systemic risk is driven by the relationships between different parties, it is also known as contagion risk. This can also be described as the risk that failure in one firm, sector or market will result in further failures.

There are four broad types of systemic risk:

- financial infrastructure;
- liquidity;
- common market positions; and
- exposure to a common counter-party.

7.7.1 Financial infrastructure

The risk relating to financial infrastructure arises if a commonly used system fails. This is particularly true if it relates to payment or settlement of financial transactions – such a failure can paralyse the entire financial system.

A classic example of this failure was the Herstatt crisis. In 1974, the small Hamburg-based Herstatt Bank failed due to fraud. Many US banks had made large payments to Herstatt in West German Deutschmarks earlier in the day, and were due to receive payments back in US dollars. By the time the dollar payments were due, Herstatt had been declared insolvent. This led to the paralysis of the interbank market – since the exact exposures of all banks to Herstatt was unknown, no banks wanted to be the first to make any further payments in case their counter-parties were then also declared insolvent. The effect therefore spread from the initial insolvency to affect all transactions between banks.

7.7.2 Liquidity risk

Liquidity risk has already been discussed, but it becomes a systemic risk if a run on banks occurs, or if short-term money markets become less liquid. Both result in a reduced ability for banks to raise the capital they need to remain solvent. As such this is a funding liquidity risk. The global financial crisis that started in 2007 resulted from such issues. Here, the reluctance of banks to provide short-term lending to each other, and for the wider market to provide short-term funding, led to the risk of collapse for many banks, some of which were saved by government finance and others which were left to become insolvent. The reduced solvency of banks, coupled with a desire to increase the amount of free capital held, meant that banks were less able to lend money to firms and individuals. The knock-on effect was that the economy as a whole was damaged by the system-wide fall in funding liquidity.

7.7.3 Common market positions

Exposure to common investment positions can affect individual investments or whole sectors or markets. The resulting risk is also known as feedback risk, the risk that a change in price will result in further changes in the same direction. Sometimes such movements are simply a result of sentiment and might be better characterised as behavioural risks that cause stocks either individually or in groups to trend upwards (bubbles) or downwards (crashes). These have been seen since the South Sea Bubble – a speculative investment bubble and subsequent crash of the eighteenth century – and even earlier.

However, just as important as the sentiment-driven movements are the downward risks of forced sales. Here, a fall in price of a risky asset can reduce the solvency of an investor, forcing the investor to sell the asset and buy a risk-free alternative to protect a statutory solvency position. This forced sale of the asset causes a further fall in price, resulting in further solvency problems and even more sales. If this risk extends to a significant proportion of the market, then it can threaten systemic stability. This is what happened in the case of Long Term Capital Management (LTCM), a hedge fund that was forced into near-insolvency in 1998. It was so large that if it had been obliged to close its derivative positions, the effect on prices would have been such that a number of firms with similar positions to LTCM would also have been forced into insolvency.

7.7.4 Exposure to a common counter-party

Exposure to a common counter-party is another contagion-type systemic risk. This risk requires a relatively small failure to cascade through several layers of investors – so not just those investing in the failed firm, but those investing in institutions that invested in the failed firm, and so on. To become a systemic risk, the ultimate effect must be one that damages the stability of an entire financial system.

These gains and losses might stem from financial reasons, in particular if holdings in a failed firm cause losses more widely, through some sort of direct financial relationship; however, they might simply be due to a loss of confidence in firms carrying out similar business to a failed firm. For example, if a particular country's credit rating fell, resulting in a fall in the prices of that country's debt, then this might cause the debt of similar countries to be similarly affected. This might be due to exposure to similar economic factors, or it might be simply driven by negative sentiment. Contagion could also result in wider effects being felt. For example, banks holding the affected government bonds could find their share prices falling, and any cross-holdings of one bank's shares by another could exacerbate the effect.

Such a contagion effect could also cause other types of systemic problems. For example, anything that reduced the solvency of banks could also reduce their ability to raise capital, leading to systemic liquidity issues as described above. Here too sentiment has a role to play, since a reduction in lending could be driven by fear as much as economic rationality.

7.8 Demographic risk

Demographic risk can be interpreted as covering a wide range of risks. It includes proportions married or with partners, age differences of partners, numbers of children (all for dependent benefits), lapses (for insurance products) or withdrawals (for pension schemes), pension scheme new entrant and retirement patterns, but, most importantly, mortality or longevity.

Mortality risk (the risk that a portfolio will suffer from mortality being heavier than expected) and longevity risk (the risk that a portfolio will suffer from mortality being lighter than expected) are significant factors for both pension schemes and life insurance companies. The former only suffers from the longevity risk, but both risks are present for life insurance companies: term and whole-life insurance carries mortality risk, whereas general and pension annuity business carry longevity risk. The International Actuarial Association (2004) defines four types of mortality or longevity risk:

- level;
- volatility;
- catastrophe; and
- trend.

Level risk is the risk that the underlying mortality of a particular population differs from that assumed. This is distinct from volatility risk, which is the risk that the mortality experience will differ from that assumed due to there being a finite number of lives in the population considered. Losses can be made due to volatility risk in a small population even if the underlying mortality assumption is correct and there is no level risk.

An extreme version of volatility risk is catastrophe risk. This is the risk of large losses due to some significant event increasing mortality rates beyond simple random volatility. Examples would be natural disasters, such as floods or earthquakes or pandemics. It is important to note that volatility risk only affects mortality risk. Whilst it is possible to have a sudden spike in mortality rates increasing losses due to a temporary increase in rates, it is not likely that a sudden temporary dip in mortality rates, increasing losses in annuity portfolios, will occur.

The final risk is trend risk. This is the risk that mortality rates will improve over time at a rate different to that assumed. This risk is distinct from the other three risks as it considers the development of mortality rates over the long term, whereas the other three risks consider mortality rates only in the immediate future.

Lapses, withdrawals and pension scheme new entrants and early retirements are also of particular interest because they are not necessarily independent, either from each other (for the pension scheme items) or from market and economic variables. For example, early withdrawals from a pension scheme are likely to be higher if a sponsor has to make employees redundant in the face of difficult economic conditions. This suggests that some demographic variables should be considered together with market and economic conditions.

However, it is worth noting that whilst salary increases might be allowed for in funding valuations and for other planning purposes, the firm's obligation only extends as far as accrued benefits, which are not affected by these decrements. The main exception here is if unreduced early retirement is offered as an alternative to redundancy. This can significantly increase the value of benefits a pension scheme is committed to pay.

7.9 Non-life insurance risk

This is generally the main risk faced by firms writing non-life insurance business. The shorter time horizon for most non-life insurers mean that market and economic risks are less relevant, although this is not necessarily the case, and some long-tail classes can mean that the investment and economic risks are significant.

Non-life insurance risk is the key factor in arriving at a correct premium rate for the business to be written and in arriving at the correct reserves for the business that has already been taken on. Two aspects need to be considered: the incidence of claims, and their intensity. In a way, incidence is not dissimilar to mortality risk, except that it can be assessed over a shorter time horizon, is often at a higher rate (for some classes of insurance) and can be much less stable from year to year.

Unlike most mortality risks, the intensity of each claim is not necessarily the same from one claim to another. In some cases the maximum possible claim is known (for example, buildings insurance), whereas for others the maximum potential claim amount is unlimited (for example, employer liability insurance). Because the risks differ significantly from class to class, a variety of approaches is needed to model them correctly.

Another similarity with mortality risk is that it too can be considered as four separate risks:

- underwriting;
- volatility;
- catastrophe; and
- trend.

Underwriting risk is the analogue of the level of risk in life insurance: it is the risk that the average level of claims in the portfolio as measured by incidence and intensity is different from that assumed.

Volatility risk is a risk that remains even if risks are correctly underwritten, and reflects uncertainty in the incidence and intensity of claims resulting from the fact that only a finite number of policies exist.

Catastrophes can occur when high-intensity low-probability events occur; however, they can also occur as a combination of a smaller event combined with a high concentration of claims by frequency, perhaps with an unusually high average claim amount. In all cases, catastrophes can be caused by some natural disaster such as a hurricane, flood or earthquake, or by something less direct such as a legal judgement affecting a particular class.

Trend risk is again the odd one out as it relates to future changes. It refers to the risk of unexpected changes from current levels in the incidence and intensity of claims. The incidence part of this risk relates to the change in the number of claims per policy. The intensity aspect is distinct from claims inflation, which is effectively an economic risk; rather it is about the type of claim. It is important to note that the trends here might be fairly short lived, and might be better described as cycles since they often follow (or lead) economic or underwriting cycles.

The final three risks are often referred to as reserving risk.

Although this risk is greater than market or economic risk for an insurer, in many cases it should be considered together with these risks. In common with some of the demographic risks, non-life insurance risk changes over the economic cycle with claims in certain classes being higher in economic downturns. Considering claim levels together with economic and market variables would seem to be sensible here as well.

7.10 Operational risks

Operational risks are a group of risks which impact on the way in which a firm carries on business. They include a wide number of different risks, which often overlap each other to a significant degree. This means that any classification is necessarily arbitrary; however, the classifications described do cover the majority of risks faced.

If not correctly managed, these risks can be the biggest risks faced by any organisation. Operational failures have led to the ultimate demise of more than one firm. This is because poor control of operational risk allows other types of risks, such as market or credit risk, to be excessive. On a less extreme level, operational failures or inadequacies can result in mistakes and inefficiencies that result in fines or lost business. Similarly, poor project implementation has been a source of shareholder value destruction in many firms across many industries, as has strategic mismanagement.

7.10.1 Business continuity risk

This is the risk that an external event will affect the physical ability of a firm to carry on business at its normal place of work. This might be a major event affecting a whole city or country, such as an earthquake or hurricane, or it might be an event affecting only the firm, such as power failure or arson. In the case of the former, it is important to consider the extent to which these risks are linked to other risks faced by a firm. For an insurer, the link is clear, but as seen following the Kobe earthquake, natural disasters can also have an effect on financial markets.

It is also important to consider the extent to which suppliers and business partners might also be affected by such risks, and any concentrations of risk arising across suppliers or between one's own organisation and a supplier.

7.10.2 Regulatory risk

Regulatory risk covers the risk that an organisation will be negatively impacted by a change in legislation or regulation, or will fall foul of legislation or regulations that are already in place. Such changes might result in additional compliance costs being faced, existing activities being prohibited or sales of business units being required. A failure to comply with existing rules might bring fines or even expensive litigation. Even if this does not occur, there might be a loss of business due to a failure of confidence.

As well as regulations and legislation from governments, any firms quoted on stock exchanges must also follow the listing rules of those markets, or face censure from the exchange.

The large number of regulatory issues was discussed earlier under the external risk management environment, and a lack of compliance in any of the areas covered can be costly.

7.10.3 Technology risk

This is the risk of a failures in technology, including unintended loss or disclosure of confidential information, data corruption and computer system failure. The latter is particularly important if a business transacts a significant proportion of business electronically, or if a large number of employees work remotely. There is clearly an overlap between technology and crime risk, discussed later, if the failure in technology is deliberate. For example, hacking and electronic data theft are criminal acts, as is damage caused by a virus or a loss of business caused by a denial-of-service attack (a deliberate attempt to overwhelm the bandwidth of a web server).

Another aspect of technology risk is the risk that there are undiscovered errors in software used in an organisation. Such errors might result in losses from mis-pricing, or in incorrect payments being made. The results could be direct financial loss together with a loss of business resulting from a lack of confidence.

Technology risks often increase exponentially with the number of systems an organisation has. Getting different systems to be able to communicate effectively and consistently can be difficult, and data errors can occur. This issue can arise particularly when firms using different systems merge.

7.10.4 Crime risk

Crime risks result from the dishonest behaviour of individuals in relation to a firm. This includes the theft of money or intellectual property by an employee (fraud) and the unauthorised access of systems by an outside party with the same aims (hacking). As discussed above, computer-related crime could also be regarded as a technology risk. Similarly, the risks described here could be included under people risks, although these risks differ because they relate specifically to crimes. Furthermore, crime risk could include aspects of moral hazard or adverse selection if there is deliberate non-disclosure in obtaining insurance or loans, or fraudulent claims are made; however, those are risks involved in carrying out particular types of business, whereas crime risks are directed against the firm rather than one of its business lines.

Crime risks do include risks such as arson, which disrupt a firm's business. Even though the effects are the same as any other business continuity risk, the measures taken to guard against criminal acts and the circumstances in which such acts might occur are quite different.

Crime risks are not necessarily multi-million dollar fraudulent enterprises – claiming expenses for taxis taken for personal business is still fraud, and can be significant in aggregate. Much of this risk relates to the culture of an organisation, an industry and a country; however, it can also be affected by the economic climate. When times are harder, fraud might be more likely – although anti-fraud measures might be more stringent if companies are also feeling the pinch.

7.10.5 People risk

People are a factor in a large number of risks faced by organisations, including of course in the risk of criminal actions. However, the term 'people risk' is reserved for non-criminal actions that can adversely affect an enterprise.

Employment-related risks

People risks start with the risk that the wrong people are employed. It is important that the people employed have the skills an organisation needs to run its business. Once employees have been recruited, it is important that the right ones are promoted, and that such promotions are good for the organisation. Similarly, it is important that the right employees are retained. Losing employees can result in a loss of valuable intellectual capital and can damage the morale of remaining employees. It can also be expensive – recruitment costs time and money, and every time a new recruit is taken on, there is the risk that the employee is not right for the role or the organisation. At its most extreme, this can be another case of adverse selection against an organisation by an employee.

Another aspect of people risk relates to the risk of disruption caused by employees. This can be a result of absenteeism, including through sickness, and on a wider basis through industrial action. Whilst the negative publicity and widespread disruption caused by the latter make it an important issue, the long-term damage to an institution caused by persistently absent employees can also be significant – as well as the financial cost involved, morale can suffer.

If an employee must be dismissed, then the legal implications need to be considered. Employment itself involves a number of legal aspects. The terms of employment contracts must be considered carefully and legislation relating to issues such as discrimination and statutory leave for maternity and paternity must be complied with.

Adverse selection

Adverse selection is a particular issue relating to underwriting risk in both life and non-life insurance. It is the risk that the demand for insurance is positively correlated with the risk of loss. For example, unhealthy people might be more likely to buy life insurance if they are charged the same premiums as healthy people. Adverse selection arises as a result of asymmetry of information and the inability to differentiate between different risks when pricing. In extreme cases, it can lead to market failure, as with 'Akerlof's lemons' (Akerlof, 1970).[1]

Adverse selection is also an issue for banks, where those with poor credit ratings will be more likely to apply for loans with banks that do not charge higher rates to reflect the higher risks. It can even be an issue for defined benefit pension schemes if the pension can be commuted to a tax-free cash lump sum

---

[1] This article shows that if a buyer cannot distinguish between good cars ('peaches') and bad cars ('lemons'), then those owning peaches will not wish to sell at the price offered, so only lemons will be sold.

at an actuarially calculated rate, with those having shorter expectations of life being more likely to commute pension.

Moral hazard

This is the risk that behaviour will depend on the level of their exposure to a particular risk. In particular, if there is insurance in place, the incentive to avoid risk is reduced. An example of this is the potential incentive for pension scheme trustees to take more investment risk after the introduction of an industry-wide insurance scheme for pension scheme members. As with adverse selection, moral hazard is linked to the asymmetry of information, but it is more about the inability of an insurer to control the behaviour of the insured once the insurance is in place. In simplistic terms, if someone is more likely to juggle a set of lead crystal glasses because he has household contents insurance in place, then this is moral hazard; if someone who enjoys juggling lead crystal glasses is more likely to buy household contents insurance, then this is adverse selection.

Agency risk

Agency risk is the risk that one party appointed to act on behalf of another will instead act on its own behalf. Company managers acting for themselves rather than the shareholders whose interests they are supposed to protect are the prime example. In banks a key agency risk occurs if bonus systems create perverse incentives for traders – for example, if good results can give unlimited bonus potential but the downside from poor results is limited, then this can create an incentive for traders to take too much risk. Within insurance companies, the fact that the actuaries responsible for regulatory reporting are remunerated by the firms, which might be more focussed on shareholder value than policyholder security, gives another example of agency risk. For pension schemes, conflicts of interest are the main sources of agency risk, examples being company-appointed trustees and actuaries acting on behalf of both the employer and the trustees. However, another key agency risk for pension schemes relates to the views of company management on investment policy. There is a risk that managers will aim to increase pension scheme equity weightings in order to improve apparent profitability (through the effect of the impact on the expected return on assets) and to reduce transparency (through the opportunity to use opaque actuarial techniques).

The costs arising from agency risks are agency costs. There are two main sources for these costs. The first is the loss associated with the action of the agents, whilst the second is the cost of any action taken to modify the behaviour of agents. A clear principal here is that the cost of any action should not exceed any savings made – in other words, action should only be taken if it reduces the total agency cost.

✓ 7.10.6 Bias

A systemic risk which can be deliberate or subconscious is bias. This is often the manifestation of a form of agency risk, where a project will be given too optimistic an appraisal because approval will result in greater rewards for a proponent. Similarly, insurance or pension reserves might be understated in order to increase apparent profits, or to improve the standing (and maintain the appointment) of the professional advisor providing the valuation.

Deliberate bias can arise if key risks are intentionally omitted or down-played, or their consequences misrepresented. Similarly, the links between different risks might be understated, as might the impact of the business or underwriting cycles. There might also be deliberate optimism around positive outcomes, such as growth in future business or returns on assets, or simply a failure to allow for the true level of uncertainty. These events can be compounded if the assumptions underlying the down-playing of downside risks are inconsistent with those underlying the over-statement of upside potential.

Many of the above biases can also arise unintentionally. Risks can be forgotten accidentally, or underestimated due to a lack of data. However, it is difficult to determine the extent to which many of these accidents are true oversights.

A particular unintentional bias to which those working in finance are susceptible is overconfidence. In particular, it has been said that overconfidence is greatest for difficult tasks with low predictability, which lack fast clear feedback (Jones *et al.*, 2006). These criteria could be applied to most financial work. Other aspects of overconfidence such as the illusion of knowledge (the belief that more information improves forecast accuracy) or the illusion of control (the belief that greater control improves results) have wide-ranging implications for all areas of finance, particularly as the volume of information that is readily available is growing rapidly all the time.

Anchoring is another behavioural bias with clear implications in the world of finance. This occurs when decisions are made relative to an existing position rather than based solely on the relevant facts – the question asked is 'given where we are, where should we be?'; it should be 'given the relevant facts, where should we be?'. This bias can clearly be seen when, for example, insurance reserves change only gradually in response to rapidly changing information.

Representativeness (making the assumption that things with similar properties are alike) and heuristic simplification (using rules-of-thumb) can also
be a source of problems in all financial organisations where the eventual
level of risk might turn out to be very different to an initial estimation or
approximation.

7.10.7 Legal risk

Legal risk is sometimes used to describe the regulatory risks covered above;
however, here it is used to describe the risk arising from poorly drafted legal
documents within an organisation. This extends to policy documents, which
form legal agreements between firms and policyholders. Legal risk can also be
linked to regulatory risk, since ambiguities in legal contract may ultimately be
dealt with by courts.

7.10.8 Process risk

A key component of operational risk is the risk inherent in the processes used
by firms. The range of processes used by institutions is huge – some examples
are given below:

- credit checks on bank loan and mortgage applicants;
- bank payment clearing;
- bank collateral management;
- bank trading and settlement;
- dividend and coupon payment;
- employee remuneration;
- policy underwriting;
- claim handling;
- benefit payment;
- premium and contribution collection;
- external investment manager monitoring; and
- risk management.

This list is not exhaustive, but it gives an idea of the range of systems and processes that are in place. A failure in any one might lead at
best to embarrassment, and at worst to litigation. Even if processes do not
fail, inefficient processes can damage the competitiveness of an organisation,
resulting in too much time being taken or money being spent to complete particular tasks. In this sense, process risk is clearly closely linked to
technology risk.

7.10.9 Model risk

This can be thought of as a type of process risk; however, because of its importance to financial institutions, it is worth considering separately. Model risk is the risk that financial models used to assess risk, to determine trades or otherwise to help make financial decisions are flawed. The flaws can be in the structure of a model, which may be overly simplistic or otherwise unrealistic, or it can be in the choice of parameters used for an otherwise sound model.

Model risk might also relate to the incorrect translation of a model from theory into code, although this is more of a technology risk, since it assumes that the model itself is sound.

Model risk also occurs if models are put to uses other than those for which they were intended. For example, a model may give reasonable estimates of the expected returns from a particular strategy, and the range of results that might be expected in normal market conditions, but it might be very poor at predicting the range of adverse outcomes that might occur in stressed markets. In other words, model risk is present if models are put to inappropriate uses. An example is the Black–Scholes (Black and Scholes, 1973) model for option pricing. This is good for giving the approximate value of a financial option for, say, accounting for stock options granted to directors, but is entirely inappropriate for determining tactical options trades.

7.10.10 Data risk

Another sub-type of process risk is the risk of using poor data. This is a particular issue in relation to personal data. Even if there is no deliberate misreporting, data can be entered incorrectly, or fill-in codes can be used when information is not available. A separate issue arises when data are being analysed, in that a single individual may have a number of records in his or her name. This can skew any analysis carried out if duplicates are not removed or consolidated.

7.10.11 Reputational risk

Reputational risk is essentially a risk that arises from other operational risks. For example, the loss of data – potentially a technology risk – can result in a loss of confidence in an organisation due to reputational damage. Similarly, repeated delays in claim payments by an insurance company is likely to be a process risk, but the subsequent loss of business due to a loss of confidence in the firm is a reputational issue.

What this means is that when considering the direct cost that might arise from particular operational risks, it is important also to consider any potential subsequent costs arising from loss of business due to reputational damage.

7.10.12 Project risk

Project risk is an umbrella term covering all of the various operational risks in the context of a particular project. In the case of financial institutions such projects may include the creation of physical assets, such as property development for investment purposes, or a new head-office building or computer system for the institution itself. However, they may also include projects of a less tangible nature associated with the launch of a new product, expansion overseas, winding up or downsizing. The inclusion of this term is really a reminder that operational risks occur not just in the day-to-day running of an organisation, but also in the approach to each project carried out.

7.10.13 Strategic risk

Strategic risk is similar to project risk, in that it includes many of the operational risks covered previously. However, it covers a more fundamental subject: the achievement of the organisation's core objectives.

The most basic strategic risk is that no coherent strategy for future development exists; however, assuming that this risk is overcome, it is important that an organisation makes a conscious decision of what its strategy is and how it intends to implement it.

7.11 Residual risks

Residual risks are those risks that remain once any action has been taken to treat the risks. It is important that once risks are dealt with, any risks that remain are recognised and correctly allowed for.

There are a number of distinct types of residual risk that exist in the financial services sector. The first has already been mentioned – credit risk. This occurs in the form of counter-party risk if, for example, derivatives have been used to reduce risk. Specifically, a pension scheme might use interest rate and inflation swaps to reduce its exposure to changes in nominal and real interest rate risks. However, in entering into these swaps, it is taking on an additional (though residual) risk, namely the risk that the bank with whom it has traded in unable to make its payments on the swap. Similarly, a pension scheme buying annuities in respect of its pensioner liabilities is exposed to the residual risk that the insurer providing the annuities might become insolvent.

In the interest and inflation swaps example, other residual risks also remain. There is the risk that the life expectancy of the pension scheme members will be different to that expected. This is a function of the fact that the swaps do not deal with this risk. However, there is also the risk that the change in the value of the liabilities as a result of interest rate changes will not be exactly matched by changes in the value of the swaps. This might occur if only a few swaps have been used to try and match the liabilities. This particular residual risk is known as basis risk, the risk arising from an imperfect hedge.

7.12 Further reading

There are a large number of books that seek to define risks, and the way in which risks are defined is not necessarily consistent. Chapman (2006) defines a wide range of risks in a broad context, and considering the risks faced by firms that are not necessarily in the financial industry can be helpful. Lam (2003) looks at fewer risks, but these are considered in the context of the financial services industry. Mandatory risk frameworks also use very precise definitions of risk for the purpose of calculating capital requirements, so it is important to be familiar with the terminology used here.

8

Risk identification

8.1 Introduction

Once the context within which risks are being analysed is clear, and full risk taxonomy available, it is time to start identifying risks. The point of the risk identification process is to decide which of the many risks that might affect an organisation are currently doing so, or may do so in future. Part of the risk identification process also involves determining the way in which risks will then be analysed, in particular whether a qualitative or quantitative approach will be used. These, and other factors, are included in a risk register, discussed later in this chapter.

Risk identification should be done as part of a well-defined process. This ensures not only that as many risks as possible are identified, but also that they are properly recorded.

There are four broad areas to risk identification. The first concerns the tools that can be used, whilst the second concerns the ways in which the tools are employed. Identification also includes an initial assessment of the nature of the risk, and also the way in which the risk is recorded. Each of these aspects is discussed in turn.

8.2 Risk identification tools

In this section, a range of potential risk identification tools are discussed. These can generally by used in a number of ways and simply describe the starting point for the generation of ideas. Some common tools are described below.

8.2.1 SWOT analysis

SWOT – standing for strengths, weaknesses, opportunities and threats – analysis is one of the best-known techniques for strategy development.

112

Table 8.1. *Potential factors in SWOT analysis*

Strengths	Weaknesses
Market dominance	Low market share
Economies of scale	Extensive specialism
Low cost base	High cost base
Effective leadership	Lack of direction
Strong balance sheet	Financial weakness
Good product innovation	Reliance on contracting markets
Strong brand	Limited recognition
Differentiated products	Differentiation by price alone
Opportunities	**Threats**
Innovation	New entrants
Additional demand	Price pressure
Opportunities for diversification	Contraction of key markets
Positive demographic change	Damaging demographic change
Cheap funding	Falling liquidity
Economic liberalisation	Increased regulation

Source: Based on Chapman, R.J.: *Simple Tools and Techniques for Enterprise Risk Management* (2006).

However, it can also be used to identify risks. Having said this, its scope is much broader, covering not just the negative aspects of the risks but the positive prospects for future strategies.

Strengths and weaknesses are internal to the organisation, whilst opportunities and threats are external. In this way, SWOT analysis ensures that both the internal and external risk management contexts of an organisation are considered.

It is important to recognise what constitutes a strength or a weakness. In particular, strengths only matter if they can be used to take advantage of an opportunity or to counter a weakness; conversely, weaknesses are important only if they result in exposure to a threat.

Some broad categories for SWOT analysis are given in Table 8.1.

8.2.2 Risk check lists

Risk check lists are lists of risks that are used as a reference for identifying risks in a particular organisation of situation. There are two main sources for such check lists: *experiential knowledge* is the collection of information that a person or group has obtained through their experiences, whilst *documented knowledge* is the collection of information or data that has been documented about a particular subject by some external source. Documented knowledge is also sometimes referred to as *historical information* if the risks concerned are widely accepted as fact.

Caution must be used when using any knowledge-based information to ensure it is relevant and applicable to the current situation. It is also important to understand any caveats that may accompany the documented information.

8.2.3 Risk prompt lists

Similar to check lists are prompt lists. However, rather than seeking to pre-identify every risk, prompt lists simply identify the various categories of risk that should be considered. These categories are then intended to prompt a broader and more specific range of risks for the institution being analysed.

The classic prompt list categories where political, economic, social and technological, giving rise to PEST analysis. However, environmental, legal and industry risks are now also commonly cited, giving the acronym PESTELI.

8.2.4 Risk taxonomy

Part-way between the check list and the prompt list falls the risk taxonomy. This is a more detailed list than the prompt list, containing a full list and description of all risks that might be faced, with these risks also being fully categorised. However, it is not as specific as the check list, containing both a wider range of risks – some of which may be irrelevant – and less focussed than an institution- or project-specific check list.

8.2.5 Risk trigger questions

Risk trigger questions are lists of situations or events in a particular area of an organisation that can lead to risk for that organisation. They are derived from situations or areas where risks have emerged previously.

8.2.6 Case studies

Case studies can perform a number of uses in risk identification. First, they can suggest specific risks, particularly if there are clear parallels between the organisation in question and that in the case study. However, even if the case study concerns a very different type of organisation, it might suggest areas where similar risks might occur in future. Case studies are particularly useful as they do not detail risks in isolation, but show the contexts in which risks are allowed to develop and the links between various different risks.

8.2.7 Risk-focussed process analysis

This approach to risk identification involves constructing flow charts for every process used by an organisation and analysing the points at which risks can occur. Every broad process should be listed and described in detail, taking into account who and what is involved and, therefore, where failures can occur. Ideally, the links between different processes should also be considered.

In order to establish what the processes are, it is important to have input from all key areas of an organisation to establish how it does what it does. The areas for a financial services firm might include:

- advertising products;
- selling products;
- collecting premiums;
- investing assets;
- making payments;
- raising capital;
- placing contracts (core and incidental);
- hiring staff;
- paying salaries . . .

. . . and so on.

8.3 Risk identification techniques

There are a number of ways in which risks can be identified. Each have their advantages and disadvantages, but all should take information from as wide a range of contributors as possible. This means that employees and directors from all departments should be involved, and from all levels of seniority. There should also be a mix between those who have been with an organisation for some time with a depth of experience, and recent joiners with the advantage of fresh views. Finally, the contributors should not necessarily be confined to people within an organisation – the perceptions of external stakeholders are important and worth considering.

8.3.1 Brainstorming

Brainstorming is the term used to describe an unrestrained or unstructured group discussion. Such a discussion should be led by an experienced facilitator in order to draw out as many different points as possible, to ensure that as broad a range of points as possible is investigated and that each point is discussed in sufficient depth.

When brainstorming, it is important that ideas are not initially censored – all ideas should be recorded, no matter how relevant they initially appear to be. This is because even bad ideas may trigger good suggestions from other members of the group.

Once a detailed list of risks has been compiled, the facilitator can organise the risks into appropriate groups, removing any which are irrelevant.

It is not necessary for the facilitator to be an expert in the business for which risks are being investigated; however, it is helpful if members of the group identifying the risks are collectively familiar with all aspects of the organisation. This does not mean that all members of the group must have familiarity with all aspects. In fact, new perspectives on potential risks are often helpful; however there must be sufficient knowledge within the group of the way in which an organisation works.

A potential drawback of brainstorming is that the potential exists for 'free riders' to exist – individuals may attend brainstorming sessions but fail to contribute. Whilst good facilitation can avoid this to an extent, an alternative is to require each group member to lead the discussion on a particular risk category, producing the first set of ideas around which other suggestions can be allowed to form.

However, brainstorming also has other disadvantages, such as the need to get all the participants together in a single location. Having all participants together might also lead to convergent thinking, with participants' ideas being influenced by prior contributions. There is also the risk that the open nature of brainstorming can lead to a lack of completeness, even with a good facilitator. Other approaches address these shortcomings.

8.3.2 Independent group analysis

This is another technique for group analysis which attempts to avoid some of the problems that working in groups can cause. In this approach, all participants write down in silence and without collaboration ideas on the risks that might arise. These ideas are aggregated by a facilitator after which there is a discussion. The primary purpose of the discussion is to determine the exact nature of the various risks and the extent to which the risks identified are genuinely distinct from one another. However, it also serves to draw out justifications for the relevance of the risks identified, with each risk being defended by someone who has proposed it. Finally, there is a discussion of the relative importance of the risks. However, the ranking of risks is also done independently and, this time, anonymously. The ranks are then combined mathematically to give an objective ranking of risks.

The approach described here is designed in particular to avoid convergent thinking. However, it is heavily dependent on the constitution of the group. If there is a lack of balance, then the results will be biased. For example, if too many participants are from the finance department, then corporate finance risks will be ranked too highly.

8.3.3 Surveys

A way of ensuring wider participation in the process is to carry out a survey of risks instead, either by post or by email. A survey would include a list of questions about different aspects of an organisation and its place in the industry to try to draw out the risk faced.

As well as allowing the views of a much larger group to be canvassed, this approach can ensure that a wide range of risks is covered and avoids the risk of participants influencing each other. However, the responses can be heavily influenced by the way in which questions are asked – the problem of framing. There is also a risk that people will not respond to the survey, and a low response rate could invalidate a risk identification exercise, particularly if key business units fail to produce any responses. Furthermore, if the results are used to rank the results, bias could occur in the rankings unless some sort of weighting is applied.

Surveys also pose a problem because of the way in which information can be collected. The only way to ensure that the results can be analysed quantitatively is to use a multiple choice approach. However, this will clearly have the effect of limiting the possible responses, so the only risks analysed will be those initially suggested. Alternatively, responses can be given in free text. However, these can be difficult to analyse. In particular, it can be difficult to work out the extent to which the same risk is being raised several times or several different risks are being identified. This is particularly true if the survey does not allow any subsequent questioning of participants to clarify the initial responses.

If a survey is used, then it is important that a pilot survey is carried out first. This can help to ensure that the questions asked are as unambiguous as possible and that the full survey gives results that are as useful as possible.

8.3.4 Gap analysis

One particular type of survey that can be used in risk identification is gap analysis. This involves asking two types of question, to identify both the desired and actual levels of risk exposure. It is important to note that the two types of question will not necessarily be asked of the same people. Whilst senior management might have strong views on the desired levels of risk exposures, it is more likely that more junior employees from around the firm will have clearer ideas of the actual levels of risk to which the firm is exposed.

If gap analysis is carried out by survey, then it potentially suffers from the same shortcomings as any other survey-based approach; however, there are other ways of gathering knowledge.

8.3.5 Delphi technique

This is another type of survey, where acknowledged experts are asked to comment on risks anonymously and independently. In order to make best use of expert knowledge, and time is taken to properly analyse the results rather than the answers simply being aggregated, the questionnaires used here generally allow much more flexibility than surveys otherwise might. The Delphi technique starts with an initial survey being sent out. This is followed up by subsequent surveys which are based the responses to the initial survey. This process continues until there is a consensus (or stalemate) on the nature and importance of the risks faced, meaning that the technique is used for assessment as well as identification.

The design of the initial questionnaire is important here, but not as important as subsequent revisions based on new information.

8.3.6 Interviews

Interviewing individuals is another way to identify the risks present in an organisation. This has the advantages of structure and independence of view that come with a survey, but also with the advantage that, if an answer is unclear, clarification can be sought immediately. The potential framing of questions is again an issue here, as is the time that would be taken to carry out all of the interviews. This is perhaps the most time-consuming approach of all those discussed. As a result, several interviewers might be used. However, if this is the case, it is important that the different interviewers' results are treated consistently.

8.3.7 Working groups

The approaches discussed so far are suitable for identifying which risks might be important for an organisation to consider. However, once a risk has been identified – for example, the risk of payment systems failure – it may be appropriate to investigate more thoroughly the exact nature of this risk. Working groups, which are groups comprised of a small number of individuals who have familiarity with the issue concerned, provide a good way to analyse a particular area or topic. Such groups can discover additional details about the risks that exist beyond the level of detail that might be expected to arise from the initial risk identification exercise.

The remit of the working party may extend beyond the task of identification and into analysis. This is particularly true for unquantifiable risks.

8.4 Assessment of risk nature

The identification of risks should also include an initial assessment of the nature of those risks, in particular whether they are quantifiable or unquantifiable. The process for analysing quantifiable risks is quite involved and modelling of these risks will typically be done by a specific group within the organisation. However, unquantifiable risks can often be analysed by the groups that identify them. Unquantifiable risks are discussed in more detail later.

8.5 Risk register

Once identified, risks should be put onto a risk register. This is a central document that details all of the risks faced by an organisation. It should be a living document which in constantly updated to reflect the changing nature of risks and the evolving environment in which an organisation operates.

Each entry in a risk register should ideally include a number of factors:

- a unique identifier;
- the category within which the risk falls;
- the date of assessment for the risk;
- a clear description of the risk;
- whether the risk is quantifiable;
- information on likelihood of the risk;
- information on the severity of the risk;
- the period of exposure to the risk;
- the current status of risk;
- details of scenarios where the risk is likely to occur;
- details of other risks to which this risk is linked;

- the risk responses implemented;
- the cost of the responses;
- details of residual risks;
- the timetable and process for review of the risk;
- the risk owner;
- the entry author.

8.6 Further reading

The advisory risk frameworks describe useful approaches to risk identification, and Chapman (2006) covers this area in some detail. The Delphi technique is discussed in detail by Linstone and Turoff (2002).

9

Some useful statistics

Many of the measures here will be familiar to readers from early studies in statistics. However, it is important that the basic statistics are fully understood, as they form an important basis for subsequent work.

9.1 Location

A measure of location gives an indication of the point around which observations are based. It can refer to one of two points. The first is a parameter used in a statistical distribution to locate it; the second is a statistic calculated from the data. The focus here is on the calculation of the second item from the data. This can be used to estimate the first item for some distributions, but this is not necessarily the case.

9.1.1 Mean

The mean is often the most useful – and used – measure of location. The sample mean, $\bar{X}$, of a set of observations is given as:

$$\bar{X} = \frac{1}{T} \sum_{t=1}^{T} X_t.$$
(9.1)

This is the most commonly used measure of central tendency in modelling: summing the observations and dividing by their number. The mode of a distribution (the most popular observation, or the maximum value of the function) does not have a clear application in stochastic modelling, and the median (the observation that is greater than one half of the sample and less than the other half) is of more interest in the risk assessment phase.

The population mean, μ, is calculated in the same way as sample mean, $\bar{X}$, so it is also true to say that:

$$\mu = \frac{1}{T} \sum_{t=1}^{T} X_t. \tag{9.2}$$

However, the population mean is frequently unobservable. Whilst it might be assumed that, say, asset returns are drawn from a distribution with a defined mean, it is impossible to know with complete certainty what that mean is. This lack of knowledge does not cause any issues for the estimation of the mean, but it does have an impact on the way in which higher moments are determined.

9.1.2 Median

The mean is frequently used to help parameterise a distribution; however, the median is more commonly seen in the analysis of simulated data. It is a measure of the mid-point in that half of the distribution lies above the median and half below. It is therefore helpful in considering the most likely outcome – the 50th percentile – rather than the most likely weighted by the size point of the distribution, which is given by the mean.

9.1.3 Mode

The mode of the distribution is the most common observation. For a discrete distribution, this can be determined by counting the observations; for a continuous distribution, it is the point at which the first derivative or gradient of the probability density function is zero – the maximum value of the density function, or the point at which the gradient of the distribution ceases to increase.

9.2 Spread

Knowing where an observation is most likely to occur is a useful part of risk management – but it is at least as important to know how far away from this an observation could fall. The first aspect to consider is the spread of a distribution. This can be used to give a general idea of the uncertainty implicit in a particular estimate, so helping to establish the level of confidence that an estimate merits.

9.2.1 Variance

The variance is the most popular measure of the spread of a distribution. The population variance, σ^2, is calculated as:

$$\sigma^2 = \frac{1}{T} \sum_{t=1}^{T} (X_t - \mu)^2. \tag{9.3}$$

Density functions for normal distributions with high and low variances are shown in Figure 9.1.

This measure is appropriate if the dataset represents all possible observations. However, in many risk management problems this will not be the case. In particular, some of the possible observations exist in the future and cannot be known. This means that this statistic is not a good estimate of the true population variance, and is biased downwards in finite samples, with the bias increasing as the sample size falls.

In order to mitigate the level of bias, an adjustment is frequently made to the calculation of the variance to give a more robust sample measure. The sample variance, s^2, is therefore usually calculated as:

$$s^2 = \frac{1}{T-1} \sum_{t=1}^{T} (X_t - \bar{X})^2. \tag{9.4}$$

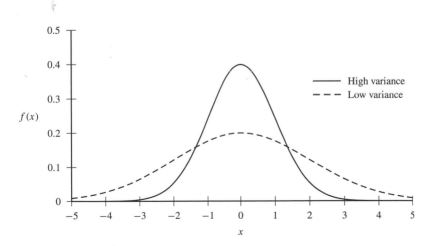

Figure 9.1 High and low variance density functions

9.2.2 Range

Whilst the variance is the most common measure of spread, it is not the only one. A simple alternative measure is to take the range of a set of observations, being the difference between the largest and smallest value. This measure can capture information about the effect of potential extreme events.

The range is therefore straightforward to calculate from a series of observations; however, it cannot necessarily be calculated for a parametric distribution to give the potential difference between highest and lowest outcomes. This will be the case for the distribution that is unbounded on at least one side. For example, if a distribution can take any value from zero to infinity, then the theoretical range will be infinite.

The solution to this is to consider a limited version of the range. The most common is the inter-quartile range, which is the difference between the 75th and 25th centiles, below which 75% and 25% of observations respectively lie; however, the 95th and 5th centiles, the 90th and 10th centiles or any other combination can be used.

9.3 Skew

The previous two measures are adequate for simple analysis; however, they ignore the possibility of skewed distributions. It is important to consider skew, as otherwise risk might be underestimated. If a distribution is assumed to be symmetric, then variance calculated might understate the likelihood of loss if the distribution is skewed. This could lead to a higher-than-anticipated level of risk being taken. Similarly, ignoring skew could lead to potentially profitable projects being rejected if the likelihood of large profits is understated.

negative skew means that the left-hand tail of the distribution is longer than the right-hand tail; the opposite is naturally true for positive skew. This means that if returns are negatively skewed, the chance of a large loss (relative to the expected return) is greater than the chance of a large gain. Density functions for distributions with positive, zero and negative skew are shown in Figure 9.2.

The population skew, ω, is given as:

$$\omega = \frac{1}{T} \frac{\sum_{t=1}^{T} (X_t - \mu)^3}{\sigma^3}. \tag{9.5}$$

This is the appropriate statistic if the full distribution is available; however, this is not usually the case, so the statistic will again be biased and a separate sample measure is needed.

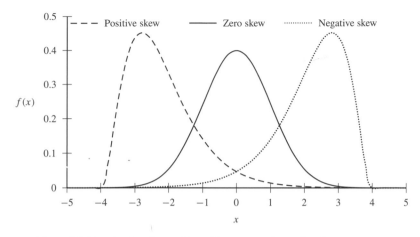

Figure 9.2 Density functions with positive, zero and negative skews

The adjustment needed to give the sample skew, w, is similar to that made for the variance, and the expression typically used for the sample skew is:

$$w = \frac{T}{(T-1)(T-2)} \frac{\sum_{t=1}^{T}(X_t - \bar{X})^3}{s^3}. \tag{9.6}$$

9.4 Kurtosis

The mean, standard deviation and skewness of a distribution are based on the first, second and third moments of a distribution. Considering the fourth moment leads us to the issue of kurtosis. This gives an indication of the likelihood of extreme observations relative to those that would be expected with the normal distribution.

Kurtosis is most commonly measured relative to the normal distribution. This has kurtosis of 3, and is described as a mesokurtic distribution. If a distribution has thin tails relative to the normal distribution, its kurtosis will be less than 3, or, relative to the normal distribution, it will have negative excess kurtosis. Such distributions are known as platykurtic. If a distribution has fat tails relative to the normal distribution, its kurtosis will be greater than 3, or, relative to the normal distribution, it will have positive excess kurtosis. Such distributions are known as leptokurtic. Leptokurtic, mesokurtic and platykurtic density functions are shown in Figure 9.3.

Leptokurtosis is an important issue when trying to quantify risk. If it is present – and not properly allowed for – then the probability of extreme events

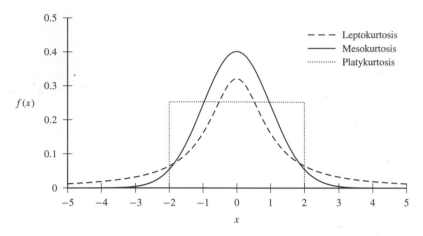

Figure 9.3 Leptokurtic, mesokurtic and platykurtic density functions

will be underestimated. It is therefore important to pay attention to the tails of a distribution when considering which statistical distribution to use. However, this can be difficult since, by definition, there will be fewer observations in the tail of the distribution than there will in the body.

The population measure of excess kurtosis, κ, is:

$$\kappa = \frac{1}{T} \frac{\sum_{t=1}^{T}(X_t - \mu)^4}{\sigma^4} - 3. \tag{9.7}$$

There is a deduction of 3 to reflect the fact that the kurtosis of the normal distribution is 3, and it is the normal distribution against which excess kurtosis is measured.

As with the sample standard deviation and sample skew, an adjustment is needed to reduce bias if the excess kurtosis is being calculated from a sample. The sample excess kurtosis, k, is given as:

$$k = \frac{T(T+1)}{(T-1)(T-2)(T-3)} \cdot \frac{\sum_{t=1}^{T}(X_t - \bar{X})^4}{s^4} - \frac{3(T-1)^2}{(T-2)(T-3)}. \tag{9.8}$$

9.5 Correlation

As well as considering the analysis of individual variables, it is also worth looking at some basic relationships between two variables, X and Y. Correlation is an important concept in ERM, as a core part of the process involves aggregating risks. If two risks have a strong positive correlation, then the risk of both occurring simultaneously is high; if the correlation is low, then the risks can diversify one another; and if the correlation is strongly negative, then there is an incentive to increase the level of one risk taken in order to offset the second.

As well as helping to establish the total amount of risk that an enterprise holds, correlation can also be used to help determine how much business should be taken on in different areas after taking into account the returns available, the risks taken on and the amount of diversification.

Three measures of correlation are discussed below: Pearson's rho, Spearman's rho and Kendall's tau. Whilst Pearson's rho is calculated directly from the two data series, the other two measures are rank correlation coefficients. This means that they are calculated from the position of the variables, or their rank, in each series. As a result, changing the value of an individual observation will change the value of Pearson's rho, but so long as the position of the observation in a data series does not change nor will a rank correlation coefficient.

Pearson's rho is attractive as it is widely used and easy to calculate. However, it is only a valid measure of association when the data series on which it is being calculated are jointly elliptical, a property described more fully in Chapter 2. Because rank correlation coefficients to not depend on the underlying shape of data series, only the relative position of observations, their results are always valid. However, whereas Pearson's rho can be used directly in some common multivariate distributions, such as the normal and t, the rank correlation coefficients are more usually combined with copulas. Kendall's tau in particular has simple relationships with the parameters of a number of copula functions.

Whichever measure of correlation is used, it must be understood that it describes only one aspect of the relationship between two variables. The choice of copula, either explicitly or implicitly through the use of a particular multivariate distribution, also helps describe the shape of this relationship beyond the broad measure of association described by the correlation.

9.5.1 Pearson's rho

The most basic is the correlation coefficient, $\rho_{X,Y}$, also known as the linear correlation coefficient. It is given as:

$$\rho_{X,Y} = \frac{\sigma_{X,Y}}{\sigma_X \sigma_Y}, \tag{9.9}$$

where $\sigma_{X,Y}$ is the population covariance between X and Y, and σ_X and σ_Y are the population standard deviations of those variables. The population covariance is calculated as:

$$\sigma_{X,Y} = \frac{1}{T} \sum_{t=1}^{T} (X_t - \mu_X)(Y_t - \mu_Y), \tag{9.10}$$

where μ_X and μ_Y are the population means for X and Y respectively. The calculation for the sample correlation, $r_{X,Y}$ is exactly the same, but this is only because the bias in the calculation of the standard deviations is balanced by that in the calculation of the covariance. This means that if sample standard deviations are used, then a sample covariance, $s_{X,Y}$, must be calculated as:

$$s_{X,Y} = \frac{1}{T-1} \sum_{t=1}^{T} (X_t - \bar{X})(Y_t - \bar{Y}), \tag{9.11}$$

where $\bar{X}$ and $\bar{Y}$ are the sample means for X and Y, and that the sample correlation coefficient must be calculated as:

$$r_{X,Y} = \frac{s_{X,Y}}{s_X s_Y}, \tag{9.12}$$

where s_X and s_Y are the sample standard deviations for X and Y. The sample covariance is also used in statistical techniques discussed later.

Pearson's rho is only a valid measure of correlation if the marginal distributions are jointly elliptical. This essentially means that the distributions are related to the multivariate normal distribution. This is important because it means that these measures can only be used appropriately in stochastic modelling if one of these distributions is used to model the data. If the marginal distributions are not jointly elliptical, then a Pearson's rho of zero does not necessarily imply that two variables are independent. Elliptical distributions are discussed in more detail under multivariate models.

9.5.2 Spearman's rho

Spearman's rank correlation coefficient, also known as Spearman's rho, $_s\rho$. For two variables, X and Y, Spearman's sample rho, $_s r_{X,Y}$, is defined as:

$$_s r_{X,Y} = 1 - 6 \frac{\sum_{t=1}^{T} (V_t - W_t)^2}{T(T^2 - 1)}, \tag{9.13}$$

where V_t and W_t are the rankings of X_t and Y_t respectively. Because the differences between the ranks are squared, it does not matter whether the ranks are in ascending or descending order, so long as the same system is used for each series.

Spearman's rho is linked to Pearson's rho in that the measures are equal if the underlying distribution used is uniformly distributed. In fact, if there are tied ranks, then one approach is to calculate the ranks from the data and then use Pearson's sample rho instead. However, unlike Pearson's rho it is independent of the statistical distribution of the data – only the order of the observations matters.

3 9.5.3 Kendall's tau

Another rank correlation coefficient is Kendall's tau, τ. It is calculated by comparing pairs of data points. Consider two variables, X and Y, each of which contain T data points, so we have $X_1, X_2, \ldots, X_T$ and $Y_1, Y_2 \ldots Y_T$. The combination (X_t, Y_t) is referred to as an observation. Now consider two observations, (X_1, Y_1) and (X_2, Y_2). If $X_2 - X_1$ and $Y_2 - Y_1$ have the same sign, then these observations are concordant; if they have different signs, then they are discordant. Concordant and discordant pairs are shown in Figure 9.4.

For T observations, the total number of pairs that can be considered is $T(T-1)/2$. This fact can be used to normalise any statistic calculated based on the numbers of concordant and discordant pairings. The calculation of concordant and discordant pairings, normalised by the total number of pairings, forms the basis of Kendall's tau, and the sample tau, $t_{X,Y}$, is calculated as follows:

$$t_{X,Y} = \frac{2(p_c - p_d)}{T(T-1)}, \tag{9.14}$$

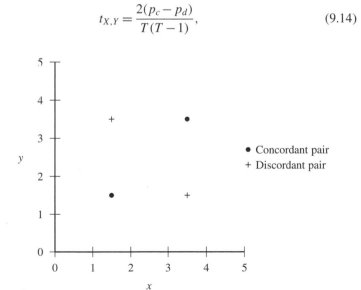

Figure 9.4 Concordant and discordant pairs

where p_c is the number of concordant pairs and p_d is the number of discordant pairs. Spearman's rho and Kendall's tau are related in the following way:

$$
\begin{cases}
\dfrac{3}{2}\tau - \dfrac{1}{2} \leq_s \rho \leq \dfrac{1}{2} + \tau - \dfrac{1}{2}\tau^2 & \text{if } \tau \geq 0 \\[2mm]
-\dfrac{1}{2} + \tau + \dfrac{1}{2}\tau^2 \leq_s \rho \leq \dfrac{3}{2}\tau + \dfrac{1}{2} & \text{if } \tau < 0.
\end{cases}
\tag{9.15}
$$

Example 9.1 An insurance company has the following total claim values from two portfolios, X and Y, over a five-year period, with claims X_t and Y_t in each year t:

t	X_t	Y_t
1	10	20
2	95	25
3	15	10
4	35	15
5	45	30

What are the correlations of these two series, as measured by Pearson's rho, Spearman's rho and Kendall's tau?

First, consider Pearson's sample rho, defined as:

$$
r_{X,Y} = \frac{s_{X,Y}}{s_X s_Y}.
$$

This means that standard deviations s_X and s_Y need to be found as does the covariance $s_{X,Y}$. The sample means $\bar{X}$ and $\bar{Y}$ are therefore also required. These are calculated as:

$$
\bar{X} = \frac{1}{T}\sum_{t=1}^{T} X_t \text{ and } \bar{Y} = \frac{1}{T}\sum_{t=1}^{T} Y_t.
$$

Adding this detail to the table above gives:

t	X_t	Y_t
1	10	20
2	95	25
3	15	10
4	35	15
5	45	30
Total	200	100
$\bar{X}, \bar{Y}$	40	20

This information allows the calculation of s_X, s_Y and $s_{X,Y}$, defined as:

$$s_X = \sqrt{\frac{1}{T-1}\sum_{t=1}^{T}(X_t - \bar{X})^2}, \; s_Y = \sqrt{\frac{1}{T-1}\sum_{t=1}^{T}(Y_t - \bar{Y})^2}$$

$$\text{and } s_{X,Y} = \frac{1}{T-1}\sum_{t=1}^{T}(X_t - \bar{X})(Y_t - \bar{Y}).$$

The summations can be calculated from the table above as:

t	X_t	Y_t	$X_t - \bar{X}$	$(X_t - \bar{X})^2$	$Y_t - \bar{Y}$	$(Y_t - \bar{Y})^2$	$(X_t - \bar{X}) \times (Y_t - \bar{Y})$
1	10	20	-30	900	0	0	0
2	95	25	55	3,025	5	25	275
3	15	10	-25	625	-10	100	250
4	35	15	-5	25	-5	25	25
5	45	30	5	25	10	100	50
Total	200	100		4,600		250	600
$\bar{X}, \bar{Y}$	40	20					

This means that $s_X = \sqrt{4600/4} = 33.91$, $s_Y = \sqrt{250/4} = 7.91$ and $s_{X,Y} = 600/4 = 150$, so $r_{X,Y} = 150/(33.91 \times 7.91) = 0.5595$.

Spearman's sample rho is defined as:

$$_s r_{X,Y} = 1 - 6\frac{\sum_{t=1}^{T}(V_t - W_t)^2}{T(T^2 - 1)},$$

where V_t and W_t are the rankings of X_t and Y_t respectively. This means that the data need to be ranked. The differences between the ranks are then taken and the results squared and summed. This information can added to the original data as follows:

t	X_t	Y_t	V_t	W_t	$V_t - W_t$	$(V_t - W_t)^2$
1	10	20	5	3	2	4
2	95	25	1	2	-1	1
3	15	10	4	5	-1	1
4	35	15	3	4	-1	1
5	45	30	2	1	1	1
Total						8

The number of observations, T, is 5 so Spearman's sample rho is calculated as $_s r_{X,Y} = 1 - (6 \times 8)/(5 \times (5^2 - 1)) = 0.6$.

Kendall's sample tau is defined as:

$$t_{X,Y} = \frac{2(p_c - p_d)}{T(T-1)},$$

where p_c is the number of concordant pairs and p_d is the number of discordant pairs. To calculate whether one pair is concordant with another, consider observations from two periods, s and t. If $X_s - X_t$ has the same sign as $Y_s - Y_t$, then the pairs are concordant, otherwise they are discordant. The table below shows the results of these calculations for each possible pair:

t	X_t	Y_t	vs $t=1$	vs $t=2$	vs $t=3$	vs $t=4$
1	10	20				
2	95	25	C (+,+)			
3	15	10	D (+,−)	C (−,−)		
4	35	15	D (+,−)	C (−,−)	C (+,+)	
5	45	30	C (+,+)	D (−,+)	C (+,+)	C (+,+)

The number of concordant pairs, p_c, is 7, whilst the number of discordant pairs, p_d, is 3. Since T is still equal to 5, Kendall's sample tau is calculated as $t_{X,Y} = 2 \times (7-3)/(5 \times (5-1)) = 0.4$.

9.5.4 Tail correlation

All of these measures imply the same level of association whatever the values of X and Y. However, it is often helpful to consider the relationship between these variables in extreme situations.

One approach to dealing with this is to consider some measure of correlation applied only to the tails of two variables, such as a correlation coefficient between X and Y for the lowest and highest 10% of observations for X. However, it is difficult to determine at which point the tail of the joint distribution between X and Y starts: the data points need to be far enough from the centre of the distribution to be regarded as extreme tail events, but not so far that there are too few to analyse. It is also important that the choice of tail does not result in too much instability in the parametrisation.

9.6 Further reading

There is a significant volume of academic literature around the characteristics of distributions and, in particular, the links between sets of data. Most of the papers concentrate on very specific aspects of measures. Malevergne and Sornette (2006) provide some interesting analysis of conditional rank correlation coefficients, as does Venter (2002). These papers consider the correlation between sub-sets of a group of observations. Blomqvist (1950) discusses a simple alternative measure, whilst a broad summary of the different measures is given in Sweeting and Fotiou (2011).

10

Statistical distributions

10.1 Univariate discrete distributions

The univariate statistical distribution of each variable on its own – also known as its marginal distribution – is an important factor in the risk it poses. Many of the features above can be modelled directly by the appropriate choice of marginal distribution, or they can be added to a more 'basic' marginal distribution.

Univariate discrete distributions are generally only used when the number of observations is small, as they quickly become difficult to deal with as the numbers involved increase. However, even if continuous approximations are used, it is important to recognise the nature of whatever is being approximated.

10.1.1 The binomial and negative binomial distributions

The binomial distribution is fundamental to many risks faced. In particular, it reflects the risk of a binary event – one which may or may not occur. Such an event could be the payment of a claim, the default of a creditor or the survival of a policyholder.

The binomial distribution is parameterised by the number of trials (or observations), n, the number of successes (or claims, defaults or other events), x, and the probability that an event will occur, p. The probability must be constant for each trial.

The probability that in n independent trials there will be x successes followed by $n - x$ failures if p is the probability of a success in each case is $p^x (1 - p)^{(n-x)}$. However, if the successes are allowed to occur in any order, the probability increases. The number of possible combinations of x successes in n trials is given by the binomial coefficient, which is itself calculated using the factorial function, $x! = x \times (x-1) \times \ldots \times 2 \times 1$. The binomial coefficient, describing the number of possible ways in which there can be x successes from n trials, is therefore given by:

$$\binom{n}{x} = \frac{n!}{x!(n-x)!}. \tag{10.1}$$

This means that the probability that the number of successes, X, will be a particular integer number, x, is:

$$f(x) = \Pr(X = x) = \binom{n}{x} p^x (1-p)^{n-x}$$

$$= \frac{n!}{x!(n-x)!} p^x (1-p)^{n-x}. \tag{10.2}$$

The mean of this distribution is np and the variance is $np(1-p)$.

Related to the binomial distribution is the negative binomial distribution. This gives the probability that $X = x$ trials will be needed until there have been r successes. If the probability of a success is p, then this probability is:

$$f(x) = \Pr(X = x) = \binom{x-1}{r-1} p^r (1-p)^{x-r}$$

$$= \frac{(x-1)!}{(r-1)!(x-r)!} p^r (1-p)^{x-r}. \tag{10.3}$$

The mean of this distribution is $r(1-p)/p$ and the variance is $r(1-p)/(p^2)$.

There are two practical issues with the binomial distribution. The first is that a commonly needed result is the cumulative distribution function, which is $f(1) + f(2) + \ldots + f(x)$. This is laborious to calculate. More importantly, as n increases, the value of $n!$ becomes enormous – for example, $100!$, or $100 \times 99 \times \ldots \times 2 \times 1$ is equal to 9.33×10^{157}. Given that the number of loans in a bank (for example) would be many times this number, the results would be impossible to calculate in any reasonable time scale. More importantly, the level of accuracy given by this calculation is spurious given the likely uncertainty in the parameters, so it makes sense to use some sort of approximation.

Example 10.1 An insurance company has a small portfolio of twenty identical policies. If the probability that any policyholder will make a claim in the following year is 0.25 and all claims are independent, what is the probability that there will be exactly four claims?

If the number of claims is X, the probability of a claim is p and the number of policies is n, then the probability that $X = x$ is given by:

$$\Pr(X = x) = \frac{n!}{x!(n-x)!} p^x (1-p)^{n-x}.$$

Substituting $p = 0.25$, $n = 20$ and $x = 4$ into this expression gives:

$$\Pr(X = 4) = \frac{20!}{4!(16)!} 0.25^4 \times 0.75^{16} = 0.1897.$$

10.1.2 The Poisson distribution

The Poisson distribution is derived from the binomial distribution. It gives the probability of a number of independent events occurring in a specified time. In this distribution, the rate of occurrence – the expected number of occurrences in any given period – is λ. In terms of the parameters of the binomial distribution, this means that with n trials and a probability of success p, $\lambda = np$. Substituting λ/n for p in Equation (10.2) gives:

$$
\begin{aligned}
f(x) = \Pr(X = x) &= \frac{n!}{x!(n-x)!} \left(\frac{\lambda}{n}\right)^x \left(1 - \frac{\lambda}{n}\right)^{n-x} \\
&= \frac{n!}{x!(n-x)!} \left(\frac{\lambda}{n}\right)^x \left(1 - \frac{\lambda}{n}\right)^n \left(1 - \frac{\lambda}{n}\right)^{-x} \\
&= \frac{n!}{n^x(n-x)!} \left(\frac{\lambda^x}{x!}\right) \left(1 - \frac{\lambda}{n}\right)^n \left(1 - \frac{\lambda}{n}\right)^{-x}. \quad (10.4)
\end{aligned}
$$

In this formulation, as n tends to infinity, $\lambda^x/x!$ is unaffected, $(1 - \lambda/n)^n$ tends to $e^{-\lambda}$ and all other terms tend to one. The result is that the probability that the actual number of occurrences, X, will be equal to some number, x, is:

$$f(x) = \Pr(X = x) = \frac{\lambda^x e^{-\lambda}}{x!}. \quad (10.5)$$

Both the mean and variance of the Poisson distribution are equal to λ. An important assumption of this distribution is that the rate of occurrences is low. This means that it can be used as an approximation to the binomial distribution with $\lambda = np$ if the probability is sufficiently small. This is often the case when mortality rates or bond defaults are being considered.

The fact that λ must be small helps limit the problem arising from large factorial calculations as seen with the binomial distribution; however, summations are still needed to give a cumulative Poisson distribution.

Example 10.2 An insurance company has a large portfolio of 1,000 identical policies. If the probability that any policyholder will make a claim in the following year is 0.005 and all claims are independent, what is the probability that there will be exactly four claims?

If the number of claims is X, the mean number of claims under the Poisson distribution is λ, then the probability that $X = x$ is given by:

$$\Pr(X = x) = \frac{\lambda^x e^{-\lambda}}{x!}.$$

The Poisson mean here is $\lambda = 1000 \times 0.005 = 5$ and $x = 4$. Substituting these values into the above expression gives:

$$\Pr(X = 4) = \frac{5^4 e^{-5}}{4!} = 0.1755.$$

10.2 Univariate continuous distributions

Univariate continuous distributions are more commonly seen than discrete ones in financial modelling. This is because the variables being measured are almost always either continuous or based on such large numbers that they can be regarded as such.

Whilst the probability density function for a continuous distribution, $f(x)$, gives an instantaneous measure of the likelihood of an event under a particular distribution, the actual probability of an event happening at any particular point is zero. This means that probabilities can only be evaluated between different values using a distribution function. If a probability is calculated from the minimum value of a distribution to some other specified value, then the distribution function is known as the cumulative distribution function, $F(x)$. This gives the probability that a random variable X is below a certain level x, denoted $\Pr(X \leq x)$. In other words:

$$F(x) = \Pr(X \leq x) = \int_{-\infty}^{x} f(s)\,ds. \qquad (10.6)$$

In order to make comparison between the various distributions more straightforward, the following conventions are adopted:

- the location parameter for a distribution is denoted α – an increase in α shifts the distribution to the right, a decrease to the left;
- the scale parameter for a distribution is denoted β – an increase in β increases the spread of the distribution; and
- the shape parameter for a distribution is denoted γ – this can have a variety of impacts on shape.

Location parameters have generally been used only for the unbounded distributions, with the lower-bounded distributions always having a minimum value of zero. It is straightforward to shift many of these distributions simply by replacing x in the formulation with $x - \alpha$. This will not generally work when the distribution is the exponential or the square of a function that ranges from $-\infty$ to ∞, but shifted distributions are frequently used with the gamma distribution (and the exponential as a special case), whilst a common alternative parametrisation of the Pareto distribution uses a non-zero lower bound.

For distributions that are lower- and upper-bounded, the bounds are generally set at zero and one, since these are the most useful cases due to their relevance to rates of claim, default, mortality and so on.

For most cases random variables can be obtained with the distribution required simply by using a spreadsheet or statistical package to apply an inverse distribution function to a series of random variables between zero and one; however, in some cases there are straightforward alternative approaches. Where appropriate, these are described with the distributions.

The distributions below are considered in the following order:

- unbounded distributions;
- lower-bounded distributions (at zero); and
- lower- and upper-bounded distributions (at zero and one).

10.2.1 The normal distribution

In modelling terms, the most basic continuous distribution is the normal or Gaussian distribution, which has the following probability density function, $f(x)$:

$$f(x) = \frac{1}{\sqrt{2\pi}} e^{-\frac{1}{2}\left(\frac{x-\alpha}{\beta}\right)^2}, \tag{10.7}$$

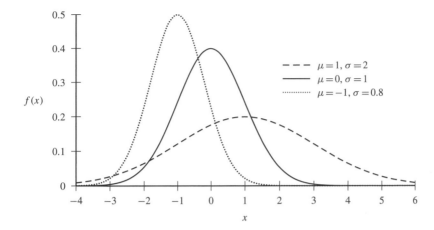

Figure 10.1 Various normal density functions

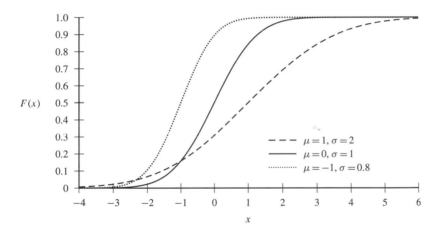

Figure 10.2 Various normal distribution functions

where α and β are the location and scale parameters. For the normal distribution, α is more commonly referred to as μ, which is also the mean of the distribution; similarly, β is more commonly referred to as σ, which is also the standard deviation of the distribution. Any real value of x can be used.

The probability density function cannot be integrated analytically to give the more useful cumulative probability distribution function, $F(x)$. However, this can be obtained from standard tables and with most spreadsheet applications. Various normal density and distribution functions are shown in Figures 10.1 and 10.2 respectively.

The normal distribution is a particularly popular choice for many models for two reasons. First, the central limit theorem says that if you have enough independent and identically distributed random variables with finite mean and variance, then their distribution will be approximately normal. This makes it the distribution of choice if there is any doubt over the true distribution, or as a large sample approximation to discrete distributions such as the binomial (approximated as a normal distribution with a mean of np and a variance of $np(1 - p)$) or the Poisson (approximated as normal distribution with a mean and a variance of λ). However, even if it is known that variables are not normally distributed, the normal distribution will still sometimes be adopted as it is analytically tractable – in other words, it can be used to give neat solutions to initially complex problems. This is fine if it is understood that this is the reason for using the normal distribution, and the results are treated with sufficient care; however this is not always the case, and using the normal distribution might be inappropriate. To understand why, it is important to recognise the characteristics of the normal distribution:

- it can take values from $-\infty$ to ∞;
- it is a symmetrical distribution (its measure of skew is 0); and
- it is mesokurtic, having neither a sharp peak and fat tails (leptokurtosis) or a rounded peak and thin tails (platykurtosis) when measured relative to itself (its kurtosis is 3), although this is clearly only helpful when considering other distributions.

The normal distribution is used in a key area of financial modelling, the random walk with drift. The standard formulation for this process is:

$$X_t = \mu + X_{t-1} + \epsilon_t, \tag{10.8}$$

where X_t is the observation of variable X at time t, ϵ_t is a normal random variable with zero mean and a variance of σ_t and μ is the rate of drift. For this to be a random walk, ϵ_t can have no correlation with ϵ_s, the error term in any other period.

This parametrisation of the normal distribution can be adjusted to reflect different means and standard deviations in the data. However, the normal distribution given in most statistical tables is the standard normal distribution, which has a mean of zero and a unit standard deviation. This has the density function $\phi(x)$, and is given by a simplified version of Equation (10.7):

$$\phi(x) = \frac{1}{\sqrt{2\pi}} e^{-\frac{1}{2}x^2}. \tag{10.9}$$

The cumulative distribution function for the standard normal distribution evaluated at x is referred to as $\Phi(x)$, which is defined as:

$$\Phi(x) = \int_{-\infty}^{x} \phi(s)\mathrm{d}s. \tag{10.10}$$

It is $\Phi(x)$ that is given in most standard tables.

Example 10.3 It is claimed that the average annual return for a particular investment strategy is normally distributed with a mean of 8% per annum with a standard deviation of 4%. In the past year, the return was 1%. Is this significantly different from the mean return at the 95% level of confidence? Is it significantly lower at the same level of confidence?

The test statistic here is:

$$Z = \frac{X - \mu}{\sigma},$$

where μ is equal to 8%, σ is equal to 4% and X is equal to 1%. This means that the test statistic is $Z = (0.01 - 0.08)/0.04 = -1.75$. From the standard normal distribution, $\Phi(-1.75) = 0.0401$. For the return to be significantly different from the mean at the 95% level of confidence, a number less than 0.025 or greater than 0.975 would be needed. The return is therefore not significantly different from the mean at this level of confidence. However, for the return to be significantly lower than the mean at the 95% level of confidence, a number less than 0.05 is be needed. The return is significantly lower than the mean at this level of confidence.

Alternatively, it is possible to calculate the inverse cumulative normal distribution function at the required levels of confidence. For the two-tailed test, $\Phi^{-1}(0.025) = -1.96$, whilst $\Phi^{-1}(0.975) = 1.96$, Φ^{-1}, being the inverse cumulative standard normal distribution. Since Z lies between these values, the observation is not significantly different from the mean at this level of confidence. For the one-tailed test, $\Phi^{-1}(0.05) = -1.645$. Since Z is lower than this value, the observation is significantly lower than the mean at this level of confidence.

The standard normal distribution is also used to determine whether an observation, X, is significantly different to an assumed mean, μ, if the standard deviation, σ, is known. The test statistic here is:

$$Z = \frac{X - \mu}{\sigma}, \tag{10.11}$$

which has a normal distribution with a mean of zero and a standard deviation of one, so it can be evaluated from the standard normal tables. The normal distribution can also be used to determine whether the sample mean, $\bar{X}$, is significantly different from the mean, with the test statistic being calculated as:

$$Z = \frac{\bar{X} - \mu}{\sigma/\sqrt{T}}, \qquad (10.12)$$

with T being the number of observations. This statistic too has a standard normal distribution.

Example 10.4 The investment strategy in Example 10.3 continues for another ten years. Over this period, the average return has been 5.75% per annum. Using the data from the previous example, does this suggest that the mean is significantly different or significantly lower than the assumed mean at the 95% level of confidence?

The test statistic here is:

$$Z = \frac{X - \mu}{\sigma/\sqrt{T}},$$

where T is equal to ten. This means that the test statistic is $Z = (0.0575 - 0.08)/(0.04 \times \sqrt{10}) = -1.78$. From the standard normal distribution, $\Phi(-1.78) = 0.0376$. For the calculated mean to be significantly different from the assumed mean at the 95% level of confidence, a number less than 0.025 or greater than 0.975 would be needed. The calculated mean is therefore not significantly different from the assumed mean at this level of confidence. However, for the calculated mean to be significantly lower than the assumed mean at the 95% level of confidence, a number less than 0.05 is be needed. The calculated mean is significantly lower than the assumed mean at this level of confidence.

Alternatively, it is possible to calculate the inverse cumulative normal distribution function at the required levels of confidence. For the two-tailed test, $\Phi^{-1}(0.025) = -1.96$, whilst $\Phi^{-1}(0.975) = 1.96$ Since Z lies between these values, the observation is not significantly different from the assumed mean at this level of confidence. For the one-tailed test, $\Phi^{-1}(0.05) = -1.645$. Since Z is lower than this value, the observation is significantly lower than the assumed mean at this level of confidence.

There are a number of ways in which a dataset can be tested to determine whether it is normally distributed. A graphical approach is to use a Q-Q ('quantile-quantile') plot. This involves plotting each observation, X_t, where $t = 1, 2, \ldots, T$ on the vertical axis against the inverse normal distribution function of the position of that variable on the horizontal axis. If the position of the variable is defined as $G(X_t)$, the item plotted is therefore $\Phi^{-1}(G(X_t))$.

There are a number of ways in which $G(X_t)$ can be calculated. The starting point is to order the data from lowest to highest such that for a data point X_t, X_1 would be the lowest observation and X_T the largest. One approach for calculating the position of X_t is to set $G(X_t) = t/(T+1)$. This means that the smallest observation is $1/(T+1)$ and the largest is $T/(T+1)$. Another option is to define $G(X_t) = (t - 0.5)/T$, which ranges from $1/2T$ to $(T - 1/2)/T$. The important point is that the smallest observation should be greater than zero and the largest less than one, so that the inverse normal distribution function can be calculated.

Once a plot has been created, it can be analysed visually. If the observations are normally distributed, then they should lie on or close to the diagonal line running between the bottom left and top right of the chart. If there are any systematic deviations, then the implication is that the observations are not normally distributed.

It should be clear that this approach can be used to test the extent to which observations fit any distribution, not just the normal – all that is needed is to substitute another inverse distribution function for $\Phi^{-1}(G(X_t))$.

Example 10.5 Are the monthly returns on index-linked gilts from the end of 1999 to the end of 2009 normally distributed?

This question can be addressed using a Q-Q plot. First, rank the monthly returns, X_t, from the lowest to the highest, or $t = 1, 2, \ldots, 120$. The lowest return, calculated as the difference between the natural logarithms of the total return indices, is -0.0683, followed by -0.0418. These have ranks of 1 and 2 respectively. The largest monthly return is 0.0875, which has a rank of 120. If $G(X_t)$ is taken to be $(t - 0.5)/T$, then the cumulative distribution functions calculated from these ranks become $0.0042, 0.0125, \ldots, 0.9958$. The standard normal quantile for each value is given by $\Phi^{-1}[G(X_t)]$, meaning that these quantiles are $-2.6383, -2.2414, \ldots, 2.6383$. These figures are shown in the table below:

t	X_t	$G(X_t)$	$\Phi^{-1}[G(X_t)]$
1	−0.0683	0.0042	−2.6383
2	−0.0418	0.0125	−2.2414
⋮	⋮	⋮	⋮
120	0.0875	0.9958	2.6383

The next stage is to plot X_t against $\Phi^{-1}(G(X_t))$, as shown below:

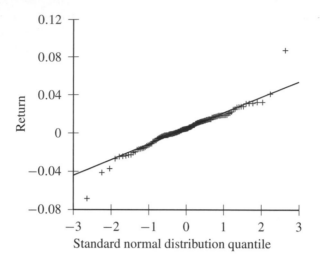

Comparing the points plotted with a diagonal line drawn through the bulk of the observations, it is clear that the very low returns are lower than would be implied by the standard normal quantiles, whilst the very high returns are higher. This suggests that the normal distribution does not describe the monthly returns on this dataset very well, at least for extreme observations.

A common numerical test of normality is the Jarque–Bera test (Jarque and Bera, 1980a,1980b). The test statistic, JB, is calculated as:

$$JB = \frac{T}{6}\left(\omega^2 + \frac{\kappa^2}{4}\right),\tag{10.13}$$

where ω is the skew and κ the excess kurtosis for the data, both being calculated with no adjustment for sample bias. The variable T is the number of observations. The distribution of this statistic tends towards χ^2_2 as T tends to ∞. A description of the χ^2 distribution is given below.

10.2.2 Normal mean–variance mixture distributions

The normal distribution can also be used as a building block to create more flexible distributions known as normal mean-variance mixture distributions. In many cases, the results are well-known distributions in their own right. However, the fact that they can be described as normal mean-variance mixture distributions is helpful when random variables are to be generated, since their relationship to the normal distribution makes simulation more straightforward.

A normal mean-variance mixture distributions is one where some variable X is defined in relation to a standard normal random variable, Z, such that:

$$X = m(W) + \sqrt{W}\beta Z, \qquad (10.14)$$

where β is a scaling factor so that, essentially, βZ is a random normal variable from a distribution with a standard deviation of β, W is a positive random variable that is independent of Z and $m(W)$ is some function of W. This means that if W is equal to one and $m(W)$ is equal to some constant μ, whilst β is equal to σ, then X is simply a normally distributed variable with a standard deviation of σ and a mean of μ.

The most general case is where W has a generalised inverse Gaussian (GIG) distribution, described later, with parameters β_1, β_2 and γ_{GIG}. In this case, if $m(W) = \alpha + \delta W$, where α is a location parameter and δ is a non-centrality or skewness parameter, then the result is a generalised hyperbolic distribution.

A special case of the GIG distribution is obtained by setting $\beta_2 = 0$: the inverse gamma distribution with $\beta_{I\Gamma} = \beta_1/2$ and $\gamma_{\Gamma} = -\gamma_{GIG}$, where $\beta_{I\Gamma}$ and $\gamma_{I\Gamma}$ are the β and γ parameters for the inverse gamma distribution. This means that if W now has a inverse gamma distribution, then $1/W$ now has a gamma distribution. Setting the two remaining parameters equal to $\gamma/2$ means that γ/W now has a chi-squared distribution with γ degrees of freedom. As discussed later, a chi-squared distribution with γ degrees of freedom is simply the sum of γ squared, independent, standard normal variables. This means that $1/W$ is equal to a chi-squared variable with γ degrees of freedom divided by the number of degrees of freedom.

There are a number of ways in which this inverse gamma distribution approach can be used to generate normal mixture distributions. In particular:

- if $m(W) = \alpha$, then the result is a t-distribution with γ degrees of freedom; and
- if $m(W) = \alpha + \delta W$, then the result is a skewed t-distribution with γ degrees of freedom.

These two distributions are discussed in more detail below.

10.2.3 Student's t-distribution

Student's t-distribution, more commonly known as just the t-distribution, can be regarded as a generalisation of the normal distribution. It is, like the normal distribution, a symmetric distribution but the degrees of freedom in the distribution determine the fatness of the tails. The probability density function for the general t-distribution is:

$$f(x) = \frac{\Gamma\left(\frac{\gamma+1}{2}\right)}{\beta\sqrt{\pi\gamma}\,\Gamma\left(\frac{\gamma}{2}\right)}\left[1 + \frac{1}{\gamma}\left(\frac{x-\alpha}{\beta}\right)^2\right]^{-\frac{\gamma+1}{2}}, \tag{10.15}$$

where:

$$\Gamma(y) = \int_0^\infty s^{y-1}e^{-s}\,ds, \tag{10.16}$$

and α is a location parameter, β is a scale parameter and the number of degrees of freedom – which determines the shape – is γ. Like the normal distribution, the t-distribution can take any real value of x. Note that, whilst α is the mean of the distribution, the variance of the distribution is actually $\beta^2\gamma/(\gamma-2)$. As γ tends to infinity, the distribution tends to the normal distribution; however, as γ falls, the degree of leptokurtosis increases. In fact, the excess kurtosis can be calculated only for values of $\gamma > 4$, for which it is $3(\gamma-2)/(\gamma-4)$, and skew can only be calculated if $\gamma > 3$, although for these values of γ it is zero. If $\gamma > 2$, the variance is finite, but it is infinite if $\gamma = 2$ and undefined if $\gamma = 1$. Even the mean only exists for $\gamma > 1$.

The special case of the t-distribution where $\gamma = 1$ is also known as the Cauchy distribution. This has tails so fat that it has no defined mean, variance or higher moments.

As with the normal distribution, the cumulative probability distribution function for the t-distribution cannot be calculated by integrating the density function analytically – except for the special case of the Cauchy distribution, where the integral reduces to:

$$F(x) = \Pr(X \le x) = \frac{1}{\pi}\arctan\left(\frac{x-\alpha}{\beta}\right) + \frac{1}{2}, \tag{10.17}$$

where α and β are again the measures of location and scale respectively. Various t density and distribution functions are shown in Figures 10.3 and 10.4 respectively.

For finite values of γ, the tail of the t-distribution follows what is known as a 'power law'. This means that the probability of an event falls as the magnitude of the event increases, with the probability being proportional to the magnitude raised to a fixed power. In particular, for the t-distribution the probability is proportional to the size of the event raised to the power of $1/(\gamma+1)$. This also means that for the tail of the Cauchy distribution, the probability of an event is proportional to the square root of its size. The increased importance of the tails

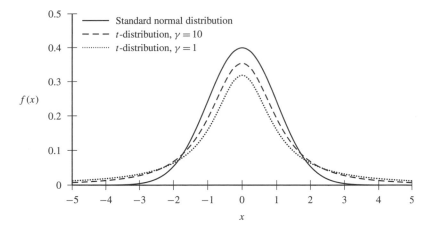

Figure 10.3 Various t-density functions

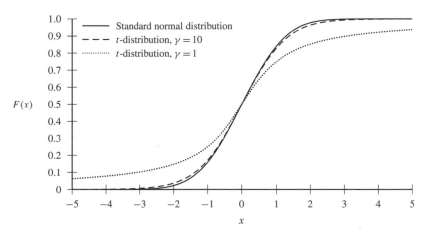

Figure 10.4 Various t-distribution functions

can be seen in the charts of various standard t-distributions when compared with the standard normal distribution.

As discussed above, the t-distribution is a normal mixture distribution. This means that random variables with a standard t-distribution with γ degrees of freedom can be simulated easily. First, a random normal variable from a standard normal distribution, Z, must be simulated. Then a random variable from a χ^2 distribution with γ degrees of freedom, X_γ^2, is taken, square-rooted and divided into the normal variable. This can then be converted into a random variable with a general t-distribution. The first element of this part of the calculation is to adjust the variable by the scale parameter required, β. This can either be an assumed value or one calculated from the sample standard deviation, s, and the degrees of freedom, γ. Finally, the distribution is re-centred by adding the location parameter, α, which is also the required mean. This means that it can be calculated from the data as $\bar{x}$ or specified as some other value. This means the total process can be summarised as:

$$\frac{Z}{\sqrt{X_\gamma^2/\gamma}}\beta+\alpha. \tag{10.18}$$

The t-distribution was not designed as a distribution to project leptokurtic time series variables; it was designed to test the whether a statistic was significantly different from the hypothesised population mean, μ, when the population variance was unknown and only the sample variance, s^2, was available. The lower the number of degrees of freedom, the higher the test statistic, reflecting the fact that having fewer observations reduces the certainty over the distribution of the observations. This gives a distribution with varying levels of kurtosis, which is useful for time series projections.

The test statistic uses the standard t-distribution, t_γ. This is a special case of the general t-distribution where $\alpha = 0$ and $\beta = 1$, so it is defined only by the degrees of freedom, γ. This has the following cumulative distribution function:

$$t_\gamma(x) = \int_{-\infty}^{x} \tau_\gamma(s)ds, \tag{10.19}$$

where:

$$\tau_\gamma(x) = \frac{\Gamma\left(\frac{\gamma+1}{2}\right)}{\sqrt{\pi\gamma}\Gamma\left(\frac{\gamma}{2}\right)}\left(1+\frac{x^2}{\gamma}\right)^{-\frac{\gamma+1}{2}}, \tag{10.20}$$

$\tau_\gamma(x)$ being the probability density function at x for a t-distribution with γ degrees of freedom. If the question relates to a single observation, X, then the test statistic is:

$$Z = \frac{X-\mu}{s}, \tag{10.21}$$

which has a standard t-distribution with $\gamma = T - 1$ degrees of freedom, where T is the number of observations. The t-distribution can also be used to determine whether the sample mean, $\bar{X}$, is significantly different from the mean, with the test statistic being calculated as:

$$Z = \frac{\bar{X}-\mu}{s/\sqrt{T}}, \tag{10.22}$$

This statistic too has a standard t-distribution with $\gamma = T - 1$ degrees of freedom. The standard t-distribution is given in statistical tables. The mean and variance and kurtosis of a dataset can be used to derive the parameters of the dataset. The parameter α can be estimated as the mean of the dataset, and the number of degrees of freedom, γ, can be derived by calculating the sample excess kurtosis, setting the result equal to $3(\gamma - 2)/(\gamma - 4)$ and rearranging for γ. A value of β can then be derived by calculating the sample standard variance, setting this equal to $\beta^2\gamma/(\gamma - 2)$, substituting the derived value for γ and rearranging for β.

Example 10.6 The investment strategy in Examples 10.3 and 10.4 are still under analysis. A further calculation of the previous ten years is needed, this time using the observed sample standard deviation of 4% over that period. In this case, is the mean significantly different or significantly lower than the assumed mean at the 95% level of confidence?

The test statistic here is:

$$Z = \frac{X - \mu}{s/\sqrt{T}},$$

where s is the estimated standard deviation, 4%. This means that the test statistic is still $Z = (0.0575 - 0.08)/(0.04 \times \sqrt{10}) = -1.78$. However, because the standard deviation is unknown, a t-test is instead used. The test statistic has a t-distribution with $T - 1 = 9$ degrees of freedom. From the standard t-distribution, $t_9(-1.78) = 0.0545$. For the calculated mean to be significantly different from the assumed mean at the 95% level of confidence, a number less than 0.025 or greater than 0.975 would be needed. The calculated mean is therefore not significantly different from the assumed mean at this level of confidence. For the calculated mean to be significantly lower than the assumed mean at the 95% level of confidence, a number less than 0.05 is be needed. Again, the calculated mean is not significantly lower than the assumed mean at this level of confidence.

Alternatively, it is possible to calculate the inverse cumulative t-distribution function at the required levels of confidence. For the two-tailed test, $t_9^{-1}(0.025) = -2.262$, whilst $\Phi^{-1}(0.975) = 2.262$, where t^{-1} is the inverse standard cumulative t-distribution function. Since Z lies between these values, the observation is not significantly different from the assumed mean at this level of confidence. For the one-tailed test, $\Phi^{-1}(0.05) = -1.833$. Since Z is higher than this value, the observation is not significantly lower than the assumed mean at this level of confidence.[1]

---

[1] In many tables the t-distribution returns only positive numbers, so in the example here $\Phi^{-1}(0.025)$ would equal 2.262

10.2.4 The skewed t-distribution

An extension of the t-distribution that is also a normal mixture distribution is the skewed t-distribution. It is important to note, however, that this is not the only distribution with this name – there are in fact a number of different skew or skewed t-distributions, each with a different form. The density function for this version is:

$$f(x) = cK_{\frac{\gamma+1}{2}}\left(\frac{|\delta|}{\beta}\sqrt{\gamma + \left(\frac{x-\alpha}{\beta}\right)^2}\right)\left(\frac{|\delta|/\beta\sqrt{\gamma+[(x-\alpha)/\beta]^2}}{1+[(x-\alpha)/\beta]^2/\gamma}\right)^{\frac{\gamma+1}{2}} e^{\frac{(x-\alpha)\delta}{\beta^2}},$$

$$(10.23)$$

where:

$$c = \frac{2^{\left(1-\frac{\gamma+1}{2}\right)}}{\beta\sqrt{\pi\gamma}\Gamma\left(\frac{\gamma}{2}\right)},$$

$$(10.24)$$

and $K_\zeta()$ is a modified Bessel function of the second kind with index ζ. Various skewed t density and distribution functions are shown in Figures 10.5 and 10.6 respectively. The parameters α, β and γ alter the location, scale and shape – in particular, the degree of leptokurtosis – of the distribution, whilst δ alters the amount of skewness. Whilst β and γ must be positive, α and δ can take any real value.

As this is a normal mixture distribution, it can be understood in terms of the normal and χ^2 distributions. From Equation (10.18), it can be seen that observations with a t-distribution with a mean of α, a scale parameter of β and γ degrees of freedom can be constructed from a standard normal random variable, Z, and a random variable from an χ^2 distribution with γ degrees of freedom, X_γ^2, as follows:

$$\alpha + \frac{1}{\sqrt{X_\gamma^2/\gamma}}\beta Z.$$

$$(10.25)$$

However, if a skewness parameter, δ, is scaled by the χ^2 variable and then added to the scaled normal term, the result is a skewed t-distribution:

$$\alpha + \frac{1}{\sqrt{X_\gamma^2/\gamma}}\beta Z + \frac{1}{X_\gamma^2/\gamma}\delta.$$

$$(10.26)$$

The mean of this distribution – which only exists if $\gamma > 2$ – is:

$$\mu = \alpha + \frac{\gamma\delta}{\gamma-2}.$$

$$(10.27)$$

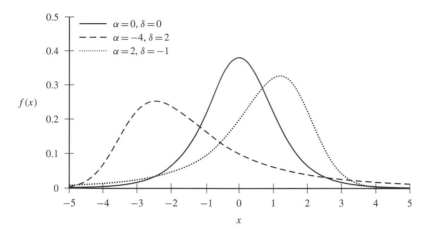

Figure 10.5 Various skewed t-density functions ($\beta=1, \gamma=5$)

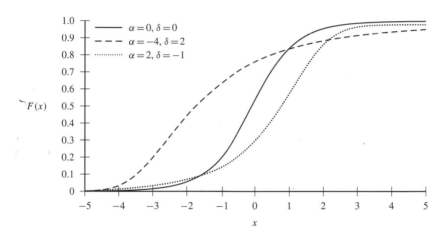

Figure 10.6 Various skewed t-distribution functions ($\beta=1, \gamma=5$)

The variance only exists if $\gamma > 4$, in which case it is given by:

$$\sigma^2 = \frac{\beta^2\gamma}{\gamma-2} + \frac{2\gamma^2\delta^2}{(\gamma-2)^2(\gamma-4)}. \tag{10.28}$$

The fact that the mean and variance exist only for twice the degrees of freedom than are needed for the standard t-distribution are a measure of how fat the tails of this distribution are. The distribution is skewed to the left if $\delta < 0$ and to the right if $\delta > 0$. If $\delta = 0$, the result is the t-distribution.

10.2.5 The Gumbel distribution

This distribution is an unbounded one, but it is skewed to the right. The focus also tends to be on the right-hand tail as it was designed with extreme values in mind. It is a straightforward two-parameter model with the following cumulative distribution function:

$$F(x) = \Pr(X \le x) = e^{-e^{-\frac{x-\alpha}{\beta}}}. \tag{10.29}$$

This shows one of the attractions of the Gumbel distribution – that the distribution rather than the density function is given. This means that cumulative probabilities can be calculated easily without the need to resort to numerical methods or standard tables.

The mean of the Gumbel distribution is $\alpha + \beta\eta$ and its variance is $\pi^2\beta^2/6$. The term η is the Euler–Mascheroni constant (which is equal to around 0.557).

10.2.6 The lognormal distribution

The distributions above can take values of x from $-\infty$ to ∞. However, in many cases, this range of values is not appropriate. In particular, if many variables can take only non-negative values. Examples include the price of an asset, the size of a population or the number of claims. Sometimes, the volatility in the distribution is so small that the probability of a negative observation is trivial when an unbounded distribution is used. In this case, a symmetrical distribution such as the normal might give an adequate approximation. However, if it does not, there are ways of manipulating the normal distribution to give only positive results.

A common manipulation of the normal distribution is to apply it to log-transformed data in time series analysis. If (and only if) the data take positive values, then natural logarithms can be taken and the result treated as being normally distributed, since the exponential (or inverse-logarithm) of any variable will always be positive. Not only is the lognormal distribution lower-bounded at zero, it also has positive skew. Various lognormal density and distribution functions are shown in Figures 10.7 and 10.8 respectively.

The logarithmic transformation of a dataset together with the assumption that the natural logarithm of an asset value will follow a random walk with drift is frequently used to model financial variables, as shown below:

$$\ln X_t = \mu + \ln X_{t-1} + \epsilon_t. \tag{10.30}$$

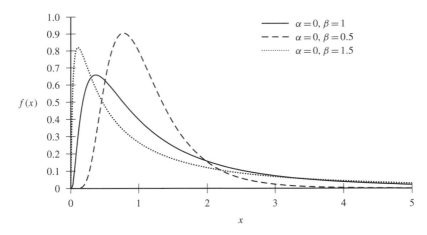

Figure 10.7 Various lognormal density functions

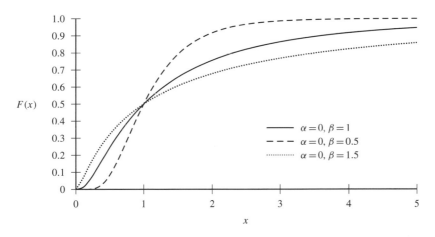

Figure 10.8 Various lognormal distribution functions

The easiest way to generate lognormal random variables is to take the natural logarithm of each data point, and then treat the logarithms of the data as being normally distributed.

The lognormal distribution itself has a density function identical to the normal distribution but with $\ln x$ substituted for x:

$$f(x) = \frac{1}{\sqrt{2\pi}} e^{-\frac{1}{2}\left(\frac{\ln x - \alpha}{\beta}\right)^2}. \tag{10.31}$$

10.2.7　The Wald distribution

The Wald Distribution is also known as the inverse normal or inverse Gaussian distribution. It describes the time it takes for a random walk with drift to reach a particular level, x, so takes only positive values. It also has positive skew, and the following probability density function:

$$f(x) = \sqrt{\frac{\gamma}{2\pi x^3}} e^{-\frac{\gamma(x-\alpha)^2}{2\alpha^2 x}}.\qquad(10.32)$$

The mean of the distribution is α, the location parameter, whilst the shape parameter is γ. The variance is α^3/γ. Both α and γ must be greater than zero.

The Wald distribution has some useful properties in terms of aggregation. In particular:

- if $X_n \sim \text{Wald}(\alpha_0 w_n, \gamma_0 w_n^2)$ and all X_n are independent, then $\sum_{n=1}^{N} X_n \sim \text{Wald}(\alpha_0 \sum_{n=1}^{N} w_n, \gamma_0 [\sum_{n=1}^{N} w_n]^2)$; and
- if $X \sim \text{Wald}(\alpha, \gamma)$, then for $n > 0$, $nX \sim \text{Wald}(n\alpha, n\gamma)$.

Various Wald density and distribution functions are shown in Figures 10.9 and 10.10 respectively.

10.2.8　The chi-squared distribution

Another approach to modelling variables bounded at zero is to treat the observations as being from a chi-squared distribution with γ degrees of freedom, χ_γ^2.

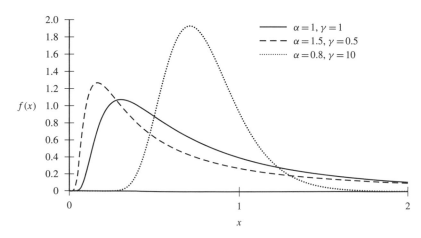

Figure 10.9　Various Wald density functions

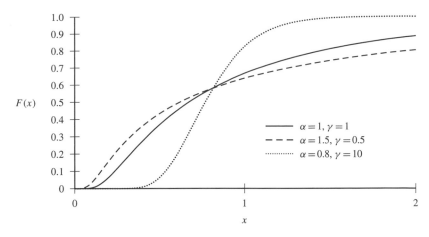

Figure 10.10 Various Wald distribution functions

The chances are that this is not the case. An χ^2_γ distribution represents the distribution of the sum of γ squared, independent variables drawn from a standard normal distribution. However, the shape of the distribution means that it is also used to simulate time series. The cumulative distribution function for the χ^2 distribution with γ degrees of freedom, $\chi^2_\gamma(x)$ is:

$$\chi^2_\gamma(x) = \int_0^x k_\gamma(s)\,ds, \tag{10.33}$$

where:

$$k_\gamma(x) = \frac{1}{2^{\gamma/?}\Gamma(\gamma/2)} x^{\frac{\gamma}{2}-1} e^{-\frac{x}{2}}, \tag{10.34}$$

γ being a positive integer. This distribution has a mean of γ and a variance of 2γ, meaning that the range of variables that this distribution will fit is limited.

Since this distribution represents the sum of squared normal variables, simulating a χ^2 distribution with γ degrees of freedom is simply a case of generating γ normally distributed random variables for each data point, then squaring and summing them. Various χ^2 density and distribution functions are shown Figures 10.11 and 10.12 respectively.

The main use of the χ^2 distribution is not in modelling time series but in testing goodness of fit. The χ^2 test is used to compare the actual number of observations in N categories with those expected. This test works by using the normal approximation to the binomial distribution. Suppose that the probability of an observation in category n where $n = 1, 2, \ldots, N$ is p_n, where $\sum_{n=1}^N p_n = 1$ and the total number of observations is T. This means that the

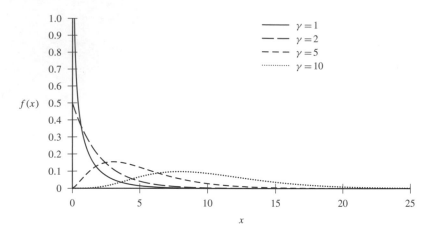

Figure 10.11 Various chi-square density functions

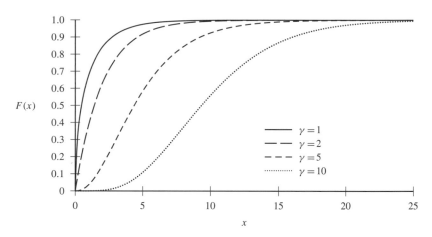

Figure 10.12 Various chi-square distribution functions

expected number of observations in category N is Tp_n and the variance is $Tp_n(1 - p_n)$. If the actual number of observations in category n is T_n, then for large values of T the difference between the actual and expected number of observations can be assumed to have a normal distribution. This means that $X_n = (T_n - Tp_n)/\sqrt{Tp_n(1 - p_n)}$ can be assumed to have a normal distribution. Since the sum of squared normal distributions has a χ^2 distribution, the χ^2 test statistic, k is:

$$k = \sum_{n=1}^{N} X_n^2 \sim \chi_{N-1}^2. \qquad (10.35)$$

Since large deviations suggest that the assumed probabilities are incorrect, this statistic needs to be tested against the upper tail test statistics of the χ^2 distribution with $N - 1$ degrees of freedom. The '-1' is because the total probability is always one, so a degree of freedom is lost.

Example 10.7 An insurance company has designed its pricing structure to target a particular mix of business, both between classes and across regions, as described below:

	Expected proportion of policies by type (t)			
Country (c)	Household buildings (1)	Household contents (2)	Car (3)	Total
England (1)	0.20	0.20	0.10	0.50
Scotland (2)	0.10	0.10	0.05	0.25
Wales (3)	0.06	0.06	0.03	0.15
N. Ireland (4)	0.04	0.04	0.02	0.10
Total	0.40	0.40	0.20	1.00

After a month, 1,000 policies have been sold, distributed as follows:

	Number of policies sold by type (t)			
Country (c)	Household buildings (1)	Household contents (2)	Car (3)	Total
England (1)	242	231	93	566
Scotland (2)	99	94	38	231
Wales (3)	52	50	20	122
N. Ireland (4)	35	33	13	81
Total	428	408	164	1,000

Is the pricing structure bringing in levels of business in line with those required?

This can be determined using an χ^2 test. If the number of policies in country c of of type t is defined as $N_{t,c}$ with the expected proportion of policies defined as $p_{t,c}$, then the expected number of policies sold is $1000p_{t,c}$, and the variance of this amount is $1000p_{t,c}(1 - p_{t,c})$. This means that $X_{t,c} = (N_{t,c} - 1000p_{t,c})/\sqrt{1000p_{t,c}(1 - p_{t,c})}$ is normally distributed and $\sum_{t=1}^{3}\sum_{c=1}^{4} X_{t,c}^2$ has a χ^2 distribution with $(3 \times 4) - 1 = 11$ degrees of freedom. The value of each $X_{t,c}^2$ is shown below:

| | $X^2_{t,c}$ | | | |
Country (c)	Household buildings (1)	Household contents (2)	Car (3)	Total
England (1)	11.03	6.01	0.54	17.58
Scotland (2)	0.01	0.40	3.03	3.44
Wales (3)	1.13	1.77	3.44	6.34
N. Ireland (4)	0.65	1.28	2.50	4.43
Total	12.82	9.46	9.51	31.79

The sum of these values is therefore 31.79. The critical value for the upper tail of the χ^2 distribution with 11 degrees of freedom is 26.76 at the 0.5% level, suggesting that number of policies sold is significantly different from that intended with at least a 99.5% level of confidence.

10.2.9 The F-distribution

The F-distribution is another lower-bounded statistical distribution; this time based on the χ^2 distribution. It comes from the ratio of two χ^2 variables, such that if X_1 and X_2 are two independent variables, each having a χ^2 distribution with γ_1 and γ_2 degrees of freedom respectively, then $(X_1/\gamma_1)/(X_2/\gamma_2)$ has an F-distribution with γ_1 and γ_2 degrees of freedom. It has the following probability density function:

$$f(x) = \frac{1}{x B(\gamma_1/2, \gamma_2/2)} \sqrt{\left(\frac{\gamma_1}{\gamma_2}\right)^{\gamma_1} \frac{x^{(\gamma_1-2)}}{[1+(\gamma_1 x/\gamma_2)]^{(\gamma_1+\gamma_2)}}}, \qquad (10.36)$$

where $B(y_1, y_2)$ is the beta function, defined as:

$$B(y_1, y_2) = \frac{\Gamma(y_1)\Gamma(y_2)}{\Gamma(y_1+y_2)}. \qquad (10.37)$$

The parameters γ_1 and γ_2 must be positive, and the distribution is defined only for positive values. Whilst the F-distribution could be used to model lower-bounded data, it is more frequently used to test the differences between two statistics, often in relation to model selection. One such test, the Chow test, is described later in Chapter 13. Various F density and distribution functions are shown in Figures 10.13 and 10.14 respectively.

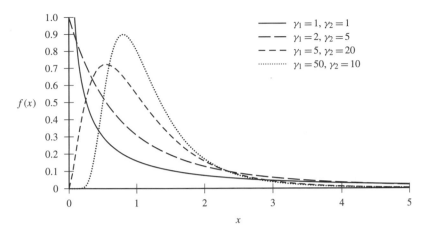

Figure 10.13 Various F-density functions

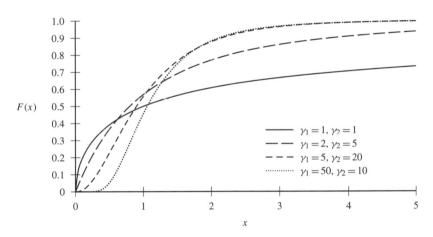

Figure 10.14 Various F-distribution functions

10.2.10 The Weibull distribution

The Weibull distribution is a flexible two-parameter distribution that is defined for positive values of x. Its distribution function is:

$$F(x) = \Pr(X \le x) = 1 - e^{-\left(\frac{x}{\gamma}\right)^{\beta}}, \qquad (10.38)$$

with a mean of $\beta\Gamma(1 + 1/\gamma)$ and a variance of $\beta^2(\Gamma(1 + 2/\gamma) - \Gamma^2(1 + 1/\gamma))$.

The distribution can be used to simulate failure or mortality rates, with a γ less than one implying a reducing rate of failure, a γ greater than one implying an increasing rate of failure and a γ equal to one implying a constant rate of failure. It was also used in the past as a proxy for the normal distribution (and others) as the distribution function can be expressed without the need for integrals.

10.2.11 The Burr distribution

Another distribution that can be used to model failure rates and is also analytically tractable is the Burr distribution. There are a number of versions of this distribution that exist, but the following is a good example:

$$F(x) = \Pr(X \leq x) = 1 - (1 + x^{\beta})^{-\gamma}. \qquad (10.39)$$

The expressions for the mean and variance of the Burr distribution are quite involved, so are not given here.

10.2.12 The Lévy distribution

A frequently used distribution in asset pricing is the Lévy distribution. This is another skewed distribution with the tails that follow a power law. This makes the Lévy distribution a good leptokurtic alternative to distributions such as the lognormal distribution, meaning that it is of interest when asset returns are being modelled and the risk of extreme results is being considered. Like the symmetrical Cauchy distribution, the Lévy distribution has no defined mean, variance or higher moments.

The probability density function for the Lévy distribution is:

$$f(x) = \sqrt{\frac{\beta}{2\pi}} \frac{e^{-\frac{\beta}{2x}}}{x^{\frac{3}{2}}}, \qquad (10.40)$$

where β is a scale parameter. The distribution is defined for all values of $x > 0$ and β must be greater than zero. Various Lévy density and distribution functions are shown in Figures 10.15 and 10.16 respectively.

The Lévy distribution is closely related to the normal distribution, to the extent that the cumulative distribution function can be expressed as:

$$F(x) = 2\Phi\left(-\sqrt{\frac{\beta}{x}}\right). \qquad (10.41)$$

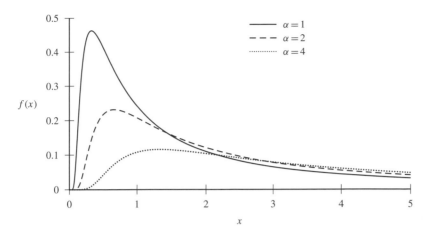

Figure 10.15 Various Lévy density functions

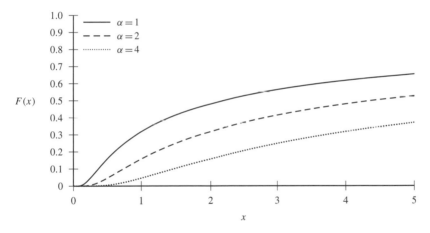

Figure 10.16 Various Lévy distribution functions

10.2.13 The gamma and inverse gamma distributions

Two very flexible distributions are the gamma and inverse gamma distributions, which again give only non-negative values. All gamma distributions have a scale or rate parameter, β, and a shape parameter, γ, both of which must also be greater than zero. The probability density function is:

$$f(x) = \frac{1}{\beta^{\gamma}\Gamma(\gamma)} x^{\gamma-1} e^{-\frac{x}{\beta}}. \tag{10.42}$$

The mean of this distribution is $\beta\gamma$ and the variance is $\beta^2\gamma$. Various gamma density and distribution functions are shown in Figures 10.17 and 10.18 respectively.

There are a number of special cases of the gamma distribution. In particular:

- if $\gamma = 1$, the result is the exponential distribution, discussed below; and
- if $\beta = 2$, the result is a χ^2 distribution with 2γ degrees of freedom.

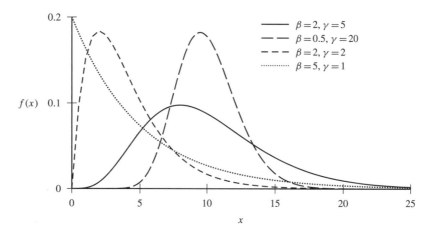

Figure 10.17 Various gamma density functions

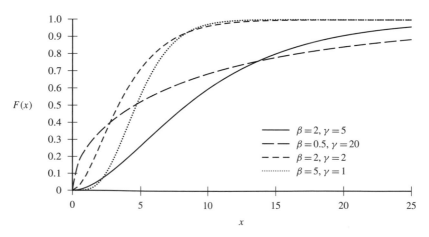

Figure 10.18 Various gamma distribution functions

The gamma distribution has some useful properties in terms of aggregation. In particular:

- if $X_n \sim$ gamma(β, γ_n) and all X_n are independent, then $\sum_{n=1}^{N} X_n \sim$ gamma$(\beta, \sum_{n=1}^{N} \gamma_n)$; and
- if $X \sim$ gamma(β, γ), then for $n > 0$, $nX \sim$ gamma$(n\beta, \gamma)$.

These properties mean that it is possible to calculate the probability under a gamma distribution by converting it to a χ^2 distribution and using the standard values from that table. In other words, if $X \sim$ gamma(β, γ), then $2X/\beta \sim \chi^2_{2\gamma}$.

If Y has a gamma distribution, then $X = 1/Y$ has an inverse gamma distribution. This has a similar probability density function to the gamma distribution:

$$f(x) = \frac{\beta^\gamma}{\Gamma(\gamma)} \frac{1}{x^{\gamma+1}} e^{-\frac{\beta}{x}}. \tag{10.43}$$

As with the gamma distribution, it is defined only for values of x greater than zero, whilst β and γ must also be greater than zero. The distribution has a mean of $\beta/(\gamma - 1)$ and a variance of $\beta^2/(\gamma - 1)^2(\gamma - 2)$. The mean is only defined for $\gamma > 1$, and the variance for $\gamma > 2$. Various inverse gamma density and distribution functions are shown in Figures 10.19 and 10.20 respectively.

The Lévy distribution is a special case of the inverse gamma distribution with $\beta = 1/2$ and $\gamma_{I\Gamma} = \gamma_L/2$, where $\gamma_{I\Gamma}$ is the γ parameter from the inverse gamma distribution and γ_L is the γ parameter from the Lévy distribution.

The gamma and inverse gamma distributions can be fitted by calculating the sample mean and variance of a dataset and rearranging the expressions for the mean and variance to find β and γ.

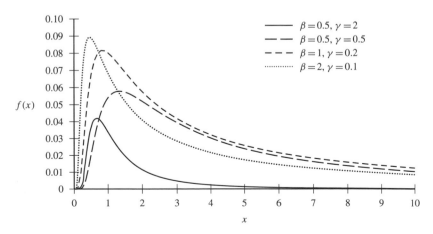

Figure 10.19 Various inverse gamma density functions

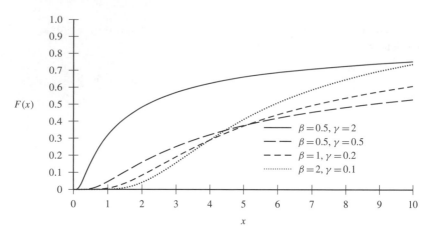

Figure 10.20 Various inverse gamma distribution functions

10.2.14 The generalised inverse Gaussian (GIG) distribution

An even more flexible distribution is the GIG distribution, which is capable of delivering a wide range of shapes. Its probability density function is:

$$f(x) = kx^{\gamma-1}e^{-\frac{1}{2}\left(\frac{\beta_1}{x}+\frac{x}{\beta_2}\right)}, \tag{10.44}$$

where k is defined as:

$$k = \frac{(\beta_1\beta_2)^{-\frac{\gamma}{2}}}{2K_\gamma(\sqrt{\beta_1/\beta_2})}, \tag{10.45}$$

and $K_\zeta()$ is a modified Bessel function of the second kind with index ζ; however, for practical purposes, k can be regarded as a constant that ensures that the integral of $F(x)$ between zero and ∞ is equal to one. The parameter γ can take any real value, whilst β_1 and β_2 must generally be positive.

The GIG distribution has a number of special cases. In particular:

- if $\beta_1 = 0$, then the result is a gamma distribution with $\beta_\Gamma = 2\beta_2$ and $\gamma_\Gamma = \gamma$, where β_Γ and γ_Γ are the β and γ parameters for the gamma distribution;
- if $\beta_2 = 0$, then the result is an inverse gamma distribution with $\beta_{I\Gamma} = \beta_1/2$ and $\gamma_\Gamma = -\gamma$, where $\beta_{I\Gamma}$ and $\gamma_{I\Gamma}$ are the β and γ parameters for the inverse gamma distribution;
- if $\gamma = -1/2$, then the result is a Wald (inverse Gaussian) distribution.

10.2.15 The exponential distribution

As mentioned above, the exponential distribution is a special case of the gamma distribution where $\gamma = 1$. It is a monotonically decreasing distribution with a very straightforward parametrisation:

$$f(x) = \frac{1}{\beta} e^{-\frac{x}{\beta}}. \tag{10.46}$$

This distribution has a mean of β and a variance of β^2. The term β is essentially a scale parameter.

The exponential distribution is linked to the discrete Poisson distribution in that it gives the expected time between observations under a Poisson distribution.

The shape of the distribution is, clearly, exponential, which means that it will not give high probabilities of extreme values. Furthermore, the limited parametrisation means that it is unlikely to provide a good fit to data.

10.2.16 The Fréchet distribution

The Fréchet distribution is another monotonically decreasing statistical distribution. It has a single parameter, β, which determines the distribution's scale, and the distribution function is:

$$F(x) = \Pr(X \leq x) = e^{-x^{-\beta}}. \tag{10.47}$$

10.2.17 The Pareto distribution

This entire distribution follows a power law. As such it is useful for modelling variables such as the distribution of wealth (for which the distribution was derived by Pareto) or the population of cities. The Pareto distribution function is:

$$F(x) = \Pr(X \leq x) = 1 - \left(\frac{\beta}{\beta + x}\right)^{\gamma}, \tag{10.48}$$

where γ determines the shape (and power) of the distribution and β determines the scale, both taking only positive values. The mean of this distribution is $\beta/(\gamma - 1)$ and the variance is $\beta^2 \gamma/(\gamma - 1)^2(\gamma - 2)$. Simulation is then simply a case of generating a uniform random variable, U, and calculating the following statistic:

$$\frac{\beta}{U^{(1/\gamma)}}. \tag{10.49}$$

10.2.18 The generalised Pareto distribution

Whilst the exponential, Fréchet and Pareto distributions can be used to model monotonically decreasing distirbutions, the range of shapes and scales available is limited. For this reason, the generalised Pareto distribution can be used instead. However, this distribution is of more fundamental importance given its use in extreme value theory, as discussed later. The cumulative distribution function for this distribution is:

$$F(x) = \begin{cases} 1 - \left(1 + \dfrac{x}{\beta\gamma}\right)^{-\gamma} & \text{if } \gamma \neq 0; \\[2ex] 1 - e^{-\frac{x}{\beta}} & \text{if } \gamma = 0. \end{cases} \tag{10.50}$$

As with the basic Pareto distribution, γ and β are the shape and scale parameters. Whilst γ has the same meaning in each, $\beta_P = \beta_{GP}\gamma_{GP}$, where the subscripts P and GP refer to the parameters from the Pareto and generalised Pareto distributions respectively. The generalised Pareto distribution has a mean of $\beta\gamma/(\gamma-1)$, providing $\gamma < 1$. It also has the property that $E(X^k)=\infty$ is undefined if $k \geq \gamma$. So, for example, if $\gamma = 2$, then $E(X^2) = \infty$, so the variance is undefined, as are all higher moments; if $\gamma = 3.5$, then $E(X^4) = \infty$, so the kurtosis is undefined, but the variance and skew of the distribution exist.

Whilst β must be positive as with the Pareto distribution, γ can take any value. If $\gamma = 0$, the formula reduces to the exponential distribution; if $\gamma > 0$, the result is the Pareto distribution, which follows a power law. However, if $\gamma < 0$, x not only has a lower bound of zero, but also an upper bound of $-\beta\gamma$.

10.2.19 The uniform distribution

A distribution that is relevant to all of those that follow is the continuous uniform distribution. For a variable with an equal probability of landing between β_1 and β_2, the probability density function is:

$$f(x) = \begin{cases} \dfrac{1}{\beta_2 - \beta_1} & \text{if } \beta_1 \leq x \leq \beta_2; \\[2ex] 0 & \text{otherwise,} \end{cases} \tag{10.51}$$

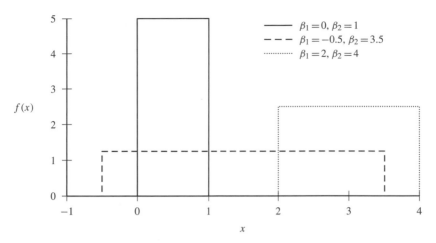

Figure 10.21 Various uniform density functions

and the cumulative distribution function is:

$$F(x) = \Pr(X \leq x) = \begin{cases} 0 & \text{if } x < \beta_1; \\ \dfrac{x - \beta_1}{\beta_2 - \beta_1} & \text{if } \beta_1 \leq x \leq \beta_2; \\ 1 & \text{if } x > \beta_2. \end{cases} \tag{10.52}$$

The parameters β_1 and β_2 are the lower and upper bounds of the distribution, effectively making them scale parameters. This distribution does not reflect many real-life variables; however, setting β_1 to zero and β_2 to one gives the distribution of uniform random variables in the range zero to one – a $U(x)$ distribution. This forms the building block of many other approaches for simulating random variables, as it represents a series of random probabilities. Various uniform density and distribution functions are shown in Figures 10.21 and 10.22 respectively.

10.2.20 The triangular distribution

Another bounded distribution, but one that allows for higher probabilities in the centre of its range than at the tails, is the triangular distribution. This can be used when as well as the maximum and minimum values, the most likely value – the mode of the distribution – is also known. It has one location parameter, α, representing this maximum, and two scale parameters, β_1

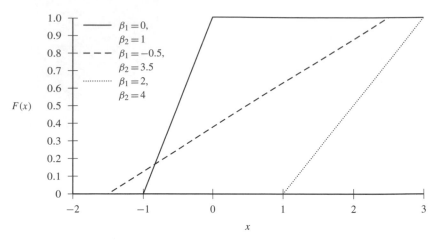

Figure 10.22 Various uniform distribution functions

and β_2, which again are the lower and upper bounds of the distribution. The probability density function here is:

$$
f(x) = \begin{cases} \dfrac{2(x - \beta_1)}{(\beta_2 - \beta_1)(\alpha - \beta_1)} & \text{if } \beta_1 \leq x \leq \alpha; \\[3mm] \dfrac{2(\beta_2 - x)}{(\beta_2 - \beta_1)(\beta_2 - \alpha)} & \text{if } \alpha \leq x \leq \beta_2; \\[3mm] 0 & \text{otherwise,} \end{cases} \tag{10.53}
$$

and the cumulative distribution function is:

$$
F(x) = \Pr(X \leq x) = \begin{cases} 0 & \text{if } x < \beta_1; \\[3mm] \dfrac{(x - \beta_1)^2}{(\beta_2 - \beta_1)(\alpha - \beta_1)} & \text{if } \beta_1 \leq x \leq \alpha; \\[3mm] 1 - \dfrac{(\beta_2 - x)^2}{(\beta_2 - \beta_1)(\beta_2 - \alpha)} & \text{if } \alpha \leq x \leq \beta_2; \\[3mm] 1 & \text{if } x > \beta_2. \end{cases} \tag{10.54}
$$

Various triangular density and distribution functions are shown in Figures 10.23 and 10.24 respectively.

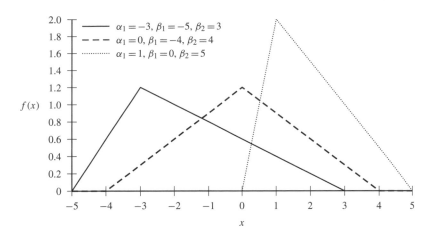

Figure 10.23 Various triangular density functions

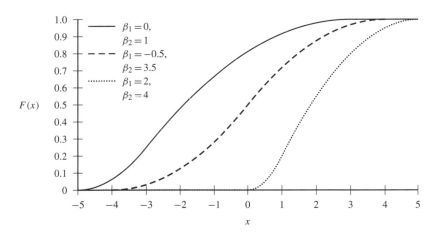

Figure 10.24 Various triangular distribution functions

10.2.21 The beta distribution

Many sets of observations are bounded by zero and one, particularly when probabilities such as mortality rates are involved. For these, a popular distribution to use is the beta distribution. This has two parameters, γ_1 and γ_2, both of which must be positive. The probability density function for the beta distribution is given in Equation (10.55):

$$f(x) = \frac{1}{B(\gamma_1, \gamma_2)} x^{\gamma_1 - 1} (1 - x)^{\gamma_2 - 1}, \tag{10.55}$$

where B is the beta function and $0 \le x \le 1$. If $\gamma_1 = \gamma_2$, then the distribution is symmetrical, and if $\gamma_1 = \gamma_2 = 1$, then the result is the standard uniform distribution. The mean of the beta distribution is $\gamma_1/(\gamma_1 + \gamma_2)$ and the variance is $\gamma_1\gamma_2/(\gamma_1 + \gamma_2)^2(\gamma_1 + \gamma_2 + 1)$. Various beta density and distribution functions are shown in Figures 10.25 and 10.26 respectively.

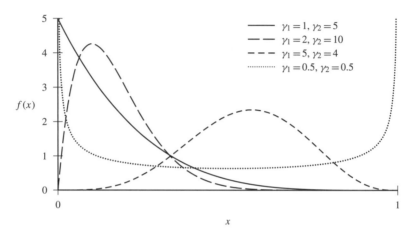

Figure 10.25 Various beta density functions

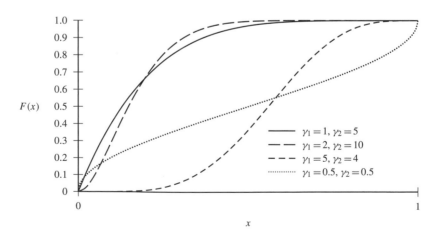

Figure 10.26 Various beta distribution functions

10.3 Multivariate distributions

A simple way of modelling several random variables at once is to use a multivariate distribution. This is a distribution which simultaneously defines the values of more than one variable. It is a slightly restrictive approach, as it involves modelling the marginal distributions and their relationships at the same time. In practice, it might be desirable to separate these, using copulas to link the fitted marginal distributions. However, multivariate distributions offer a simple way of modelling a group of variables that might be appropriate for an approximate modelling exercise, or if only limited data are available.

The multivariate distributions discussed here are all related to the normal distribution, so are defined in terms of linear correlation coefficients. This also means that they are all jointly elliptical distributions.

Multivariate distributions can also defined in terms of location, scale and shape parameters. However, there can also be interactions between these parameters. In particular, there are measures of co-scale that combine individual measures of spread from the marginal distributions with the linear correlation coefficients. These are sometimes – but not always – equal to the covariances between the variables.

Strictly speaking, whilst a univariate distribution deals with a single variable, a distribution that deals with two variables is bivariate and a multivariate distribution deals with more than two. Both bivariate and multivariate distributions are joint distributions; however, I use the term multivariate to include bivariate distributions. In most cases, bivariate distributions will be discussed first before the concepts are generalised to multivariate cases.

10.3.1 Matrix algebra

When dealing with multivariate data, it is often easier to work with matrices rather than with linear algebra. Some simple concepts in matrix algebra are set out here.

A matrix with S rows and T columns has $S \times T$ 'elements'. In an $S \times T$ matrix, A, each element is denoted $A_{s,t}$ where $s = 1 \ldots S$ and $t = 1 \ldots T$. The matrix is arranged as:

$$A = \begin{pmatrix} A_{1,1} & A_{1,2} & \cdots & A_{1,T} \\ A_{2,1} & A_{2,2} & \cdots & A_{2,T} \\ \vdots & \vdots & \ddots & \vdots \\ A_{S,1} & A_{S,2} & \cdots & A_{S,T} \end{pmatrix}. \tag{10.56}$$

If $S = 1$, then this reduces to a row vector; if instead $T = 1$, then the result is a column vector. The transpose of matrix A is denoted A'. This is obtained by transposing each row of the matrix with each column, so A' is a $T \times S$ matrix with T rows and S columns:

$$A' = \begin{pmatrix} A_{1,1} & A_{2,1} & \cdots & A_{S,1} \\ A_{1,2} & A_{2,2} & \cdots & A_{S,2} \\ \vdots & \vdots & \ddots & \vdots \\ A_{1,T} & A_{2,T} & \cdots & A_{S,T} \end{pmatrix}. \tag{10.57}$$

If matrices are added or subtracted, corresponding elements are added, so:

$$A = B + C \tag{10.58}$$

can be expressed as:

$$\begin{pmatrix} A_{1,1} & A_{1,2} & \cdots & A_{1,T} \\ A_{2,1} & A_{2,2} & \cdots & A_{2,T} \\ \vdots & \vdots & \ddots & \vdots \\ A_{S,1} & A_{S,2} & \cdots & A_{S,T} \end{pmatrix} = \begin{pmatrix} B_{1,1} & B_{1,2} & \cdots & B_{1,T} \\ B_{2,1} & B_{2,2} & \cdots & B_{2,T} \\ \vdots & \vdots & \ddots & \vdots \\ B_{S,1} & B_{S,2} & \cdots & B_{S,T} \end{pmatrix}$$

$$+ \begin{pmatrix} C_{1,1} & C_{1,2} & \cdots & C_{1,T} \\ C_{2,1} & C_{2,2} & \cdots & C_{2,T} \\ \vdots & \vdots & \ddots & \vdots \\ C_{S,1} & C_{S,2} & \cdots & C_{S,T} \end{pmatrix},$$

$$\tag{10.59}$$

and $A_{s,t} = B_{s,t} + C_{s,t}$ for all values of s and t in the matrix. This means that matrices can only be added and subtracted if they have the same number of rows and columns or dimensions. It also means that the order of addition or subtraction does not matter:

$$B + C = C + B. \tag{10.60}$$

In other words, matrix addition is commutative. Matrix addition is also associative: the order of calculation does not matter. In other words, if another matrix, D, is added:

$$(B + C) + D = B + (C + D). \tag{10.61}$$

The same principle applies to matrix transposition:

$$(B + C)' = B' + C'. \tag{10.62}$$

However, the order of multiplication does usually matter, due to the way that matrix multiplication is carried out. Matrices can only be multiplied if the number of columns of the first matrix equals the number of rows of the second. For example, if B is an $M \times T$ matrix and C is a $T \times N$ matrix, then it is possible to 'pre-multiply' C by B (or, equivalently to 'post-multiply' B by C) to give:

$$A = BC, \tag{10.63}$$

where A is a $M \times N$ matrix with elements $a_{m,n}$, $m = 1 \ldots M$, $n = 1 \ldots N$, $s, t = 1 \ldots T$ and:

$$A_{m,n} = \sum_{s=1}^{T} \sum_{t=1}^{T} B_{m,s} C_{t,n}. \tag{10.64}$$

In other words, $A_{1,2}$ would be calculated using the elements highlighted below:

$$
\begin{pmatrix} \square & A_{1,2} & \cdots & \square \\ \square & \square & \cdots & \square \\ \vdots & \vdots & \ddots & \vdots \\ \square & \square & \cdots & \square \end{pmatrix} = \begin{pmatrix} B_{1,1} & B_{1,2} & \cdots & B_{1,T} \\ \square & \square & \cdots & \square \\ \vdots & \vdots & \ddots & \vdots \\ \square & \square & \cdots & \square \end{pmatrix}
$$

$$
\times \begin{pmatrix} \square & C_{1,2} & \cdots & \square \\ \square & C_{2,2} & \cdots & \square \\ \vdots & \vdots & \ddots & \vdots \\ \square & C_{T,2} & \cdots & \square \end{pmatrix}. \tag{10.65}
$$

Although it is not commutative, matrix multiplication is associative:

$$(AB)C = A(BC); \tag{10.66}$$

it is distributive:

$$A(B+C) = AB + AC; \tag{10.67}$$

and the transpose of the product of matrices is equal to the reversed product of their transposes:

$$(ABC)' = C'B'A'. \tag{10.68}$$

It is also possible simply to multiply all elements of a matrix by the same number, a process known as 'scalar multiplication'. If matrix A is multiplied by a scalar, D, to give DA, then each element of A, $A_{s,t}$ would be scaled to $D \times A_{s,t}$.

Another useful aspect of matrix algebra is the determinant of a matrix, denoted $|A|$. The determinant can be calculated only if a matrix is square. However, as the size of a matrix increases, the calculation becomes increasingly complex. for a 2×2 matrix, the determinant is:

$$|A| = \begin{vmatrix} A_{1,1} & A_{1,2} \\ A_{2,1} & A_{2,2} \end{vmatrix} = A_{1,1}A_{2,2} - A_{1,2}A_{2,1}. \tag{10.69}$$

For a 3×3 matrix, it is:

$$
\begin{aligned}
|A| = \begin{vmatrix} A_{1,1} & A_{1,2} & A_{1,3} \\ A_{2,1} & A_{2,2} & A_{2,3} \\ A_{3,1} & A_{3,2} & A_{3,3} \end{vmatrix} &= A_{1,1} \begin{vmatrix} A_{2,2} & A_{2,3} \\ A_{3,2} & A_{3,3} \end{vmatrix} \\
&\quad - A_{1,2} \begin{vmatrix} A_{2,1} & A_{2,3} \\ A_{3,1} & A_{3,3} \end{vmatrix} \\
&\quad + A_{1,3} \begin{vmatrix} A_{2,1} & A_{2,2} \\ A_{3,1} & A_{3,2} \end{vmatrix}.
\end{aligned}
\tag{10.70}
$$

Expanding this to a $T \times T$ matrix gives:

$$
\begin{aligned}
|A| = \begin{vmatrix} A_{1,1} & A_{1,2} & \cdots & A_{1,T} \\ A_{2,1} & A_{2,2} & \cdots & A_{2,T} \\ \vdots & \vdots & \ddots & \vdots \\ A_{T,1} & A_{T,2} & \cdots & A_{T,T} \end{vmatrix} &= A_{1,1} \begin{vmatrix} A_{2,2} & A_{2,3} & \cdots & A_{2,T} \\ \vdots & \vdots & \ddots & \vdots \\ A_{T,2} & A_{T,3} & \cdots & A_{T,T} \end{vmatrix} \\
&\quad - A_{1,2} \begin{vmatrix} A_{2,1} & A_{2,3} & \cdots & A_{2,T} \\ \vdots & \vdots & \ddots & \vdots \\ A_{T,1} & A_{T,3} & \cdots & A_{T,T} \end{vmatrix} \\
&\quad + \vdots \\
&\quad \pm \\
&\quad A_{1,T} \begin{vmatrix} A_{2,1} & A_{2,2} & \cdots & A_{2,T-1} \\ \vdots & \vdots & \ddots & \vdots \\ A_{T,1} & A_{T,2} & \cdots & A_{T,T-1} \end{vmatrix},
\end{aligned}
\tag{10.71}
$$

where each sub-determinant, or 'minor' is calculated in the same way as the determinant itself, down to the simple calculation for 2×2 matrices.

The determinant cannot always be calculated. In particular, if two rows or columns are equal, or one is a simple multiple of another, or if one row or column is a linear combination of two or more other rows or columns, then there will be a 'minor' somewhere that has a denominator of zero. This results in the determinant being undefined.

If a matrix is square and its determinant is defined, then it has an inverse. The inverse of matrix A is denoted A^{-1}. It is defined such that:

$$AA^{-1} = A^{-1}A = I, \tag{10.72}$$

where I is the identity matrix. This is a matrix whose elements are all zero except for the diagonal, where they are one:

$$A = \begin{pmatrix} 1 & 0 & \cdots & 0 \\ 0 & 1 & \cdots & 0 \\ \vdots & \vdots & \ddots & \vdots \\ 0 & 0 & \cdots & 1 \end{pmatrix}. \tag{10.73}$$

The identity matrix also has the property that it leaves a matrix unchanged, whether pre-multiplying or post-multiplying it:

$$AI = IA = A. \tag{10.74}$$

However, it is important to note that unless A is a square matrix, the identity matrix pre-multiplying A will have a different dimension from that post-multiplying it.

Matrix inversion uses the determinant as a scaling factor, but the calculation of the inverse is even more involved than calculation of the determinant. The inverse of matrix A is calculated as:

$$A^{-1} = \frac{1}{|A|} F. \tag{10.75}$$

F is the matrix of cofactors. This is a square T-dimensional matrix with elements $F_{s,t}$ where $s, t = 1 \ldots T$:

$$F = \begin{pmatrix} F_{1,1} & F_{1,2} & \cdots & F_{1,T} \\ F_{2,1} & F_{2,2} & \cdots & F_{2,T} \\ \vdots & \vdots & \ddots & \vdots \\ F_{T,1} & F_{T,2} & \cdots & F_{T,T} \end{pmatrix}. \tag{10.76}$$

Each element $F_{s,t}$ is $(-1)^{(s+t)}$ multiplied by the determinant of a matrix with row s and column t removed. For a 4×4 version of matrix F giving the cofactors of matrix A, the cofactor in row 2 and column 3 is given as:

$$F_{2,3} = (-1)^{(2+3)} \begin{vmatrix} A_{1,1} & A_{1,2} & \square & A_{1,4} \\ \square & \square & \square & \square \\ A_{3,1} & A_{3,2} & \square & A_{3,4} \\ A_{4,1} & A_{4,2} & \square & A_{4,4} \end{vmatrix}$$

$$= (-1)^{(2+3)} \begin{vmatrix} A_{1,1} & A_{1,2} & A_{1,4} \\ A_{3,1} & A_{3,2} & A_{3,4} \\ A_{4,1} & A_{4,2} & A_{4,4} \end{vmatrix}. \tag{10.77}$$

Non-square matrices can also have left- and right-inverses, but these will be different since the dimensions needed for pre- and post-multiplication are different. For this reason, they are not regarded as 'true' inverses.

A special type of square matrix, which is important in some procedures, is the orthogonal matrix. This is defined as a matrix whose transpose is equal to its inverse, so:

$$A'A = AA' = I. \tag{10.78}$$

These matrix manipulations are available in most spreadsheet and statistical packages.

10.3.2 The multivariate normal distribution

The univariate normal distribution has already been discussed. However, it is also possible to project correlated normal random variables. This is popular for the same sort of reasons that the univariate normal distribution is popular – it is easy to parameterise and project, and gives reasonable results if little is known about the data. Considering first the bivariate case, only one additional param-eter is required: the linear correlation between the two variables, Pearson's rho. The bivariate normal probability density function is then defined as:

$$f(x, y) = \frac{1}{2\pi \beta_X \beta_Y \sqrt{\left(1 - \rho_{X,Y}^2\right)}} e^{-z}, \tag{10.79}$$

where:

$$z = \frac{1}{2(1-\rho_{X,Y}^2)} \left[\left(\frac{s-\alpha_X}{\beta_X}\right)^2 + \left(\frac{t-\alpha_Y}{\beta_Y}\right)^2 - \frac{2\rho_{X,Y}(s-\alpha_X)(t-\alpha_Y)}{\beta_X\beta_Y} \right].$$

(10.80)

In this equation, α_X and α_Y correspond to μ_X and μ_Y, the means of X and Y, whilst β_X and β_Y correspond to σ_X and σ_Y, their standard deviations. The parameter $\rho_{X,Y}$ is the linear correlation between the two variables. Both x and y can take any real value. If μ_X and μ_Y are zero, and σ_X and σ_Y are one, then the result is the standard bivariate normal distribution. This is defined by the linear correlation coefficient between the two variables, and has the following probability density function:

$$\phi_{\rho_{X,Y}}(x,y) = \frac{1}{2\pi\sqrt{\left(1-\rho_{X,Y}^2\right)}} e^{-\frac{1}{2(1-\rho_{X,Y}^2)}\left(x^2+y^2-2\rho_{X,Y}xy\right)}.$$

(10.81)

The distribution function ,$\Phi_{\rho_{X,Y}}(x,y)$, is defined as:

$$\Phi_{\rho_{X,Y}}(x,y) = \int_{-\infty}^{x}\int_{-\infty}^{y} \phi_{\rho_{X,Y}}(s,t)\,ds\,dt.$$

(10.82)

The density function of the bivariate standard normal distribution can be shown graphically in two ways: as a surface chart and as a contour chart. The contour chart can be thought of as a map of the landscape shown by the surface chart. These are shown in for two different correlations in Figure 10.27, with each contour representing an increment of 0.1 in the density function. The corresponding surface and contour charts for the distribution function are shown in Figure 10.28.

It is possible to increase the number of variables and move to a truly multivariate distribution by using matrix notation. Let X be a column vector of N variables, $X_1, X_2, \ldots, X_N$, and let the measures of spread – the means of these variables – be the column vector α whose elements are $\alpha_{X_1}, \alpha_{X_2}, \ldots, \alpha_{X_N}$. In the multivariate case, it is easier to combine the correlations and standard deviations into measures of co-spread. For the multivariate distribution, these are covariances. If the $N \times N$ matrix Σ contains the covariances of the variables X, with the diagonal elements of the matrix being the variances, then the

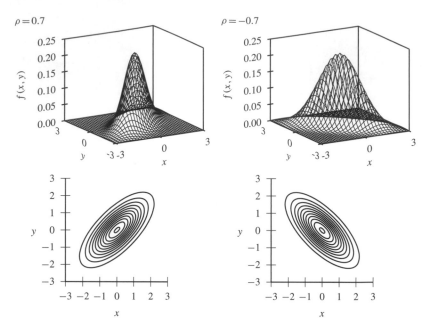

Figure 10.27 Various bivariate normal density functions

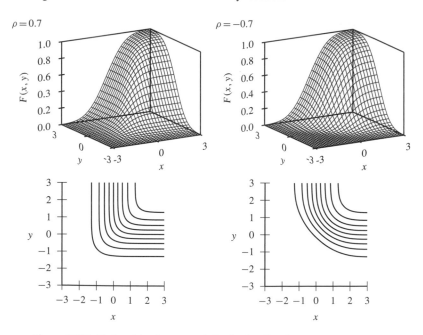

Figure 10.28 Various bivariate normal distribution functions

matrix of co-scale parameters, $\boldsymbol{B}$, is given as:

$$
\boldsymbol{B} = \begin{pmatrix}
\beta_{X_1,X_1} & \beta_{X_1,X_2} & \cdots & \beta_{X_1,X_N} \\
\beta_{X_2,X_1} & \beta_{X_2,X_2} & \cdots & \beta_{X_2,X_N} \\
\vdots & \vdots & \ddots & \vdots \\
\beta_{X_N,X_1} & \beta_{X_N,X_2} & \cdots & \beta_{X_N,X_N}
\end{pmatrix}
$$

$$
= \begin{pmatrix}
\sigma_{X_1,X_1} & \sigma_{X_1,X_2} & \cdots & \sigma_{X_1,X_N} \\
\sigma_{X_2,X_1} & \sigma_{X_2,X_2} & \cdots & \sigma_{X_2,X_N} \\
\vdots & \vdots & \ddots & \vdots \\
\sigma_{X_N,X_1} & \sigma_{X_N,X_2} & \cdots & \sigma_{X_N,X_N}
\end{pmatrix} = \Sigma, \qquad (10.83)
$$

where $\beta_{X_n,X_n} = \sigma_{X_n,X_n} = \sigma_{X_n}^2$. If $\boldsymbol{X}$ is the column vector denoting the values at which each variables is evaluated, then the joint probability that each element in $\boldsymbol{X}$ is less than its corresponding value in $\boldsymbol{x}$ is $\Pr(\boldsymbol{X} \le \boldsymbol{x})$. This is the combined probability that X_1 is less than x_1, X_2 is less than x_2 and so on. The probability density function here is:

$$
f(\boldsymbol{x}) = f(x_1, x_2, \dots, x_N)
$$

$$
= \frac{1}{(2\pi)^{(N/2)}\sqrt{|\boldsymbol{B}|}} e^{-\frac{1}{2}(\boldsymbol{x}-\boldsymbol{\alpha})'\boldsymbol{B}^{-1}(\boldsymbol{x}-\boldsymbol{\alpha})}. \qquad (10.84)
$$

If the underlying distributions are all standard normal distributions, then this function becomes the standard multivariate normal density function. Here, not only do the means disappear (being zero), but the covariance matrix becomes a correlation matrix – remember that the definition of the correlation is the covariance divided by the two standard deviations, each of which will be one in this case. This means that the standard multivariate normal distribution is defined by this correlation matrix, $\boldsymbol{R}$. Its density function is therefore denoted $\phi_{\boldsymbol{R}}$:

$$
\phi_{\boldsymbol{R}}(\boldsymbol{x}) = \frac{1}{(2\pi)^{(N/2)}\sqrt{|\boldsymbol{R}|}} e^{-\frac{1}{2}\boldsymbol{x}'\boldsymbol{R}^{-1}\boldsymbol{x}}. \qquad (10.85)
$$

The corresponding cumulative distribution function is:

$$
\Phi_{\boldsymbol{R}}(\boldsymbol{x}) = \int_{-\infty}^{s_1} \int_{-\infty}^{s_2} \dots \int_{-\infty}^{s_N} \phi_{\boldsymbol{R}}(\boldsymbol{s}) \mathrm{d}s_1 \mathrm{d}s_2 \dots \mathrm{d}s_N. \qquad (10.86)
$$

Mahalanobis distance

To test whether observations are from a multivariate normal distribution, it is important to test them jointly. Useful statistics in this regard are the Mahalanobis distance and the Mahalanobis angle (Mahalanobis, 1936).

Consider the column vector X_t which contains the observations at time t, where $t = 1, 2, \ldots, T$, for a group of N variables, so $X'_t = (X_{1,t} \, X_{2,t} \, \ldots \, X_{N,t})$. Let the column vector $\bar{X}$ contain the sample mean for each variable calculated over all $t = 1, 2, \ldots, T$, so $\bar{X}' = (\bar{X}_1 \, \bar{X}_2 \, \ldots \, \bar{X}_N)$. Then let S be an $N \times N$ matrix of the sample covariances of the N variables based on the observations from $t = 1, 2, \ldots, T$:

$$S = \begin{pmatrix} s_{X_1,X_1} & s_{X_1,X_2} & \cdots & s_{X_1,X_N} \\ s_{X_2,X_1} & s_{X_2,X_2} & \cdots & s_{X_2,X_N} \\ \vdots & \vdots & \ddots & \vdots \\ s_{X_N,X_1} & s_{X_N,X_2} & \cdots & s_{X_N,X_N} \end{pmatrix}, \qquad (10.87)$$

where s_{X_n,X_m} is the sample covariance between the observations for variables m and n calculated over all $t = 1, 2, \ldots, T$, and where $s_{X_n,X_n} = s^2_{X_n}$, the variance of the observations for variable n.

The Mahalanobis distance at time t, D_t is then calculated as:

$$D_t = \sqrt{(X_t - \bar{X})' S^{-1} (X_t - \bar{X})}. \qquad (10.88)$$

Squaring the Mahalanobis distance gives a statistic that is the sum of N normal random variables, if the variables are drawn from a multivariate normal distribution. The statistic D_t^2 therefore has a χ^2 distribution with N degrees of freedom. The multivariate normality of the data can be tested by calculating D_t^2 for each $t = 1, 2, \ldots, T$ and comparing their distribution with what would be expected if the statistics were drawn from a χ_N^2 distribution. This can be done with a Q-Q plot, where the inverse cumulative distribution function for the χ^2 distribution with N degrees of freedom is used as the comparison.

Mardia's tests

It is also possible to derive numerical tests based on measures of multivariate skew and kurtosis, known as Mardia's tests (Mardia, 1970). To carry out Mardia's multivariate test of skew, the first stage is to define the Mahalanobis angle between observations at times s and t, $D_{s,t}$:

$$D_{s,t} = (X_s - \bar{X})' S^{-1} (X_t - \bar{X}). \qquad (10.89)$$

A skew-type parameter, w_N can then be calculated:

$$w_N = \frac{1}{T^2} \sum_{s=1}^{T} \sum_{t=1}^{T} D_{s,t}^3. \tag{10.90}$$

Multiplying this by $T/6$ gives Mardia's skew test statistic, MST, that has a χ^2 distribution with $N(N+1)(N+2)/6$ degrees of freedom:

$$MST = \frac{T}{6} w_N \sim \chi^2_{N(N+1)(N+2)/6}. \tag{10.91}$$

For Mardia's test of multivariate kurtosis, the kurtosis-type parameter, k_N, is calculated from the Mahalanobis distance:

$$k_N = \frac{1}{T} \sum_{t=1}^{T} D_t^4,. \tag{10.92}$$

This can be transformed into Mardia's kurtosis test statistic, MKT, which tends to a standard normal distribution as T tends to infinity:

$$MKT = \frac{k_N - N(N+2)}{\sqrt{8N(N+2)/T}} \sim N(0,1). \tag{10.93}$$

10.3.3 Generating multivariate random normal variables

There are a number of ways that correlated random numbers can be generated. If only two variables are required, then the approach is simple. First, generate two series of normally distributed random numbers, X_1 and X_2. Then, if a correlation of ρ is required between two series, create a third variable, X_3, defined as:

$$X_3 = \rho X_1 + \sqrt{(1 - \rho^2)} X_2. \tag{10.94}$$

The variable X_3 has a correlation of ρ with variable X_1. If X_1, X_2 and (therefore) X_3 have standard normal distributions, X_1 and X_3 can be transformed to distributions with different standard deviations and means in the same way that univariate normal distributions are adjusted. This leaves the correlation unaffected. Methods for creating correlated multivariate normal variables are more involved, but there are two common approaches: Cholesky decomposition and principal components analysis.

Cholesky decomposition

The objective of Cholesky decomposition is to derive a matrix that can be multiplied by a single matrix of simulations of N random normal variables to give N simulations of N correlated normal variables. This can be repeated to give as many correlated observations as required. To do this an $N \times N$ matrix, C, must be found such that:

$$C'C = \Sigma, \tag{10.95}$$

where Σ is the $N \times N$ covariance matrix. It is assumed here that Σ has full rank. The matrix C is lower triangular – in other words, all elements above and to the right of the diagonal are zero:

$$C = \begin{pmatrix} C_{1,1} & 0 & \cdots & 0 \\ C_{2,1} & C_{2,2} & \cdots & 0 \\ \vdots & \vdots & \ddots & \vdots \\ C_{N,1} & C_{N,2} & \cdots & C_{N,N} \end{pmatrix}. \tag{10.96}$$

The transpose, C' is therefore upper triangular. Each element of C can be calculated using the following formula:

$$C_{m,n} = \begin{cases} 0 & \text{if } m < n; \\[2mm] \sqrt{\sigma_{m,m} - \sum_{u=1}^{m-1} C_{m,u}^2} & \text{if } m = n; \\[2mm] \dfrac{1}{C_{n,n}} \left(\sigma_{m,n} - \sum_{u=1}^{n-1} C_{m,u} C_{n,u} \right) & \text{if } m > n, \end{cases} \tag{10.97}$$

where $m, n = 1 \ldots N$. This means that if the elements above and to the left of a particular element are known, the element itself can be evaluated, so the matrix must be evaluated from the top left corner downwards, either by column or by row.

Once the matrix C has been found, it can be used to simulate a column vector of N correlated normal random variables, X with covariances defined by Σ and with means given by the column vector μ. A two stage process is needed to do this. First, a column vector, Z, of N matrix of independent and normally distributed random variables with means of zero and a standard deviation of one, is needed. This represents a single simulation of N variables. This is pre-multiplied by the Cholesky matrix C. However, whilst the covariances between the resulting variables are correct, each distribution is centred on zero. The column vector containing the mean values, μ, must be added to the result. The calculation required is, therefore:

$$X = CZ + \mu. \tag{10.98}$$

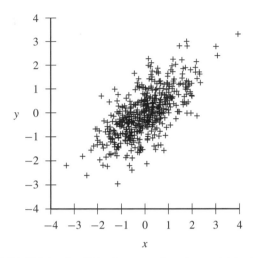

Figure 10.29 500 simulated bivariate normal random variables ($\rho = 0.7$)

Simulated bivariate data calculated using this approach are shown in Figure 10.29.

6 Principal component analysis

Another approach to modelling correlated random variables is to use principal component analysis (PCA), also known as eigenvalue decomposition. The PCA approach describes the difference from the mean for each variable as a weighted average of a number of independent volatility factors. *risk 'factors*

As for the Cholesky decomposition, the starting point is the covariance matrix, Σ, containing the covariances of N variables. For this matrix, there exists a square matrix V that can convert the covariance matrix into a diagonal matrix, Λ:

$$\Lambda = V'\Sigma V, \qquad V\Lambda = \Sigma V \qquad (10.99)$$

or, showing the elements of the matrices:

$$
\begin{pmatrix}
\Lambda_1 & 0 & \cdots & 0 \\
0 & \Lambda_2 & \cdots & 0 \\
\vdots & \vdots & \ddots & \vdots \\
0 & 0 & \cdots & \Lambda_N
\end{pmatrix}
=
\begin{pmatrix}
V_{1,1} & V_{2,1} & \cdots & V_{N,1} \\
V_{1,2} & V_{2,2} & \cdots & V_{N,2} \\
\vdots & \vdots & \ddots & \vdots \\
V_{1,N} & V_{2,N} & \cdots & V_{N,N}
\end{pmatrix}
$$

$$\times \begin{pmatrix} \sigma_{X_1,X_1} & \sigma_{X_1,X_2} & \cdots & \sigma_{X_1,X_N} \\ \sigma_{X_2,X_1} & \sigma_{X_2,X_2} & \cdots & \sigma_{X_2,X_N} \\ \vdots & \vdots & \ddots & \vdots \\ \sigma_{X_N,X_1} & \sigma_{X_N,X_2} & \cdots & \sigma_{X_N,X_N} \end{pmatrix}$$

$$\times \begin{pmatrix} V_{1,1} & V_{1,2} & \cdots & V_{1,N} \\ V_{2,1} & V_{2,2} & \cdots & V_{2,N} \\ \vdots & \vdots & \ddots & \vdots \\ V_{N,1} & V_{N,2} & \cdots & V_{N,N} \end{pmatrix}. \tag{10.100}$$

Matrix V is, like Σ, an $N \times N$ matrix. It contains N column vectors, each of length N, the N eigenvectors of the covariance matrix Σ. The diagonals of $\mathbf{3}$ are the eigenvalues of Σ. The combination of each eigenvector and eigenvalue is a principal component. This means that the first eigenvector, column vector V_1, and the first eigenvalue, Λ_1, form the first principal component of the data, such that:

$$\Lambda_1 = V_1' \Sigma V_1$$

$$= \begin{pmatrix} V_{1,1} & V_{2,1} & \cdots & V_{N,1} \end{pmatrix} \begin{pmatrix} \sigma_{X_1,X_1} & \sigma_{X_1,X_2} & \cdots & \sigma_{X_1,X_N} \\ \sigma_{X_2,X_1} & \sigma_{X_2,X_2} & \cdots & \sigma_{X_2,X_N} \\ \vdots & \vdots & \ddots & \vdots \\ \sigma_{X_N,X_1} & \sigma_{X_N,X_2} & \cdots & \sigma_{X_N,X_N} \end{pmatrix} \begin{pmatrix} V_{1,1} \\ V_{2,1} \\ \vdots \\ V_{N,1} \end{pmatrix}. \tag{10.101}$$

There are a number of methods that can be used to calculate the principal components. A simple, iterative approach is the power method. Starting with the first – and largest – principal component, this calculates successive values of $V_1(k)$ and $\Lambda_1(k)$ for $k = 0, 1, 2, \ldots$ until the changes in $V_1(k)$ and $\Lambda_1(k)$ fall below a pre-specified tolerance level. To start this process, a starting value is needed for the column vector $V_1(k)$, denoted $V_1(0)$. As good as any is $V_{1,1} = V_{2,1} = \ldots = V_{N,1} = 1$. The next stage is to calculate an interim vector $V_1^*(k)$, where:

$$V_1^*(k+1) = \Sigma V_1(k), \tag{10.102}$$

starting with $k = 0$. From $V_1^*(k+1)$, the element with the largest absolute value is taken. This is $\Lambda_1(k+1)$. These items are then used to calculate $V_1(k+1)$:

$$V_1(k+1) = \frac{1}{\Lambda_1(k+1)} V_1^*(k+1). \tag{10.103}$$

This is repeated K times, K being the number of iterations required such that the proportional changes in $V_1(k)$ and $\Lambda_1(k)$ have become sufficiently small. Then a single stage is needed to calculate the first principal component. This is to normalise the eigenvector such that $V_1'V_1 = 1$. This is done by dividing each element of $V_1(K)$ by a scalar given by $\sqrt{V_1(K)'V_1(K)}$.

The process for finding the second principal component is the same as for the first, except that the covariance matrix is replaced with a new matrix:

$$\Sigma_1 = \Sigma - \Lambda_1 V_1 V_1'. \tag{10.104}$$

This entire process is repeated until all principal components have been found. These principal components can then be used to simulate correlated normal random variables. As with the Cholesky decomposition, a two-stage process is needed if the required result is a column vector, X, of length N representing a single simulation of N correlated variables with covariances defined by Σ and with means given by the vector μ.

The starting point is again a column vector, Z, being a vector of length N of independent and normally distributed random variables with means of zero and a standard deviation of one. Next, an $N \times N$ diagonal matrix L is needed, with each diagonal being the square root of subsequent eigenvalues:

$$L = \begin{pmatrix} L_1 & 0 & \dots & 0 \\ 0 & L_2 & \dots & 0 \\ \vdots & \vdots & \ddots & \vdots \\ 0 & 0 & \dots & L_N \end{pmatrix} = \begin{pmatrix} \sqrt{\Lambda_1} & 0 & \dots & 0 \\ 0 & \sqrt{\Lambda_2} & \dots & 0 \\ \vdots & \vdots & \ddots & \vdots \\ 0 & 0 & \dots & \sqrt{\Lambda_N} \end{pmatrix}. \tag{10.105}$$

The elements are square roots because the eigenvalues represent the variances of the eigenvectors, whereas when generating random variables multiplication by standard deviations is required.

Next the $N \times N$ matrix of eigenvectors, V is needed. The random variables, the matrix of square-rooted eigenvalues and the matrix of eigenvectors are then multiplied together to give a column vector of correlated random variables with means of zero. To add non-zero means, the column vector of means, μ, must be added to the result. This means that the vector of correlated numbers is given by:

$$X = ZLV + \mu. \tag{10.106}$$

However, a key feature of PCA – at least when the principal components are derived as described above – is that the first principal component has the greatest impact on the number simulated, and the importance of the components decreases. This means that if most of the variation in a number of variables is determined only by a small number of factors, then a smaller number of simulations is needed to model a large number of variables. A classic example is changes in interest rates. Whilst it might seem desirable to model 20 different government bonds of different terms by considering the correlations of each bond with each other bond, this is unlikely to be necessary. In particular, the most common changes in bond yields generally cause the whole yield curve to rise or fall; the second most common changes cause the slope of the yield curve to change; the third most common cause it to bend around a particular term. This means that if the movements in bond yields are modelled using PCA, around 95% of the variability can be captured using only the first three principal components, significantly reducing computational complexity. In matrix terms, this means that only the first N^* eigenvalues are used where $N^* < N$, so the matrix L becomes an $N^* \times N^*$ matrix. Since only the first N^* eigenvalues are used, only the first N^* eigenvectors are needed, so V becomes an $N \times N^*$ matrix. Most importantly, only N^* uncorrelated random variables are needed, so whilst X and μ remain N-length vectors, only an N^*-length vector of uncorrelated random variables is needed for Z. Considering the order of multiplication, it should be clear how the larger matrix is generated from the smaller volume of data.

In a way, PCA is similar to the factor-based approach to modelling discussed later. However, in the factor based approach, the various factors are specifically chosen, whereas in PCA the factors 'fall out of' the model.

10.3.4 Multivariate normal mean–variance mixture distributions

In the same way that the univariate normal distribution can be generalised to give normal mean–variance mixture distributions, multivariate normal mean–variance mixture distributions also exist. These are distributions where a column vector $X' = (X_1 X_2 \ldots X_N)$, is defined in relation to a column vector $Z' = (Z_1 Z_2 \ldots Z_N)$ whose elements are drawn from a standard normal distribution, such that:

$$X = m(W) + \sqrt{W} C Z, \qquad (10.107)$$

where C is an $N \times N$ matrix, W is a strictly positive random scalar that is independent of Z and $m(W)$ is a column vector that is a function of W. The matrix C is chosen such that $C'C$ is equal to Σ, a covariance matrix. As such, it converts the uncorrelated random normal variables into correlated random normal variables with variances given by the diagonal of σ. The matrix C can be calculated by Cholesky decomposition, but this is not essential. However, using this decomposition and setting $m(W)$ equal to μ would give correlated random normal variables with means defined by the vector μ and covariances defined by the matrix Σ.

As with univariate normal mixture distributions, the most general case is where W has a generalised inverse Gaussian (GIG) distribution with parameters β_1, β_2 and γ. If $m(W) = \alpha + \delta W$, where α is a column vector of location parameters and δ is a column vector of non-centrality parameters, then the result is a multivariate generalised hyperbolic distribution.

As with the univariate distributions, using the special case where γ / W has a chi-squared distribution with γ degrees of freedom leads to three commonly used special cases:

- if $m(W) = \alpha$, then the result is a multivariate t-distribution with γ degrees of freedom; and
- if $m(W) = \alpha + \delta W$, then the result is a multivariate skewed t-distribution with γ degrees of freedom.

10.3.5 The multivariate t-distribution

The multivariate t-distribution is a useful variant of the multivariate normal distribution. It is also easy to use, but allows some flexibility in the fatness of the tails. There are, in fact, a number of versions of multivariate t-distributions. The version considered here is a simple one, where all marginal distributions have the same number of degrees of freedom. This has marginal distributions with fatter tails than for the multivariate normal distribution, but also pro-duces a larger proportion of 'jointly extreme' observations. The fatness of the marginal and joint tails is determined by the degrees of freedom, γ, assumed: the smaller γ is, the fatter the tails.

The probability density function for the bivariate version of this distribution is:

$$f(x, y) = \frac{\Gamma\left(\frac{\gamma+2}{2}\right)}{\pi \gamma \beta_X \beta_Y \sqrt{\left(1 - \rho_{X,Y}^2\right)} \Gamma\left(\frac{\gamma}{2}\right)} z^{-\left(\frac{\gamma+2}{2}\right)}, \qquad (10.108)$$

where:

$$z = 1 + \frac{1}{\gamma\left(1 - \rho_{X,Y}^2\right)} \left[\left(\frac{x - \alpha_X}{\beta_X}\right)^2 + \left(\frac{y - \alpha_Y}{\beta_Y}\right)^2 - \frac{2\rho_{X,Y}(x - \alpha_X)(y - \alpha_Y)}{\beta_X \beta_Y} \right].$$

$$(10.109)$$

If α_X and α_Y are zero, and β_X and β_Y are one, then the result is the standard bivariate t-distribution, which has the following density function:

$$\tau_{\gamma, \rho_{X,Y}}(x, y) = \frac{\Gamma\left(\frac{\gamma+2}{2}\right)}{\pi \gamma \sqrt{\left(1 - \rho_{X,Y}^2\right)} \Gamma\left(\frac{\gamma}{2}\right)} z^{-\left(\frac{\gamma+2}{2}\right)}, \qquad (10.110)$$

where:

$$z = 1 + \frac{x^2 + y^2 - 2\rho_{X,Y}xy}{\gamma(1 - \rho_{X,Y}^2)}. \tag{10.111}$$

However, $\Gamma\left(\frac{\gamma+2}{2}\right)/\gamma\Gamma\left(\frac{\gamma}{2}\right) = 1/2$, so these expressions simplify to:

$$\tau_{\gamma,\rho_{X,Y}}(x, y) = \frac{1}{2\pi\sqrt{\left(1 - \rho_{X,Y}^2\right)}} \left[1 + \frac{x^2 + y^2 - 2\rho_{X,Y}xy}{\gamma(1 - \rho_{X,Y}^2)}\right]^{-\left(\frac{\gamma+2}{2}\right)}. \tag{10.112}$$

The distribution function, $t_{\gamma,\rho_{X,Y}}(x, y)$, is given by:

$$t_{\gamma,\rho_{X,Y}}(x, y) = \int_{-\infty}^{x}\int_{-\infty}^{y} \tau_{\gamma,\rho_{X,Y}}(s, t)\mathrm{d}s\mathrm{d}t. \tag{10.113}$$

Comparing surface and contour plots of the multivariate t- and normal distributions, the more pronounced peak and greater tail dependence are clear. Density and distribution surface plots and contour lines are shown in Figures 10.30 and 10.31 respectively.

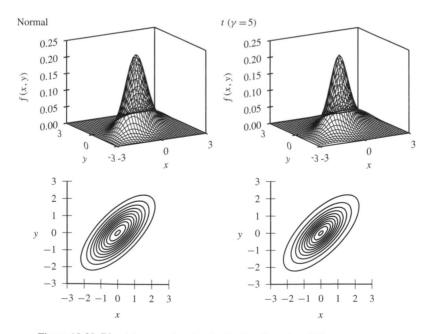

Figure 10.30 Bivariate normal and t-density functions ($\rho = 0.7$)

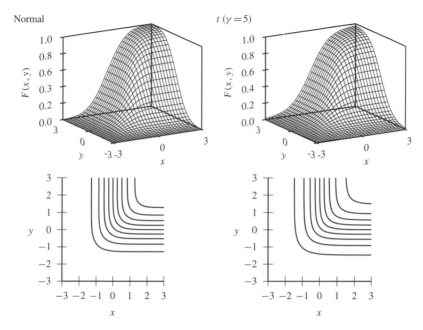

Figure 10.31 Bivariate normal and t-distribution functions ($\rho = 0.7$)

As with the univariate t-distribution, whilst α_X and α_Y represent the means of the marginal distribution, β_X and β_Y do not represent the standard deviations. This means that when moving to an N-variable multivariate version, the matrix of co-scale parameters, B, is not the same as the covariance matrix, Σ. In particular:

$$\Sigma = \frac{v}{v-2} B \qquad (10.114)$$

The multivariate t-distribution has the following probability density function:

$$f(x) = f(x_1, x_2, \ldots, x_N)$$

$$= \frac{\Gamma\left(\frac{\gamma+N}{2}\right)}{(\gamma\pi)^{(N/2)}\Gamma\left(\frac{\gamma}{2}\right)\sqrt{|B|}} \times \left(1 + \frac{(x-\alpha)'B^{-1}(x-\alpha)}{\gamma}\right)^{-\frac{\gamma+N}{2}}.$$

$$(10.115)$$

If the underlying distributions are all standard t-distributions, then this function becomes the standard multivariate t-distribution density function. Here, not only do the means disappear (being zero), but the matrix of co-scale parameters becomes a correlation matrix. This means that the standard multivariate normal distribution is defined by this correlation matrix, R, and the degrees of freedom, γ. It is therefore denoted $t_{\gamma,R}(x)$:

$$t_{\gamma,R}(x) = \int_{-\infty}^{x_1} \int_{-\infty}^{x_2} \cdots \int_{-\infty}^{x_N} \tau_{\gamma,R}(s) ds_1 ds_2 \ldots ds_N, \qquad (10.116)$$

where:

$$\tau_{\gamma,R}(x) = \frac{\Gamma\left(\frac{\gamma+N}{2}\right)}{(\gamma\pi)^{(N/2)} \Gamma\left(\frac{\gamma}{2}\right) \sqrt{|R|}} \times \left(1 + \frac{x'R^{-1}x}{\gamma}\right)^{-\frac{\gamma+N}{2}}. \qquad (10.117)$$

Random variables from a multivariate t-distribution with γ degrees of freedom can be simulated easily. First, a column vector, Z, being a series of multivariate random normal variables from standard normal distributions with a correlation matrix R must be simulated. Next, define B^D as a diagonal matrix whose diagonal elements are the scale parameters applied to each marginal distribution. By definition, the non-diagonal elements will be zero. The matrix B^D is post-multiplied by Z to give a vector whose elements are scaled to give random normal variables from distributions whose standard deviations are the diagonals B^D. These standard deviations can either be calculated from the data as sample standard deviations or derived for each variable from the value of the scale parameter $\beta_1, \beta_2, \ldots, \beta_N$ and the degrees of freedom, γ. Then a random variable from a χ^2 distribution with γ degrees of freedom, X_γ^2, is taken, square-rooted and divided into each element of the resulting vector. Each element is then re-centred by adding a vector containing the required location parameters for each element, α. Each mean can be calculated from the data as $\bar{X}$ or specified as a desired value for $\alpha_1, \alpha_2, \ldots, \alpha_N$. This means the total process can be summarised as:

$$\frac{1}{\sqrt{X_\gamma^2/\gamma}} B^D Z + \alpha. \qquad (10.118)$$

This process also confirms that the multivariate t-distribution can be constructed as a normal mixture distribution.

10.3.6 The multivariate skewed t-distribution

It is also possible to create a multivariate extension of the skewed t-distribution. The probability density function for the bivariate case is:

$$f(x,y) = c K_{\frac{\gamma+2}{2}} \left(\sqrt{(\gamma + z_1) z_2}\right) \left(\frac{\sqrt{(\gamma + z_1) z_2}}{1 + z_1/\gamma}\right)^{\frac{\gamma+2}{2}} e^{z_3}, \qquad (10.119)$$

where:

$$c = \frac{2^{\left(1-\frac{\gamma+2}{2}\right)}}{\pi\gamma\beta_X\beta_Y\sqrt{\left(1-\rho_{X,Y}^2\right)}\Gamma\left(\frac{\gamma}{2}\right)};$$ (10.120)

$$z_1 = \frac{1}{1-\rho^2}\left[\left(\frac{x-\alpha_X}{\beta_X}\right)^2 + \left(\frac{y-\alpha_Y}{\beta_Y}\right)^2 - \frac{2\rho_{X,Y}(x-\alpha_X)(y-\alpha_Y)}{\beta_X\beta_Y}\right];$$

(10.121)

$$z_2 = \frac{1}{1-\rho^2}\left[\left(\frac{\delta_X}{\beta_X}\right)^2 + \left(\frac{\delta_Y}{\beta_Y}\right)^2 - \frac{2\rho_{X,Y}(\delta_X)(\delta_Y)}{\beta_X\beta_Y}\right];$$ (10.122)

and

$$z_3 = \frac{1}{1-\rho^2}\left[\frac{(x-\alpha_X)\delta_X}{\beta_X^2} + \frac{(y-\alpha_Y)\delta_Y}{\beta_Y^2} - \frac{\rho_{X,Y}[(x-\alpha_X)\delta_Y + (y-\alpha_Y)\delta_X]}{\beta_X\beta_Y}\right].$$

(10.123)

This produces a bivariate skewed t-distribution with γ degrees of freedom. The parameters α_X and α_Y control the location of this distribution, whilst β_X and β_Y are responsible for scale. The parameters δ_X and δ_Y control the degree of skew, and the correlation between the variables is given by $\rho_{X,Y}$.

This distribution can also be extended to a multivariate, N-dimensional setting:

$$f(\boldsymbol{x}) = f(x_1, x_2, \ldots, x_N)$$

$$= cK_{\frac{\gamma+N}{2}}\left(\sqrt{(\gamma+z_1)z_2}\right)\left(\frac{\sqrt{(\gamma+z_1)z_2}}{1+z_1/\gamma}\right)^{\frac{\gamma+N}{2}}e^{z_3}$$ (10.124)

where:

$$c = \frac{2^{\left(1-\frac{\gamma+N}{2}\right)}}{(\pi\gamma)^{\frac{N}{2}}\sqrt{|\boldsymbol{B}|}\Gamma\left(\frac{\gamma}{2}\right)};$$ (10.125)

$$z_1 = (\boldsymbol{x}-\boldsymbol{\alpha})'\boldsymbol{B}^{-1}(\boldsymbol{x}-\boldsymbol{\alpha});$$ (10.126)

$$z_2 = \boldsymbol{\delta}'\boldsymbol{B}^{-1}\boldsymbol{\delta};$$ (10.127)

and

$$z_3 = (\boldsymbol{x}-\boldsymbol{\alpha})'\boldsymbol{B}^{-1}\boldsymbol{\delta}.$$ (10.128)

Here, $\boldsymbol{\alpha}$ is a vector of location parameters, whilst $\boldsymbol{B}$ is the matrix of co-scale parameters, which also contains information on the correlations between the variables. The vector $\boldsymbol{\delta}$ controls the degree of skew in each dimension.

As with the multivariate t-distribution, the matrix of co-scale parameters does not give the covariance matrix. In this case, the covariance matrix, Σ, is defined as follows:

$$\Sigma = \frac{\gamma}{\gamma - 2}B + \delta\delta'\frac{2\gamma^2}{(\gamma - 2)^2(\gamma - 4)}. \tag{10.129}$$

Whilst the density function is complicated, the distribution of random variables can again be understood in terms of the normal and χ^2 distributions. From Equation (10.118), it can be seen that observations with a multivariate t-distribution with means of α, scale parameters in the diagonal of B^D and γ degrees of freedom can be constructed from a vector of standard normal random variables, Z, and a random variable from a χ^2 distribution with γ degrees of freedom, X_γ^2, as follows:

$$\alpha + \frac{1}{\sqrt{X_\gamma^2/\gamma}}B^D Z. \tag{10.130}$$

However, if a vector of skew parameters, δ, is scaled by the χ^2 variable and then added to the vector of scaled normal observations, then the result is a set of observations from a skewed t-distribution:

$$\alpha + \frac{1}{\sqrt{X_\gamma^2/\gamma}}B^D Z + \frac{1}{X_\gamma^2/\gamma}\delta. \tag{10.131}$$

Density and distribution surface plots and contour lines for the bivariated skewed t-distribution are shown in Figures 10.32 and 10.33 respectively.

10.3.7 Spherical and elliptical distributions

An elliptical distribution is one where the relationship between N variables, $X_1, X_2, \ldots, X_N$ for a given joint probability density, $f(X_1, X_2, \ldots, X_N)$, is defined by an N-dimensional ellipse. So, considering a two-dimensional example, if the linear correlation coefficient between two variables X and Y is $\rho_{X,Y}$, the distribution evaluated at $X = x$ and $Y = y$ is elliptical if:

$$x^2 + y^2 - 2\rho_{X,Y}xy = c, \tag{10.132}$$

where c is a constant that is a function of the chosen value of $\rho_{X,Y}$ and the chosen value of the joint distribution function. For example, consider the bivariate

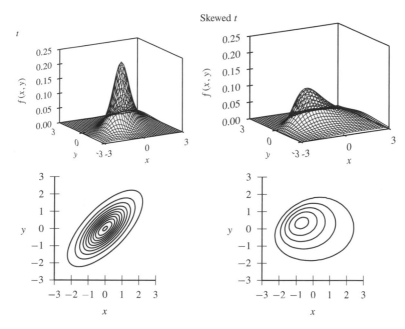

Figure 10.32 Bivariate standard and skewed t-density functions ($\rho = 0.7$, $\beta = 1$ and $\gamma = 5$ for both; $\alpha_X = -2$, $\alpha_Y = 1$, $\delta_X = 2$ and $\delta_Y = -1$ for the skewed t-distribution)

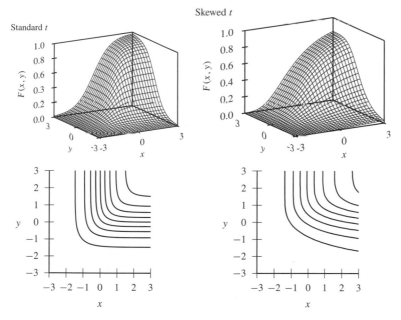

Figure 10.33 Bivariate standard and skewed t-distribution functions ($\rho = 0.7$, $\beta = 1$ and $\gamma = 5$ for both; $\alpha_X = -2$, $\alpha_Y = 1$, $\delta_X = 2$ and $\delta_Y = -1$ for the skewed t-distribution)

standard normal distribution, whose density function is:

$$\phi_{\rho_{X,Y}}(x, y) = \frac{1}{2\pi\sqrt{\left(1-\rho_{X,Y}^2\right)}}e^{-\frac{1}{2(1-\rho_{X,Y}^2)}\left(x^2+y^2-2\rho_{X,Y}xy\right)}. \qquad (10.133)$$

If $\phi_{\rho_{X,Y}}(x, y)$ is replaced with a constant, ϕ, representing the probability level of interest, then rearranging this equation and taking logarithms gives the following:

$$x^2 + y^2 - 2\rho_{X,Y}xy = -2(1-\rho_{X,Y}^2)\ln\left[2\pi\phi\sqrt{\left(1-\rho_{X,Y}^2\right)}\right]. \qquad (10.134)$$

The right-hand side of this expression is a constant for a fixed value of $\rho_{X,Y}$, so any probability can be described by an elliptical relationship between the two variables.

Returning to the more general case of elliptical distributions, the case where ρ is zero is a special one, resulting in a spherical distribution. This is so called because $x^2 + y^2 = c$ is the formula for a circle, which when generalised for N variables, $x_1, x_2, \ldots, x_N$ becomes the formula for an N-dimensional sphere:

$$x_1^2 + x_2^2 + \ldots + x_N^2 = c. \qquad (10.135)$$

The formal definition of a spherical distribution is one where the marginal distributions are:

- symmetric;
- identically distributed; and
- uncorrelated with one another.

These criteria cover (but are not restricted to) uncorrelated multivariate normal and normal mixture distributions; however, the lack of correlation implies independence for the multivariate normal distribution alone.

The surface and contour charts of two-dimensional elliptical and spherical distributions in Figure 10.34 also indicate why they are so called.

However, it is important to recognise that these properties can – and do – extend beyond two dimensions, and that N-dimensional spherical and elliptical distributions can be defined in a similar fashion.

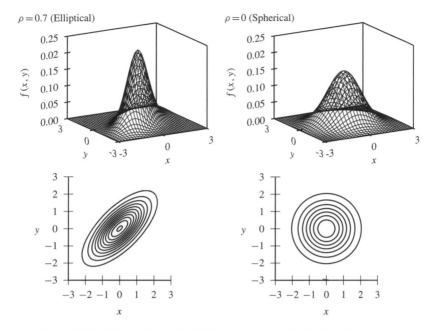

Figure 10.34 Spherical and elliptical bivariate normal density functions

🖉 10.4 Copulas

The marginal distributions are clearly important, but so are the links between them. The multivariate distributions described above provide one way of modelling these linkages, but the approach has limitations. First, the relationship modelled is assumed to be constant for all values of the marginal distribution. More importantly, though, modelling the linkages and the marginal distributions together limits the extent to which patterns in the data can be captured. An approach which can solve these problems is to use copulas.

In fact, multivariate distributions already contain copulas. However, these copulas are implicit in the distributions. In this section, the focus is on explicit copulas, which do not depend on the nature of the marginal distribution.

A copula defines the relationship between two or more variables. It is therefore a joint cumulative distribution function. However, the inputs to the function are themselves individual cumulative distribution functions rather than the raw values. This means that it is the order of the raw data that is important rather than the shape of the marginal distribution functions. It also means that if the marginal distribution of any data series changes, so long as the order of the observations remains the same in relation to the other series then so does the copula linking this distribution to the others. This is known as the property of invariance, and it can be particularly useful in modelling. Consider, for example, a situation where the returns on a number of asset classes are joined by a copula. If the distribution of returns for a particular asset class changes, then this suggests a change in the marginal distribution for this asset class is needed. However, if it is believed that the asset's relationship with other asset classes remains unchanged, then the copula need not be adjusted. However, it should be borne in mind that the change in the nature of a single asset may well be associated with a change in the relationship between that asset and others.

If the distribution function of each of N variables is defined as $F(x_n)$ where $n = 1, 2, \ldots, N$) and $C(F(x_1), F(x_2), \ldots, F(x_N))$ is the copula linking these functions, then this copula must have three properties:

- it must be an increasing function of each of its inputs, so if $F(x_n^*) > F(x_n)$, then $C(F(x_1), F(x_2), \ldots, F(x_n^*), \ldots, F(x_N))$ must be greater than $C(F(x_1), F(x_2), \ldots, F(x_n), \ldots, F(x_N))$;
- if all but one of the marginal distribution functions are equal to one, then the copula must be equal to the value of the remaining marginal distributions, so $C(1, 1, \ldots, F(x_n), 1) = F(x_n)$; and
- the copula must always return a non-negative probability, which happens if $\sum_{i_1=1}^{2} \sum_{i_2=1}^{2} \cdots \sum_{i_N=1}^{2} (-1)^{\sum_{n=1}^{N} i_n} C(F(x_{1,i}), F(x_{2,i}), \ldots, F(x_{N,i}))$, where $F(x_{n,1}) = a_n$ and $F(x_{n,2}) = b_n$, each $a_n \leq b_n$, both which are between zero and one.

The final item essentially decomposes the copula into all combinations of a_n and b_n to define the total probability in terms of the various joint distribution functions – and to ensure that the result is positive. This can best be appreciated using only two variables. Instead of using working in terms of $F(x_n)$ where $n = 1, 2, \ldots, N$), consider two variables, X and Y. In this case, the formula above reduces to $C(b_X, b_Y) - C(a_X, b_Y) - C(b_X, a_Y) + C(a_X, a_Y)$. This can be visualised in terms Figure 10.35. If the rectangle with sides $b_X - a_X$ and $b_Y - a_Y$ represents the probability that is sought, then this can be calculated by starting with the probability represented by a rectangle with sides b_X and b_Y. Subtracting a rectangle with sides b_X and a_Y removes some of the 'excess probability', as does subtracting a rectangle with sides b_Y and a_X. However, the rectangle with sides a_X and a_Y has then been subtracted twice, so needs to be added back. This is shown graphically in Figure 10.35.

Whilst copulas can be described in many dimensions, two-dimension examples using the variables X and Y will generally be discussed first in order to demonstrate the basic principles. However, multivariate copulas are important

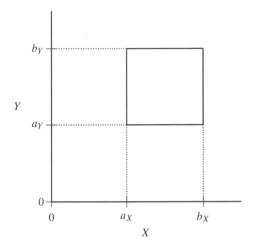

Figure 10.35 Copula verification

tools, since many risks generally need to be modelled together on a consistent basis.

10.4.1 Sklar's theorem

An important aspect of the relationship between variables is given by Sklar's theorem (Sklar, 1959). First, consider the case of two variables, X and Y. These have marginal cumulative distribution functions of $F(x)$ and $F(y)$ respectively, where $F(x) = \Pr(X \leq x)$ and $F(y) = \Pr(Y \leq y)$; however, they can also be defined in terms of a joint cumulative distribution function, $F(x, y) = \Pr(X \leq x | Y \leq y)$. Sklar's theorem says that $F(x, y)$ is linked to the marginal cumulative distributions, $F(x)$ and $F(y)$, through the use of a copula, $C(F(x), F(y))$, where:

$$F(x, y) = C(F(x), F(y)). \tag{10.136}$$

Furthermore, Sklar's theorem says that if the marginal distributions are continuous, then this copula is unique for this combination of marginal and joint distributions: there is only one way in which their link can be described. A copula could therefore be described as a joint cumulative distribution function expressed in terms of the marginal cumulative distribution functions.

It is worth thinking some more about what this means. The marginal cumulative distributions are the inputs to the copula functions. These are essentially uniform distributions for each of the marginal variables. They do not rely on the marginal distributions, which could have any form, since they are the probabilities derived from these distributions. The copula function takes these inputs and combines them somehow to arrive at a joint cumulative distribution function – another probability.

A copula need not necessarily be continuous, and an important example of a discrete copula is seen with an empirical copula function. This describes the relationship between variables in terms of their ranks. Consider two variables X and Y, each with T observations. For variable X, an empirical cumulative distribution function based on observations X_t at time t, where $t = 1, 2, \ldots, T$, is is $F(x)$. This can be calculated a number of ways, the aim being to arrive at an equally spaced series where the smallest value is greater than zero and the largest is less than one. Arriving at this series poses the same issue as exists in constructing Q-Q plots, and the methods for deriving an empirical cumulative distribution are similar. One approach gives values of $F(x)$ for observed values of X_t ranging from $1/(1+T)$ to $T/(1+T)$. This involves defining $F(x)$ as:

$$F(x) = \Pr(X_s \leq x) = \frac{1}{1+T} \sum_{t=1}^{T} I(X_t \leq x), \qquad (10.137)$$

where $I(X_t \leq x)$ is an indicator function which is one if $X_t \leq x$ and zero otherwise, and X_s is one of X_t. Alternatively, values from $1/2T$ to $(T-1/2)/T$ can be produced using the following formulation:

$$F(x) = \Pr(X_s \leq x) = \frac{1}{T} \left[\sum_{t=1}^{T} I(X_t \leq x) - \frac{1}{2} \right]. \qquad (10.138)$$

The function $F(y)$ can be calculated in the same way. A joint distribution function can be defined similarly, in the first case as:

$$F(x, y) = \Pr(X_s \leq x \text{ and } Y_s \leq y)$$

$$= \frac{1}{1+T} \sum_{t=1}^{T} I(X_t \leq x \text{ and } Y_t \leq y), \qquad (10.139)$$

where $I(X_t \leq x \text{ and } Y_t \leq y)$ is an indicator function that is equal to one if both of the conditions in the parentheses are met and zero otherwise, and in the second case as:

$$F(x, y) = \Pr(X_s \leq x \text{ and } Y_s \leq y)$$

$$= \frac{1}{T} \left[\sum_{t=1}^{T} I(X_t \leq x \text{ and } Y_t \leq y) - \frac{1}{2} \right]. \tag{10.140}$$

However, because Equations (10.139) and (10.140) are calculated using indicator functions based on the ranks of the observations, they can also be regarded as empirical copulas.

It is also worth defining a *survival* copula. This gives the joint probability that two variables X and Y will be greater than the fixed values x and y. The probability that a variable X is greater than x is denoted $\bar{F}(x) = 1 - F(x)$, with $\bar{F}(y)$ being similarly defined. The bivariate survival copula, denoted $\bar{C}(\bar{F}(x)\bar{F}(y))$, is therefore defined as follows:

$$\bar{F}(x, y) = \bar{C}(\bar{F}(x)\bar{F}(y))$$

$$= \bar{C}(1 - F(x), 1 - F(y))$$

$$= 1 - F(x) - F(y) + C(F(x), F(y)). \tag{10.141}$$

Sklar's theorem is easily expanded from the bivariate to the multivariate case. Consider N variables, $X_1, X_2, \ldots, X_N$. These have marginal cumulative distribution functions of $F(x_1), F(x_2), \ldots, F(x_N)$ and can also be defined in terms of a joint distribution function, $F(x_1, x_2, \ldots, x_N)$. The multivariate version of Sklar's theorem links the joint distribution to the marginal distributions as follows:

$$F(x_1, x_2, \ldots, x_N) = C(F(x_1), F(x_2), \ldots, F(x_N)). \tag{10.142}$$

As noted above, the expression $C(F(x_1), F(x_2), \ldots, F(x_N))$ is a cumulative distribution function. However, it is also helpful to note the copula density function, $c(F(x_1), F(x_2), \ldots, F(x_N))$. In the same was as the probability density function for a distribution gives the gradient of the cumulative distribution function, the copula density function gives the rate of change of the copula distribution function. It is defined as follows:

$$c(F(x_1), F(x_2), \ldots, F(x_N)) = \frac{\partial^N C(x_1, x_2, \ldots, x_N)}{\partial F(x_1) \partial F(x_2) \ldots \partial F(x_N)}. \tag{10.143}$$

If the distribution functions are all continuous, then it can be calculated as:

$$c(F(x_1), F(x_2), \ldots, F(x_N)) = \frac{f(x_1, x_2, \ldots, x_N)}{f(x_1) f(x_2) \ldots f(x_N)}. \tag{10.144}$$

In this equation, $f(x_1, x_2, \ldots, x_N)$ is the joint density function of the joint cumulative distribution function $F(x_1, x_2, \ldots, x_N)$, and $f(x_n)$ is the marginal density function of the marginal cumulative distribution function $F(x_n)$, where $n = 1, 2, \ldots, N$.

10.4.2 Dependence and concordance

Before moving on to discuss some specific copulas, it is helpful to discuss the more general issues of dependence and concordance. Whilst a measure of association such as Pearson's rho might give the association between one variable and another, it is not necessarily the case that there is any dependence; instead there may simply be some degree of concordance. The difference is important. One variable may not directly (or indirectly) influence another; rather, they may influence each other to some degree, or both may be influenced by a third factor. Even so, there is an association, or concordance, between them.

Pearson's rho, Spearman's rho and Kendall's tau can all be used to determine the degree of association between variables; however, it is important to understand their strengths and limitations. To do this, some set of criteria is needed. Scarsini (1984) defines a set of axioms for measures of concordance between X and Y. Consider a measure of association between X and Y, defined as $M_{X,Y}$, where X and Y are linked by a copula $C(F(x), F(y))$. For $M_{X,Y}$ to be a good measure of concordance, the following properties are required:

- completeness of domain – $M_{X,Y}$ must be defined for all values of X and Y, with X and Y being continuous;
- symmetry – $M_{X,Y} = M_{Y,X}$, or in other words switching X and Y should not affect the value of the measure;
- coherence – if $C(F(x), F(y)) \geq C(F(w), F(z))$, then $M_{X,Y} \geq M_{W,Z}$, or in other words if the joint probability is higher, then the measure of association should also be higher;
- unit range – $-1 \leq M_{X,Y} \leq 1$, and the extreme values in this range should be feasible;
- independence – if X and Y are independent, then $M_{X,Y} = 0$;
- consistency – if $X = -Z$, then $M_{X,Y} = -M_{Z,Y}$, so reversing the signs of one series should simply reverse the sign of the measure; and
- convergence – if $X_1, X_2, \ldots, X_T$ and $Y_1, Y_2, \ldots, Y_T$ are sequences of T observations with the joint distribution function $_T F(x, y)$ and the copula $_T C(F(x), F(y))$, and if $_T C(F(x), F(y))$ tends to $C(F(x), F(y))$ as T tends to infinity, then $_T M_{X,Y}$ must tend to $M_{X,Y}$.

Together, this list of features also implies other properties for good measures of concordance:

- if $g(X)$ and $h(Y)$ are monotonic transformations of X and Y, it is also true that $M_{g(X),h(Y)} = M_{X,Y}$; and
- if X and Y are co-monotonic, then $M_{X,Y} = 1$; if they are counter-monotonic, then $M_{X,Y} = -1$.

It has already been established that Pearson's rho is appropriate only if the marginal distributions are jointly elliptical. However, even if this is the case, the measure fails Scarsini's criteria. To see why this is, consider two variables, X and Y, that are co-monotonic – so an increase in one implies an increase in the other – but not linearly related. An example might be where $Y = \ln X$. Since the relationship between the variables is not linear, Pearson's rho will never equal one. In fact, any transformation to the data other than a linear shift can result in a change in the value of Pearson's rho. Both Spearman's rho and Kendall's Tau, on the other hand, fulfil all of Scarsini's criteria.

10.4.3 Tail dependence

If the data are parameterised – that is, expressed in terms of a statistical distribution – then, at the limit, the relationship between variables at their margins can be used to describe the tail dependence of those variables. In particular, consider $_L\lambda_{X,Y}$, the coefficient of lower tail dependence between two variables, X and Y. This is defined as:

$$_L\lambda_{X,Y} = \lim_{q \to 0^+} \Pr(X < F_q^{-1}(x)|Y < F_q^{-1}(y)), \qquad (10.145)$$

where $F_q^{-1}(x)$ and $F_q^{-1}(y)$ are the values of x and y for which the cumulative distribution functions, $F(x)$ and $F(y)$, are equal to q. Equation (10.145) can also be expressed in terms of a bivariate copula, as:

$$_L\lambda_{X,Y} = \lim_{q \to 0^+} \frac{C(F_q(x), F_q(y))}{q}, \qquad (10.146)$$

where $F_q(x)$ and $F_q(y)$ are the values of these distribution functions for $x = q$ and $y = q$. Equations (10.145) and (10.146) say that the coefficient of upper tail correlation is found as q tends to zero from above. If $0 <_L \lambda_{X,Y} \le 1$, then lower tail dependence exists; if $_L\lambda_{X,Y} = 0$, it does not.

Similarly, the coefficient of upper tail dependence is defined as:

$$_U\lambda_{X,Y} = \lim_{q \to 1^-} \Pr(X > F_q^{-1}(x)|Y > F_q^{-1}(y)). \qquad (10.147)$$

This can be expressed in terms of a bivariate survival copula as:

$$_U\lambda_{X,Y} = \lim_{q \to 1^-} \frac{\bar{C}(\bar{F}_q(x), \bar{F}_q(y))}{1-q},\qquad(10.148)$$

(handwritten margin note: shaded / whole area)

where $\bar{F}_q(x)$ and $\bar{F}_q(y)$ are the values of $1 - F(x)$ and $1 - F(y)$ for $x = q$ and $y = q$. This function is valued as q tends to one from below. If $0 <_U \lambda_{X,Y} \le 1$, then upper tail correlation exists; if $_U\lambda_{X,Y} = 0$, it does not.

The reasoning behind coefficients can be understood by looking at the calculation for finite values of q. Consider the $C(F_q(x), F_q(y))$. This gives the proportion of the observations for which both $F(x) = q$ and $F(y) = q$. If x and y are perfectly positively correlated, then not only will the cumulative marginal distributions be equal to q, but so will the cumulative joint distribution; however, perfect negative correlation will result in a cumulative joint distribution of zero. This means that, since $C(F_q(x), F_q(y))$ can range from zero to q, dividing $C(F_q(x), F_q(y))$ by q gives a measure of tail dependence in the range zero to one.

The presence of upper of lower tail dependence can help determine which copula is the most appropriate for a particular dataset. For example, if the level of association is higher for extreme values, then a copula with lower and upper tail dependence should be considered; however, if there is a greater degree of association for extremely low values alone – for example, in respect of extreme negative returns alone – then a copula with lower tail dependence is probably appropriate.

Interestingly, there is no tail dependence in the multivariate normal distribution, however high the correlations are (assuming that they have an absolute value of less than one).

Tail dependence is important in a risk management context because it describes potential concentrations of risk at the point in the distributions where this risk really matters. However, copulas can also be used to describe the full complexity of relationships across the range of values of marginal distributions.

10.4.4 Fréchet–Höffding copulas

As discussed above, a copula function $C(F(x), F(y))$ describes the joint distribution function in terms of the marginal distributions. Three of the simplest copulas are the independence copula:

$$_{ind}C(F(x), F(y)) = F(x)F(y);\qquad(10.149)$$

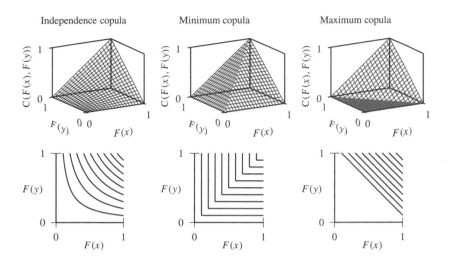

Figure 10.36 Bivariate Fréchet–Höffding copulas

the minimum, or co-monoticity copula:

$$_{\min}C(F(x), F(y)) = \min(F(x), F(y));\tag{10.150}$$

and the maximum, or counter-monoticity copula:

$$_{\max}C(F(x), F(y)) = \max(F(x) + F(y) - 1, 0).\tag{10.151}$$

The copula functions and associated contour lines are shown in Figure 10.36. However, it is worth spending a moment to examine what these formulae represent. Remember that Sklar's theorem essentially defines $C(F(x), F(y))$ as $\Pr(X \le x | Y \le y)$. The independence copula is therefore simply the product of the marginal distributions or cumulative probabilities. This comes from basic probability theory, that the probability of two independent events is the product of their probabilities.

The minimum copula is the lower of the two distribution functions. This is the copula that arises when X is simply a monotonic transformation of Y. For example, if X is the number of pounds sterling that someone earns each year whilst Y is the number of pence (just one hundred times the number of pounds), then the probability that an individual earns less then £30,000 and less than 4,000,000 pence (or £40,000) is simply the probability that an individual earns less then £30,000.

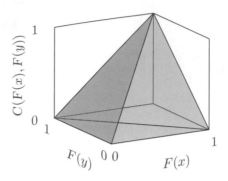

Figure 10.37 Fréchet-Höffding upper and lower bounds

Finally, the maximum copula describes the opposite scenario, where probability of X occurring is one minus the probability of Y; in other words, they are mutually exclusive. Because these copulas represent the three basic dependencies that variables can have – independence, full positive dependence and full negative dependence – they are also known as fundamental copulas.

The minimum and the maximum copulas are also interesting because they form the upper and lower boundaries for all copulas, known as the Fréchet–Höffding bounds. These are shown in Figure 10.37. In fact, these three copulas form special classes of the Fréchet–Höffding family of copulas. The more general form of this class is:

$$
\begin{aligned}
{}_FC(F(x), F(y)) = {} & p\max(F(x)+F(y)-1,0) \\
& +(1-p-q)F(x)F(y) \\
& +q\min(F(x), F(y)),
\end{aligned}
\tag{10.152}
$$

where $0 \le p \le 1$, $0 \le q \le 1$ and $p+q \le 1$. This can also be written in terms of the independence, minimum and maximum copulas:

$$
\begin{aligned}
{}_FC(F(x), F(y)) = {} & p_{\max}C(F(x), F(y)) \\
& +(1-p-q)_{\mathrm{ind}}C(F(x), F(y)) \\
& +q_{\min}C(F(x), F(y)).
\end{aligned}
\tag{10.153}
$$

For $p = 0$ and $q = 1$, this gives the Fréchet–Höffding upper bound, or the minimum copula; for $p = 1$ and $q = 0$, it gives the Fréchet–Höffding lower bound, or the maximum copula; and for $p = q = 1$, it gives the independence copula. If $p = 0$ and $q < 1$, or $q = 0$ and $p < 1$, then the result is known as a mixture copula.

For the Fréchet–Höffding family of copulas, Spearman's rho and Kendall's tau are calculated as follows:

$$\rho(s) = q - p, \tag{10.154}$$

and

$$\tau = \frac{(q - p)(2 + p + q)}{3}. \tag{10.155}$$

This means that $\tau \le \rho(s) \le -1 + \sqrt{1 + 3\tau}$ if $\tau \ge 0$, and $1 - \sqrt{1 - 3\tau} \le \rho(s) \le \tau$ if $\tau < 0$. These boundaries are stricter than the more general ones discussed earlier.

Multivariate versions of the independence and the minimum copula exist, describing the relationships between N variables. The form of the N-dimensional independence copula is:

$$_{ind}C(F(x_1), F(x_2), \dots, F(x_N)) = F(x_1)F(x_2)\dots F(x_N); \tag{10.156}$$

and for the N-dimensional minimum, or co-monoticity copula, it is:

$$_{min}C(F(x_1), F(x_2), \dots, F(x_N)) = \min(F(x_1), F(x_2), \dots, F(x_N)). \tag{10.157}$$

The maximum, or counter-monoticity copula exists in bivariate form only. This means that the only possible multivariate Fréchet–Höffding copulas apart from these are mixture copulas with $p = 0$:

$$_FC(F(x_1), F(x_2), \dots, F(x_N)) =$$
$$(1 - q)F(x_1)F(x_2)\dots F(x_N) + q\min(F(x_1), F(x_2), \dots, F(x_N)).$$

$$\tag{10.158}$$

10.4.5 Archimedean copulas

An important class of copulas is the Archimedean class. These are copulas constructed using a generator function and they have the advantage that they can be expressed in closed form. A generator function is a continuous, mono-tonically decreasing function, ψ, that transforms any number in the range zero to one to another number in the range zero to infinity, such that $\psi(1) = 0$ and

$\psi(0) = \infty$. The function ψ also has a pseudo-inverse, $\psi^{[-1]}$, such that:

$$\psi^{[-1]}(x) = \begin{cases} \psi^{-1}(x) & \text{if } 0 \leq x \leq \psi(0) \\ 0 & \text{if } \psi(0) \leq x \leq \infty. \end{cases} \qquad (10.159)$$

When applied to the generator, the pseudo-inverse gives the input to the generator if this input is between zero and one, so $\psi^{[-1]}(\psi(F(x))) = F(x)$. If the pseudo-inverse gives the same result as the 'ordinary' inverse, then ψ is referred to as a strict generator.

This combination of generator and pseudo-inverse are combined to give a bivariate Archimedean copula as:

$$_A C(F(x), F(y)) = \psi^{[-1]}[\psi(F(x)) + \psi(F(y))]. \qquad (10.160)$$

In other words:

- the generator function is applied to each of the marginal cumulative probabilities;
- the results are summed; and
- the pseudo-inverse is applied to this sum to give a joint cumulative probability.

This process shows why the pseudo inverse is needed, to ensure that whatever the sum of $\psi(F(x))$ and $\psi(F(y))$ the result can be inverted back to a number between zero and one – in other words, a probability. If the sum were greater than $\psi(0)$, then failing to define a pseudo-inverse could result in a negative probability being returned.

The bivariate example above can easily be extended to a multivariate case with N variables as follows:

$$_A C(F(x_1), F(x_2), \ldots, F(x_N)) =$$
$$\psi^{[-1]}[\psi(F(x_1)) + \psi(F(x_2)) + \ldots + \psi(F(x_N))]. \qquad (10.161)$$

The philosophy behind Archimedean copulas is quite elegant. Essentially, a number of probabilities – which can take values between zero and one – are being converted to variables that take values between zero and infinity. When combined, these will give a single value between zero and infinity, which can then be converted back into a single joint probability. This is shown graphically in Figure 10.38.

Archimedean copula are straightforward to use. They are also attractive because they provide a joint probability distribution in a closed form without the need for integration. However, since even multivariate Archimedean

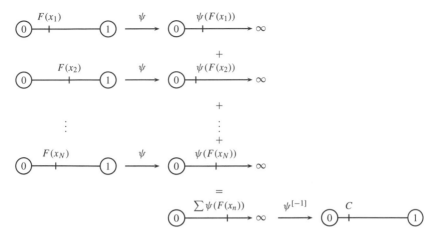

Figure 10.38 Using an Archimedean copula

copulas are defined using only one or two parameters, they are not necessarily good at modelling heterogeneous groups of variables.

So far, Archimedean copulas remain a theoretical construct, but in order to give useful copulas, specific generator functions are needed. Some important ones are given in Table 10.1. Their main features are summarised in Table 10.2. The copulas are also shown graphically in Figure 10.39. For the Gumbel and Clayton copulas, the parameter values of $\alpha = 4$ and $\alpha = 6$ correspond to a Kendall's tau of exactly 0.75, whilst parameter value of $\alpha = 14$ for the Frank copula corresponds to a Kendall's tau of approximately 0.75 (0.7479 to be precise). It is also helpful to see the copula density functions, given in Figure 10.40, so that the different levels of dependency between extreme high and low values can be appreciated.

The parameterisations highlight an important feature of these copulas – that Kendal's tau is an exact function of the parameters. This means that for all the single parameter copulas – in other words, all here except the generalised Clayton copula – the parameters can be derived to fit a bivariate dataset simply by calculating Kendall's tau from the data and rearranging the expressions in Table 10.2 to find α. For the generalised Clayton copula – and other Archimedean copulas with more than one parameter – there is no single combination of parameters that will give the required rank correlation.

In Table 10.2, the rank correlation coefficients for the Frank copula require the use of Debye functions defined as:

$$D_k(\alpha) = \frac{k}{\alpha^k} \int_0^\alpha \frac{t^k}{e^t - 1} dt. \tag{10.162}$$

Table 10.1. *Structure of Archimedean copulas*

Name	Symbol	$\psi(F(x))$	Parameter Range	Strict	Lower Limit	Upper Limit
Gumbel	$_{Gu}C_\alpha$	$(-\ln F(x))^\alpha$	$\alpha \geq 1$	Yes	$_{ind}C$	$_{min}C$
Frank	$_{Fr}C_\alpha$	$-\ln\left(\dfrac{e^{-\alpha F(x)}-1}{e^{-\alpha}-1}\right)$	$-\infty \leq \alpha \leq \infty$	Yes	$_{max}C$	$_{min}C$
Clayton	$_{Cl}C_\alpha$	$\dfrac{1}{\alpha}[(F(x))^{-\alpha}-1]$	$\alpha \geq -1$	$\alpha \geq 0$	$_{max}C$	$_{min}C$
Generalised Clayton	$_{GC}C_{\alpha,\beta}$	$\dfrac{1}{\alpha^\beta}[(F(x))^{-\alpha}-1]^\beta$	$\alpha \geq 0,\ \beta \geq 1$	Yes	–	–

Table 10.2. *Archimedean copula dependence measures*

Name	τ	$\rho(s)$	$_L\lambda$	$_U\lambda$
Gumbel	$1-\dfrac{1}{\alpha}$	no closed form	0	$2-2^{-\frac{1}{\alpha}}$
Frank	$1+\dfrac{4[D_1(\alpha)-1]}{\alpha}$	$1-\dfrac{12[D_2(-\alpha)-D_1(-\alpha)]}{\alpha}$	0	0
Clayton	$\dfrac{\alpha}{\alpha+2}$	complicated form	$\begin{cases} 2^{-\frac{1}{\alpha}} & \text{if } \alpha>0 \\ 0 & \text{if } \alpha\leq0 \end{cases}$	0
Generalised Clayton	$\dfrac{(2+\alpha)}{\alpha+2}$	complicated form	$2^{-\frac{1}{\alpha\beta}}$	$2-2^{-\frac{1}{\beta}}$

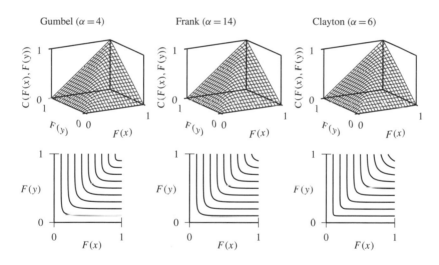

Figure 10.39 Bivariate Gumbel, Frank and Clayton copulas

10.4.6 The Gumbel copula

$\zeta = 1 - \dfrac{1}{\alpha}$

The Gumbel copula, $_{Gu}C_\alpha$, has the following generator function:

$$_{Gu}\psi_\alpha(F(x))=(-\ln F(x))^\alpha, \qquad (10.163)$$

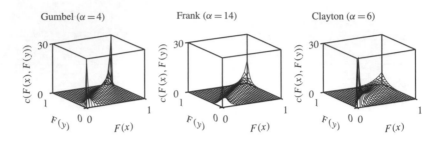

Figure 10.40 Bivariate Gumbel, Frank and Clayton copula density functions

where $1 \le \alpha < \infty$. This means that the bivariate Gumbel copula is defined as:

$$_{Gu}C_\alpha(F(x), F(y)) = e^{-[(-\ln F(x))^\alpha + (-\ln F(y))^\alpha]^{\frac{1}{\alpha}}}. \tag{10.164}$$

Example 10.8 Consider two insurance claims, X and Y. The probability that claim X is less than or equal to £50,000 is 0.873, whilst the probability that claim Y is less than or equal to £30,000 is 0.922. If the claims are linked by a Gumbel copula with a parameter α of 2.5, what is the probability that *both* X is less than or equal to £50,000 *and* Y is less than or equal to £30,000?

The generator for the Gumbel copula, $_{Gu}\psi_\alpha(F(x))$, is $(-\ln F(x))^\alpha$. Therefore, $_{Gu}\psi_\alpha(F(x)) = (-\ln 0.873)^{2.5} = 0.00680$, whilst $_{Gu}\psi_\alpha(F(y)) = (-\ln 0.922)^{2.5} = 0.00188$.

These can then be combined to given the joint probability that X is less than or equal to £50,000 *and* Y is less than or equal to £30,000 by calculating $_{Gu}C_\alpha(F(x), F(y)) = e^{-[(-\ln F(x))^\alpha + (-\ln F(y))^\alpha]^{\frac{1}{\alpha}}} = e^{-(0.00680+0.00188)^{\frac{1}{2.5}}} = 0.861$

The Gumbel copula can be generalised to the N-dimensional multivariate case as follows:

$$_{Gu}C_\alpha(F(x_1), F(x_2), \dots, F(x_N)) = e^{-\left[\sum_{n=1}^{N}(-\ln F(x_n))^\alpha\right]^{\frac{1}{\alpha}}}. \tag{10.165}$$

In both the bivariate and multivariate cases, if $\alpha = 1$, the Gumbel copula reduces to the independence copula; conversely, as α tends to ∞, it tends to the minimum copula.

The Gumbel copula has upper tail dependence, but no lower tail dependence. This means that it is particularly suitable for modelling dependency when association increases for extreme positive values. For example, losses from a credit portfolio (measured as positive) could be sensibly modelled using a Gumbel copula.

10.4.7 The Frank copula

The Frank copula, $_{Fr}C_\alpha$, has the following generator function:

$$_{Fr}\psi_\alpha(F(x)) = -\ln\left(\frac{e^{-\alpha F(x)} - 1}{e^{-\alpha} - 1}\right), \tag{10.166}$$

where α can be any real number. Some elementary algebra can be used to show that the bivariate Frank copula is defined as:

$$_{Fr}C_\alpha(F(x), F(y)) = -\frac{1}{\alpha}\ln\left[1 + \frac{(e^{-\alpha F(x)} - 1)(e^{-\alpha F(y)} - 1)}{e^{-\alpha} - 1}\right]. \tag{10.167}$$

Again, this can be generalised to the multivariate case:

$$_{Fr}C_\alpha(F(x_1), F(x_2), \ldots, F(x_N)) = -\frac{1}{\alpha}\ln\left[1 + \frac{\prod_{n=1}^{N}\left(e^{-\alpha F(x_n)} - 1\right)}{(e^{-\alpha} - 1)^{N-1}}\right].$$
$$\tag{10.168}$$

For the bivariate case, the Frank copula tends to the maximum copula as α tends to $-\infty$. However, it was established before that the maximum copula only exists in the bivariate case. In fact, the multivariate Frank copula is defined only for $\alpha > 0$ if $N > 2$. As α tends to ∞, the Frank copula tends to the minimum copula in both the bivariate and multivariate forms. The Frank copula has neither upper nor lower tail dependency

10.4.8 The Clayton copula

The Clayton copula, $_{Cl}C_\alpha$, has the following generator function:

$$_{Cl}\psi_\alpha(F(x)) = \frac{1}{\alpha}[(F(x))^{-\alpha} - 1], \tag{10.169}$$

where $\alpha \geq -1$. For values of α in this range, the bivariate Clayton copula is defined as:

$$_{Cl}C_\alpha(F(x), F(y)) = \max\{[(F(x))^{-\alpha} + (F(y))^{-\alpha} - 1]^{-(1/\alpha)}, 0\}. \tag{10.170}$$

If $\alpha = -1$, then this becomes the maximum copula. However, the generator for the Clayton copula is strict only when $\alpha > 0$. In this case, the bivariate copula can be expressed as:

$$_{Cl}C_\alpha(F(x), F(y)) = [(F(x))^{-\alpha} + (F(y))^{-\alpha} - 1]^{-\frac{1}{\alpha}}. \tag{10.171}$$

As long as $\alpha > 0$, it is possible to generalise the bivariate Clayton copula into a multivariate form:

$$_{Cl}C_\alpha(F(x_1), F(x_2), \ldots, F(x_N)) = \left[\sum_{n=1}^{N}(F(x_n))^{-\alpha} - N + 1\right]^{-\frac{1}{\alpha}}. \tag{10.172}$$

If $\alpha > 0$, then the Clayton copula has only lower tail dependency. This makes it suitable for linking portfolio returns if it is thought that extreme negative returns are likely to occur together. If $\alpha \leq 0$, then the Clayton Copula exhibits no upper or lower tail dependency.

10.4.9 The generalised Clayton copula

The generalised Clayton copula is a two-parameter Archimedean copula. Its generator function is:

$$_{GC}\psi_{\alpha,\beta}(F(x)) = \frac{1}{\alpha\beta}[(F(x))^{-\alpha} - 1]^\beta, \tag{10.173}$$

and it has the following form for a bivariate copula:

$$_{GC}C_{\alpha,\beta}(F(x), F(y)) = \left(\left\{[(F(x))^{-\alpha} - 1]^\beta + [(F(y))^{-\alpha} - 1]^\beta\right\}^{\frac{1}{\beta}} + 1\right)^{-\frac{1}{\alpha}}, \tag{10.174}$$

which can be generalised to the following multivariate form:

$$_{GC}C_{\alpha,\beta}(F(x_1), F(x_2), \ldots, F(x_N)) = \left(\sum_{n=1}^{N}\left\{[(F(x_n))^{-\alpha} - 1]^\beta\right\}^{\frac{1}{\beta}} + 1\right)^{-\frac{1}{\alpha}}. \tag{10.175}$$

From the formulae for Kendall's tau in Table 10.2, it can be seen that the generalised Clayton copula is in effect a generalisation of both the Clayton and the Gumbel copulas. In particular, it becomes the standard Clayton copula if $\beta = 0$ and the Gumbel copula if $\alpha = 1$. This formulation also means that the generalised Clayton copula has both upper and lower tail dependency, making it useful for modelling variables where jointly fat tails are thought to occur for both extreme high and extreme low values.

10.4.10 The Marshall–Olkin copula

Archimedean copulas are only one class of this type of joint distribution function. Other copulas do exist, and an interesting example is the Marshall–Olkin copula. This is driven by the desire to reflect the risk that a random shock will be fatal to one or more components, lives or companies. This means that it is a survival copula, so in bivariate form it represents the joint probability that the lifetime of X is greater than or equal to x, and that the lifetime of Y is greater than or equal to y.

The random shocks in the bivariate Marshall–Olkin copula are assumed to follow three independent Poisson processes with Poisson means λ_X, λ_Y and λ_{XY} per period, and with the subscripts denoting the parameters for the failure of component X, component Y and both X and Y together. This means that if each of these parameters is taken to describe the expected per-period frequency of failure, the total per-period frequency of failure for component X is $\lambda_X + \lambda_{XY}$, whereas for component Y it is $\lambda_Y + \lambda_{XY}$. The copula parameters α_X and α_Y are linked to these Poisson parameters as follows:

$$\alpha_X = \frac{\lambda_{XY}}{\lambda_X + \lambda_{XY}}, \tag{10.176}$$

and:

$$\alpha_Y = \frac{\lambda_{XY}}{\lambda_Y + \lambda_{XY}}. \tag{10.177}$$

The bivariate Marshall–Olkin copula for variables with random lifetimes X and Y then has the following form:

$$
\begin{aligned}
{}_{\text{MO}}\bar{C}_{\alpha_X,\alpha_Y}(\bar{F}(x), \bar{F}(y)) &= \min\left[(\bar{F}(x))^{1-\alpha_X}\bar{F}(y), \bar{F}(x)(\bar{F}(y))^{1-\alpha_Y} \right] \\
&= \begin{cases} (\bar{F}(x))^{1-\alpha_X}\bar{F}(y) & \text{if } (\bar{F}(x))^{\alpha_X} \geq (\bar{F}(y))^{\alpha_Y} \\ \bar{F}(x)(\bar{F}(y))^{1-\alpha_Y} & \text{if } (\bar{F}(x))^{\alpha_X} \leq (\bar{F}(y))^{\alpha_Y}. \end{cases}
\end{aligned}
\tag{10.178}
$$

$\lambda_X = 0.05, \lambda_Y = 0.15, \lambda_X = 0.10$

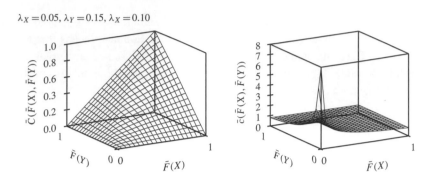

Figure 10.41 Bivariate Marshall–Olkin distribution and density functions

Both Kendall's tau and Spearman's rho are simple functions of α_X and α_Y:

$$\tau = \frac{\alpha_X \alpha_Y}{\alpha_X + \alpha_Y - \alpha_X \alpha_Y}. \tag{10.179}$$

$$\rho(s) = \frac{3\alpha_X \alpha_Y}{2\alpha_X + 2\alpha_Y - \alpha_X \alpha_Y}. \tag{10.180}$$

The bivariate Marshall–Olkin copula distribution and density functions are shown in Figure 10.41.

Example 10.9 Consider the chief executives of firms X and Y. Company-specific shocks lead to firm X replacing its chief executive on average once every 2.5 years, whilst company-specific shocks lead to firm Y replacing its chief executive on average once every five years. Furthermore, economy-wide shocks lead to firms replacing their chief executives once every ten years. Assuming these shocks occur in line with Poisson distributions, what is the probability that the chief executive of firm X stays in this post for at least a further four years and that the chief executive of firm Y stays in this post for at least a further six years?

For firm X, $\lambda_X = 1/2.5$, a rate of 0.4 per annum; for firm Y, $\lambda_Y = 1/5$, a rate of 0.2 per annum. The economy-wide parameter, λ_{XY}, is equal to $1/10$ years, a rate of 0.1 per annum.

First, consider the probabilities that the chief executives of firms X and Y will last for at least another four and six years respectively, assuming that each of these changes has a Poisson distribution. For firm X, the probability of there being no change over an x-year period is given by $\bar{F}(x) = e^{-(0.4+0.1)}x$. Evaluated at $x = 4$ for the four-year time horizon gives a probability of 0.1353.

For firm Y, the probability of there being no change over a y-year period is given by $\bar{F}(y) = e^{-(0.2+0.1)}y$. Evaluated at $y = 6$ for the six-year time horizon gives a probability of 0.1653.

Using the Marshall–Olkin copula, the parameter α_X is calculated as $\lambda_X Y/(\lambda_X \lambda_X Y) = 0.1/(0.4 + 0.1) = 0.2000$, whilst the parameter α_Y is calculated as $\lambda_X Y/(\lambda_Y \lambda_X Y) = 0.1/(0.2 + 0.1) = 0.3333$. The probability of joint survival beyond four and six years for X and Y respectively is therefore, evaluated at $x = 4$ years, $y = 6$ years, the lesser of:

$$(\bar{F}(x))^{1-\alpha_X} \bar{F}(y) = 0.1353^{1-0.2000} \times 0.1653 = 0.0334, \qquad (10.181)$$

and:

$$\bar{F}(x)(\bar{F}(y))^{1-\alpha_Y} = 0.1353 \times 0.1653^{1-0.3333} = 0.0408, \qquad (10.182)$$

The joint probability is therefore 0.0334.

If $x = y$ and the marginal distributions are assumed to be Poisson distributions, then the Marshall–Olkin copula reduces to this simple form, although since the marginal distributions and their dependence structures are not given separately, this is really a multivariate exponential distribution rather than an explicit copula:

$$
\begin{aligned}
{}_{\text{MO}}\bar{C}_{\alpha_X,\alpha_Y}(\bar{F}(x), \bar{F}(y)) &= e^{-(\lambda_X+\lambda_Y+\lambda_{XY})x} \\
&= e^{-(\lambda_X+\lambda_Y+\lambda_{XY})y}.
\end{aligned}
$$

$$(10.183)$$

It is also possible to create a multivariate extension of the Marshall–Olkin copula. Consider N components, lives or firms with lifetimes $X_1, X2, \ldots, X_N$. Assume also that there are M potential shocks that can affect some or all of these N components. It is possible to construct $\boldsymbol{A}$, an $M \times N$ matrix representing the effect of these shocks. To be specific, if an element $A_{m,n}$ is equal to one, then it means that component n fails as a result of shock m, whereas if this element is equal to zero, then component n is unaffected by the shock.

To fully specify all combinations, up to $M = 2^N - 1$ shocks must be defined. For example, for $N = 3$, the following 7×3 matrix would be needed:

$$
A = \begin{pmatrix}
1 & 0 & 0 \\
0 & 1 & 0 \\
0 & 0 & 1 \\
1 & 1 & 0 \\
1 & 0 & 1 \\
0 & 1 & 1 \\
1 & 1 & 1
\end{pmatrix},
\tag{10.184}
$$

with the individual and joint Poisson parameters being replaced with a Poisson process for each shock m, the Poisson mean for each of which is λ_{X_m}. No more parameters must be defined since any further shocks can be incorporated into the existing parameters. For example, if there are two shocks affecting all three firms, then the Poisson mean for the case where all firms are affected is simply the sum of the Poisson means for the two shocks. The multivariate copula is then given by:

$$
{}_{MO}\bar{C}_{\alpha_{X_1}, \alpha_{X_2}, \dots, \alpha_{X_N}} (\bar{F}(x_1), \bar{F}(x_2), \dots, \bar{F}(x_N)) =
$$

$$
\min \left[(\bar{F}(x_1))^{\alpha_{X_1}}, (\bar{F}(x_2))^{\alpha_{X_2}}, \dots, (\bar{F}(x_N))^{\alpha_{X_N}} \right] \prod_{n=1}^{N} (\bar{F}(x_n))^{1-\alpha_{X_N}},
$$

$$
\tag{10.185}
$$

where:

$$
\alpha_{X_n} = \frac{\sum_{m=1}^{M} \prod_{i=1}^{N} A_{m,i} \lambda_m}{\sum_{m=1}^{M} A_{m,n} \lambda_m}.
\tag{10.186}
$$

However, only one shock is needed to describe an environment-wide event that would affect all components. If this is taken to be shock M, the final shock in the list, then the numerator can be replaced with λ_M – all other terms will be zero, since at least one of $A_{m,i}$ will be zero for each $m \neq M$:

$$
\alpha_{X_n} = \frac{\lambda_M}{\sum_{m=1}^{M} A_{m,n} \lambda_m}.
\tag{10.187}
$$

This copula gives the joint probability that component X_1 will last for at least x_1 years, component X_2 will last for at least x_2 years and so on.

10.4.11 The normal copula

The Archimedean copulas in particular are limited by the small number of parameters available to describe multivariate relationships. One type of copula that does not have this restriction is the normal or Gaussian copula. The bivariate normal copula, $_{Ga}C_{\rho_{X,Y}}(F(x), F(y))$, is defined in as follows:

$$_{Ga}C_{\rho_{X,Y}}(F(x), F(y)) = \Phi_{\rho_{X,Y}}(\Phi^{-1}(F(x)), \Phi^{-1}(F(y))). \qquad (10.188)$$

In this equation, $\Phi^{-1}(F(x))$ and $\Phi^{-1}(F(y))$ are the inverse cumulative distribution functions for the standard normal distribution evaluated at the probabilities given by $F(x)$ and $F(y)$. The term $\Phi_{\rho_{X,Y}}$ is the joint cumulative normal distribution evaluated at these values for a correlation of $\rho_{X,Y}$. This copula can also be expressed as an integral:

$$_{Ga}C_{\rho_{X,Y}}(F(x), F(y)) = \frac{1}{2\pi\sqrt{1-\rho_{X,Y}^2}} \int_{-\infty}^{\Phi^{-1}(F(x))} \int_{-\infty}^{\Phi^{-1}(F(y))} e^{-z} ds\, dt,$$

$$(10.189)$$

where:

$$z = \frac{1}{2(1-\rho_{X,Y}^2)}\left(s^2 + t^2 - 2\rho_{X,Y}st\right). \qquad (10.190)$$

In this equation, $|\rho_{X,Y}| < 1$. This copula is defined by the value of $\rho_{X,Y}$. In fact, if the marginal distributions arc normal, then this distribution essentially becomes a bivariate normal distribution with zero means and unit standard deviations.

The independence, minimum and maximum copulas are special cases of the normal copula where $\rho_{X,Y} = 0$, $\rho_{X,Y} = 1$ and $\rho_{X,Y} = -1$ respectively. Another interesting feature of the normal copula is that if $|\rho_{X,Y}| < 1$, then tail dependence does not exist – as the marginal probabilities approach one, the dependence approaches zero.

The multivariate normal copula also exists. For N variables, this can be expressed as:

$$_{Ga}C_R(F(x_1), F(x_2), \dots, F(x_N)) =$$
$$\Phi_R(\Phi^{-1}(F(x_1)), \Phi^{-1}(F(x_2)), \dots, \Phi^{-1}(F(x_N))). \qquad (10.191)$$

Here, instead of a single correlation coefficient, a matrix of the $N(N-1)/2$ correlation coefficients, R, for all combinations of variables is needed. This essentially means that the parametrisation gets less robust as the number of variables, N increases. Independence and minimum copulas exist when all correlations are zero or one, but the maximum copula only exists for the bivariate normal copula (when $N = 2$). As before, if the marginal distributions are normal, then this becomes a multivariate normal distribution

10.4.12 Student's t-copula

A major drawback of the normal copula is that it is parameterised by a single variable – the linear correlation coefficient. One way of controlling the strength of the relationship between variables in the tails relative to those in the centre of the distribution is to use a copula based on the Student's t-distribution. The t-copula is based not just on the correlation coefficient but also on the degrees of freedom used. The bivariate t-copula is given by:

$$_tC_{\gamma,\rho_{X,Y}}(F(x), F(y)) = t_{\gamma,\rho_{X,Y}}(t_\gamma^{-1}(F(x)), t_\gamma^{-1}(F(y))),\qquad (10.192)$$

where $(t_\gamma^{-1}(F(x))$ and $(t_\gamma^{-1}(F(y))$ are the inverse cumulative distribution functions for Student's t-distribution with γ degrees of freedom evaluated at the probabilities given by $F(x)$ and $F(y)$. The term $t_{\gamma,\rho_{X,Y}}$ is the joint cumulative t-distribution evaluated at these values for γ degrees of freedom and a correlation of $\rho_{X,Y}$. This copula can also be expressed as an integral:

$$_tC_{\gamma,\rho_{X,Y}}(F(x), F(y)) = \frac{1}{2\pi\sqrt{1 - \rho_{X,Y}^2}}$$

$$\times \int_{-\infty}^{t_\gamma^{-1}(F(x))} \int_{-\infty}^{t_\gamma^{-1}(F(y))} \left(1 + \frac{s^2 + t^2 - 2\rho_{X,Y}st}{2\gamma(1 - \rho_{X,Y}^2)}\right)^{\left(\frac{\gamma+2}{2}\right)} ds\,dt.$$

$$(10.193)$$

The smaller the value of γ, the greater the level of association in the tails relative to that in the centre of the joint distribution. As γ tends to infinity, the form of the t-copula tends to that of the normal copula. Care is needed with low values of γ, however, as this shape of this copula implies an increasing concentration of observations in the four extreme corners of the distribution. Unlike the normal copula, the t-copula has both upper and lower tail dependency. Bivariate normal copula distribution and density functions are shown with the corresponding charts for the t-copula in Figures 10.42 and 10.43 respectively.

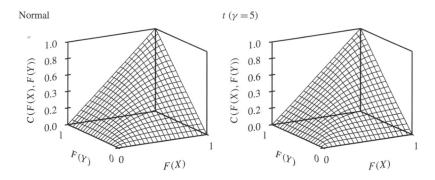

Figure 10.42 Bivariate normal and *t*-distribution functions ($\rho = 0.7$)

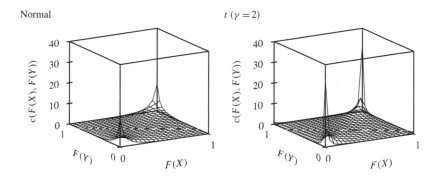

Figure 10.43 Bivariate normal and *t*-density functions ($\rho = 0.7$)

There is, of course, a multinomial version of the *t*-copula, and this is given in Equation (10.194):

$$_tC_{v,R}(F(x_1), F(x_2), \ldots, F(x_N)) =$$
$$t_{v,R}\left[t_v^{-1}(F(x_1)), t_v^{-1}(F(x_1)), \ldots, t_v^{-1}(F(x_N))\right], \tag{10.194}$$

where $t_{\gamma,R}$ is the joint cumulative *t*-distribution with γ degrees of freedom and the correlation matrix R evaluated at $t_\gamma^{-1}(F(x_1)), t_\gamma^{-1}(F(x_2)), \ldots, t_\gamma^{-1}(F(x_N))$. If all correlations are one, then the *t*-copula again becomes the minimum copula; however, if they are all zero, the result is not the independence copula, as random variables from an uncorrelated multivariate *t*-distribution are not independent.

10.5 Further reading

There are many books that give a fuller exposition of the range of statistical techniques available. A good introduction is given by Johnson and Bhattacharyya (2010), whilst Greene (2003) gives a more in-depth analysis. Information on copulas is less widely available, but there are a number of useful books on the subject. Nelsen (2006) provides a good introduction; Cherubini *et al.* (2004) gives some useful insights into the use of copulas in a financial context, whilst McNeil *et al.* (2005) gives more depth in terms of techniques and proofs. Sweeting and Fotiou (2011) give more details on the calculation of coefficients of tail dependence, together with commentary on the issues faced.

11
Modelling techniques

Model Risk

| 11.1 Introduction

One of the most common ways in which risks can be quantified is through the use of models. Models are mathematical representations of real-world processes. This does not mean that all models should attempt to exactly replicate the way in which the real world works – they are, after all, only models. However, it is important that models are appropriate for the uses to which they are put, and that any limitations of models are recognised. This is particularly important if a model designed for one purpose is being considered for another. Similarly, models calibrated using data in a particular range may not be appropriate for data outside those ranges – a model designed when asset price movements are small may break down when volatility increases. Appropriateness will also differ from organisation to organisation. A model appropriate for analysing the large annuity book of one insurer may give unrealistic answers if used with the smaller annuity book of a competitor.

Even if a model is deemed appropriate for the use to which it is put, uncertainty still remains. The structure of most models is a matter of preference, and the parameters chosen will depend on the exact period and type of data used. This uncertainty should be reflected by considering a range of structures and parameters, and analysing the extent to which changes affect the outputs of the model. This gives a guide as to how robust a model is. In particular, the structure of a model that gives significantly different outputs when calibrated using different data ranges should be reconsidered.

The complexity of models is a difficult area. In some areas, such as derivatives trading, models can grow ever more complex in order to exploit ever smaller pricing anomalies. However, in most areas of risk management greater complexity is not necessarily desirable. First, it makes checking the structure of models more difficult, and it is important that models are independently checked and are comprehensively documented. Greater complexity also makes models more difficult to explain to clients, regulators, senior management and other stakeholders, and it is important that these stakeholders do understand exactly what is going on rather than relying on the output from a 'black box'. This leads to a third concern, that greater complexity can lead to greater confidence in the ability of a model to reflect the exact nature of risks. Whilst more complex models might better represent the real world, they cannot replicate it exactly. Furthermore, if the volume of data does not change, then it becomes increasingly difficult to calibrate increasingly complex models – and the parameters become less and less reliable. Models should be treated with a degree of scepticism, and there is a strong argument for using less complex models but recognising more clearly what they can and cannot do.

Complexity is also linked to another issue, dimensionality. If trying to model a number of variables in a consistent fashion, then the ability to do this with any confidence diminishes rapidly as the number of variables increases. For example, it is relatively straightforward to determine the relationship between two variables with a thousand joint observations. However, defining the joint relationship with a thousand joint observations of ten or one hundred variables is increasingly difficult to do with any degree of certainty. This means that generalisations need to be made to give any workable joint distribution.

Once a model has been designed, it is also important to recognise that it must develop over time. As the data develops over time, the parameters of the model will change, as might the structure. It is important that reviews are scheduled for models on a regular basis, but also that there are the provisions for *ad hoc* reviews should circumstances demand. For example, a sharp fall in liquidity or widening of credit spreads might make it clear that existing models do not work, in which case it is inappropriate to simply continue using them regardless.

There are areas where modelling is difficult or inappropriate. For example, some asset classes are too new for there to be sufficient data to model. Even where data might exist, it might be difficult or expensive to collect. There are also risks that are so idiosyncratic that modelling the risks is less important than identifying, assessing and treating them. Many operational risks fall into this category.

Before considering some models in detail, it is worth considering how these models might be fitted. There are several layers to this problem. The first is the general approach used. To an extent, this depends on the model being fitted. If the data are being fitted to a statistical distribution, then the main choices are the method of moments, maximum likelihood estimation and pseudo-samples.

However, other models are linear models where one (dependent) variable is expressed in terms of a number of other (explanatory) variables. In this case, whilst maximum likelihood estimation is still used, various forms of least squares regression are also important.

Two types of generalised linear models, probit and logit regressions, can be used when the dependent variable is a category or a binary variable, as can discriminant analysis and the k-nearest neighbour method. Models can also be fitted where there are no dependent variables, or those variables are not known. In this case, principal components analysis and singular value decomposition might be appropriate.

When a model is being fitted, it is important to assess how well that model fits the data. This can be done for each of the explanatory variables and for the model as a whole.

This chapter gives a brief overview of all of these approaches and some of the issues that might be faced.

11.2 Fitting data to a distribution

As discussed above, there are two main approaches to fitting data to distributions: the method of moments and the method of maximum likelihood. Each of these approaches is discussed in turn.

11.2.1 The method of moments

The method of moments is carried out by setting as many moments of the distribution as there are parameters to statistics calculated from the data, and solving for the parameters. This is most easily demonstrated by an example.

Parametrisation of univariate distributions by the method of moments

Example 11.1 A portfolio of employer liability insurance claims has an average claim amount of £20,000, the distribution having a variance of £200,000. Working in units of £1,000, fit a gamma distribution to this dataset.

The gamma distribution has two parameters in the following formulation:

$$f(x) = \frac{1}{\beta^\gamma \Gamma(\gamma)} x^{\gamma-1} e^{-\frac{x}{\beta}}.$$

This means that the first two moments can be used to parameterise the distribution from data. The mean of this distribution is $\beta\gamma$ and the variance is $\beta^2\gamma$. Therefore the first moment, $E(X)$, is equal to the mean, $\beta\gamma$. The second moment, $E(X^2)$, can be calculated from the variance. Since the variance can be expressed as $E(X^2) - E(X)^2$, the second moment for the gamma distribution is the variance plus the mean squared, or $\beta^2\gamma + \beta^2\gamma^2$.

If for a particular set of data, $E(X) = 20$ and $E(X^2) = 600$, then the following simultaneous equations could be set up:

$$E(X) = \beta\gamma = 20,$$

and

$$E(X^2) = \beta^2\gamma + \beta^2\gamma^2 = 600.$$

Rearranging the first equation gives $\gamma = 20/\beta$. Substituting this into the second equation gives $20\beta + 400 = 600$, which means that $\beta = 10$. Substituting this back into the first equation gives $10\gamma = 20$, so $\gamma = 2$.

Parametrisation of copulas by the method of moments

This method can be also be used to fit some copulas. In particular, the parameters for Archimedean copulas are often defined exactly by one or more measures of correlation. Again, this is more easily seen from an example.

Example 11.2 The returns on two portfolios of bonds have a correlation, as measured by Kendall's tau, of 0.5. If future returns are to be simulated assuming the two series are linked by a Clayton copula, what is the single parameter of that copula?

Kendall's tau for the Clayton copula is defined as:

$$\tau = \frac{\alpha}{\alpha + 2},$$

where α is the single parameter for the Clayton copula. Rearranging this in terms of τ gives $\alpha = 2\tau/(1-\tau)$. Therefore, if Kendall's tau is calculated as 0.5 for two variables, then this means that when two variables are linked by a Clayton copula the parameter for that copula, α, is 2.

Generally speaking, the method of moments is more straightforward to implement than the other estimation approaches. However, it does not always give the most likely values for the parameters. In particular, there are instances where the values derived for the parameters are outside the acceptable ranges for the distribution. For example, negative parameters might be derived for the Gamma distribution. This becomes less likely as the number of observations increases.

11.2.2 The method of maximum likelihood

The broad principle of the method of maximum likelihood is to choose parameters that give the highest probability given the observations made. The broad approach starts with $f(x)$. If the distribution under consideration is discrete, then $f(x)$ represents $P(X = x)$ for a random variable X. However, if the distribution is continuous, then $f(x)$ is the probability density function.

Unlike the method of moments, the method of maximum likelihood only gives results that are feasible. It also has some attractive properties, such as the fact that any bias in the estimator reduces as the number of observations increases, and that as the number of observations increases the distribution of the estimates tend towards the normal distribution.

Maximum likelihood estimation for discrete distributions

The first stage of this method is to construct a likelihood function. This describes the joint probability that each $X_t = x_t$, where $t = 1, 2, \ldots, T$. The likelihood function is given by:

$$L = \prod_{t=1}^{T} f(x_t). \tag{11.1}$$

The next stage is to take the natural logarithm of each side. This is possible since the probabilities will always be positive, and because it is a monotonic transformation – if x is higher than y, then $\ln x$ will also be higher than $\ln y$. This gives:

$$\ln L = \sum_{t=1}^{T} \ln f(x_t). \tag{11.2}$$

The term $\ln L$ is referred to as the log-likelihood. Once this has been obtained, it needs to be maximised with respect to each parameter. This is done by differentiating with respect to that parameter, setting the result to zero, and solving. So, for example, if one parameter of the distribution is p:

$$\frac{\partial \ln L}{\partial p} = 0. \tag{11.3}$$

If there are several parameters, then several equations need to be derived and solved simultaneously.

Example 11.3 Consider an unfair coin which when tossed twenty times gives only five heads. Show that the maximum likelihood probability of obtaining a head is 0.25.

If the probability of obtaining a head in a single toss is p, then since coin tossing follows a binomial distribution the probability of obtaining five heads from twenty trials is:

$$L = \frac{t}{x!(t-x)!} p^x (1-p)^{t-x} = \frac{20}{5!(20-5)!} p^5 (1-p)^{(20-5)}.$$

Taking logarithms gives:

$$\ln L = \ln\left(\frac{20}{5!(20-5)!}\right) + 5\ln p + (20-5)\ln(1-p).$$

Differentiating both sides with respect to p and setting the result equal to zero gives:

$$\frac{\partial \ln L}{\partial p} = \frac{5}{p} - \frac{15}{1-p} = 0.$$

Rearranging this equation gives $p = 5/20 = 0.25$.

Maximum likelihood estimation for continuous distributions

The method of maximum likelihood can also be used to derive parameters for continuous distributions in a similar fashion.

Example 11.4 A firm believes that the time until payment, x, for work that it carries out follows an exponential distribution with:

$$f(x) = \frac{1}{\beta} e^{-\frac{x}{\beta}}.$$

If the time until payment for the last five invoices is 3, 10, 5, 6 and 16 weeks respectively, find the exponential parameter, β, using the maximum likelihood approach.

Defining the five payment times as $x_1, x_2, \ldots, x_5$, the likelihood function L is given by:

$$L = \prod_{t=1}^{5} f(x_t).$$

Given the expression for $f(x)$, this can be expanded to:

$$L = \frac{1}{\beta}e^{-\frac{3}{\beta}} \times \frac{1}{\beta}e^{-\frac{10}{\beta}} \times \frac{1}{\beta}e^{-\frac{5}{\beta}} \times \frac{1}{\beta}e^{-\frac{6}{\beta}} \times \frac{1}{\beta}e^{-\frac{17}{\beta}}$$

$$= \frac{1}{\beta^5}e^{-\frac{3+10+5+6+16}{\beta}}$$

$$= \frac{1}{\beta^5}e^{-\frac{40}{\beta}}.$$

Taking logarithms gives:

$$\ln L = -5\ln\beta - \frac{40}{\beta}.$$

Differentiating this with respect to β and setting the result equal to zero gives:

$$\frac{d\ln L}{d\beta} = -\frac{5}{\beta} + \frac{41}{\beta^2} = 0.$$

This can be rearranged to shown that $\beta = 8$.

Maximum likelihood estimation for copulas

For copulas, the approach is slightly more involved. There are a number of approaches that can be used, but in each case the first step is to derive a copula density function. This is analogous to the probability density function for a single variable, and is used in the same way in maximum likelihood estimation. Since the copula density function gives the instantaneous joint probability for the range of observations, a likelihood function can be constructed by multiplying the individual density functions together:

$$L = \prod_{t=1}^{T} c(F(x_{1,t}), F(x_{2,t}), \ldots, F(x_{N,t})). \tag{11.4}$$

This describes the joint probability that each $X_{n,t} = x_{n,t}$, where $n = 1, 2, \ldots, N$ and $t = 1, 2, \ldots, T$. The copula density function is as described earlier:

$$c(F(x_1), F(x_2), \ldots, F(x_N)) = \frac{\partial^N C(F(x_1), F(x_2), \ldots, F(x_N))}{\partial F(x_1)\partial F(x_2)\ldots\partial F(x_N)}, \tag{11.5}$$

and as noted this can be rewritten in terms of probability density functions if all distribution functions are continuous:

$$c(F(x_1), F(x_2), \ldots, F(x_N)) = \frac{f(x_1, x_2, \ldots, x_N)}{f(x_1)f(x_2)\ldots f(x_N)}. \tag{11.6}$$

If these functions are available, then parametrising the copula is straightforward – in principal at least. It is simply a case of substituting the probabilities in terms of the unknown parameters and then maximising the resulting likelihood function. However, when the number of variables is large, optimisation is not always straightforward.

There is a useful standard result for the normal or Gaussian copula. If the cumulative distribution function for variable n at time t is $F(x_{n,t})$, then the maximum likelihood estimate for the correlation matrix, $\hat{R}$, is given by:

$$\hat{R} = \frac{1}{T}\sum_{t=1}^{T} \Phi_{t-1}\Phi_{t-1}', \tag{11.7}$$

where Φ_{t-1} is a column vector, the elements of which are

$$\Phi^{-1}(F(x_{1,t})), \Phi^{-1}(F(x_{2,t})), \ldots, \Phi^{-1}(F(x_{N,t})).$$

Alternatively, the values of $F(x_{n,t})$ can be calculated empirically as inputs to calculate the densities according to candidate copulas. These copula densities can then be used to calculate likelihood functions, the copula chosen being the one whose likelihood function has the highest value. The choice of candidate copulas can be reduced by first calculating rank correlation coefficients from the raw data and restricting the choice of copulas to those with parameters reflecting the broad level of association shown in the data.

11.3 Fitting data to a model

It is common to be faced with a model, rather than a distribution, to which data must be fitted. The approach used here depends on the form of the model being fitted.

11.3.1 Least squares regression

This category of regressions is the most straightforward. Consider a variable, Y_t observed at time t where $t = 1, 2, \ldots, T$. A model to explain this dependent variable could be constructed in terms of N explanatory variables, $X_{t,n}$, where $n = 1, 2, \ldots, N$ as follows:

$$Y_t = \beta_1 X_{t,1} + \beta_2 X_{t,2} + \ldots + \beta_N X_{t,N} + \epsilon_t. \tag{11.8}$$

Here β_n determines the extent to which $X_{t,n}$ affects Y_t. The extent to which the explanatory variables fail to explain the dependent variable are captured in ϵ_t. This relationship can also be expressed in matrix form as:

$$Y = X\beta + \epsilon, \tag{11.9}$$

or more completely:

$$\begin{pmatrix} Y_1 \\ Y_2 \\ \vdots \\ Y_T \end{pmatrix} = \begin{pmatrix} X_{1,1} & X_{1,2} & \cdots & X_{1,N} \\ X_{2,1} & X_{2,2} & \cdots & X_{2,N} \\ \vdots & \vdots & \ddots & \vdots \\ X_{T,1} & X_{T,2} & \cdots & X_{T,N} \end{pmatrix} \begin{pmatrix} \beta_1 \\ \beta_2 \\ \vdots \\ \beta_N \end{pmatrix} + \begin{pmatrix} \epsilon_1 \\ \epsilon_2 \\ \vdots \\ \epsilon_T \end{pmatrix}. \tag{11.10}$$

Note that although this equation does not contain a constant term, it is straightforward to include one – all that is needed is for each element of the first column of X to be equal to one.

Ordinary least squares

Equation (11.10) can be rearranged so it is given in terms of the residual error terms:

$$\epsilon = Y - X\beta. \tag{11.11}$$

One approach to fitting this model is to choose the values of the parameters in β such that the sum of squared error terms given by $\epsilon'\epsilon$ or $\epsilon_1^2 + \epsilon_2^2 + \ldots + \epsilon_T^2$ is minimised. This is the basic principle behind ordinary least squares (OLS) regression. Fortunately, a closed-form solution exists for the vector β that gives this result, and that is:

$$b = (X'X)^{-1}X'Y, \tag{11.12}$$

where b is the vector containing estimates of the vector β. The OLS model has a number of restrictive assumptions. In particular:

- a linear relationship exists between the explanatory variables and the dependent variable – this means that variables with a non-linear relationship must be transformed first, by raising to a power, the use of logarithms or some other approach;

- the data matrix must have full column rank – in other words no columns can be linear transformations or combinations of other columns, otherwise it is not possible to calculate the matrix inverse;
- the independent variables should not be correlated with the error terms;
- the error terms should be normally distributed – this is more important for the calculation of statistics relating to the regression than for the regression itself;
- the error terms should not be correlated with each other – if they are, this suggests that there is an element of serial correlation in the model that has not been picked up by the parameters and the regression specification should be changed or the estimation procedure should be modified;
- the error terms should a constant, finite variance, σ^2 – if not, then the method of estimation must be modified.

Generalised least squares

The error terms must comply with a number of assumption if an OLS regression is to be valid. As has been discussed, a number of tests of a regression's significance requires the error terms to be normally distributed, but the validity of the regression itself requires the error terms to be uncorrelated with each other and to have a constant finite variance. This means that instead of each error term having variance of σ^2 and a covariance with any other error term of zero, the variances and covariances are given by a constant σ^2 multiplied by a matrix Ω.

If the issue with the data is that the error terms do not have a constant variance, then the matrix would simply be a diagonal one, with each diagonal element of the matrix Ω giving the weight to be applied to the constant variance σ^2 for the observation at time t. This can be restated as a diagonal matrix of the variances for each observation:

$$\sigma^2 \Omega = \sigma^2 \begin{pmatrix} \Omega_{1,1} & 0 & \dots & 0 \\ 0 & \Omega_{2,2} & \dots & 0 \\ \vdots & \vdots & \ddots & \vdots \\ 0 & 0 & \dots & \Omega_{T,T} \end{pmatrix} = \begin{pmatrix} \sigma_1^2 & 0 & \dots & 0 \\ 0 & \sigma_2^2 & \dots & 0 \\ \vdots & \vdots & \ddots & \vdots \\ 0 & 0 & \dots & \sigma_T^2 \end{pmatrix}.$$

(11.13)

If, instead, the issue is that there is serial correlation in the residuals but that the variance of the residuals is constant, then each diagonal element of the matrix Ω is a one, whilst the off-diagonal elements contain the correlations between the observations. This can be restated as a matrix of variances and covariances:

$$\sigma^2 \Omega = \sigma^2 \begin{pmatrix} 1 & \rho_{1,2} & \cdots & \rho_{1,T} \\ \rho_{2,1} & 1 & \cdots & \rho_{2,T} \\ \vdots & \vdots & \ddots & \vdots \\ \rho_{T,1} & \rho_{T,2} & \cdots & 1 \end{pmatrix} = \begin{pmatrix} \sigma^2 & \sigma_{1,2} & \cdots & \sigma_{1,T} \\ \sigma_{2,1} & \sigma^2 & \cdots & \sigma_{2,T} \\ \vdots & \vdots & \ddots & \vdots \\ \sigma_{T,1} & \sigma_{T,2} & \cdots & \sigma^2 \end{pmatrix},$$

$$(11.14)$$

where $\rho_{s,t}$, the correlation between error terms at times s and t, is equal to $\rho_{t,s}$, and $\sigma_{s,t}$, the covariance between error terms at times s and t, is equal to $\sigma_{t,s}$. If heteroskedasticity exists in addition to serial correlation, then the diagonal elements will vary as per Equation (11.13).

If any of these issues exist in the data, the parameters can be obtained using generalised least squares (GLS) rather than OLS. The matrix of coefficients can be estimated using the following formula:

$$b = (X'\Omega^{-1}X)^{-1}X'\Omega^{-1}Y. \qquad (11.15)$$

The coefficient of determination

If the average observation of Y_t where $t = 1, 2, \ldots, T$ is $\bar{Y}$, then the total sum of squares is:

$$SST = \sum_{t=1}^{T} (Y_t - \bar{Y})^2. \qquad (11.16)$$

If the predicted value of Y_t is denoted $\hat{Y}_t$ where $\hat{Y}_t = \sum_{n=1}^{N} X_{t,n} b_n$, then the sum of squares explained by the regression is given by:

$$SSR = \sum_{t=1}^{T} \left(\hat{Y}_t - \bar{Y}\right)^2. \qquad (11.17)$$

The sum of squared errors, representing the unexplained deviations in the equation, can be summarised as:

$$SSE = \sum_{t=1}^{T} \epsilon_t^2. \qquad (11.18)$$

These three items are related as follows:

$$SST = SSR + SSE. \qquad (11.19)$$

explained unexplained

They can also be used to gauge the significance of the regression as a whole, being combined to give the coefficient of determination, or R^2:

$$R^2 = \frac{SSR}{SST} = 1 - \frac{SSE}{SST}. \tag{11.20}$$

This can range from zero to one, with a higher value indicating a better fit. However, the R^2 of any regression can be increased simply by adding an extra variable. To counter this, there is an alternative to the coefficient of determination known as the *adjusted* R^2 or R_a^2:

$$R_a^2 = 1 - \frac{SSE/(T-N)}{SST/(T-1)} = 1 - \frac{T-1}{T-N}(1-R^2). \tag{11.21}$$

Testing the fit of the regression

A similar measure can be used to test the fit of the regression as a whole using an F-test. The test statistic is:

$$\frac{SSR/(N-1)}{SSE/(T-N)} = \frac{R^2/(N-1)}{(1-R^2)/(T-N)} \sim F_{T-N}^{N-1}. \tag{11.22}$$

The null hypothesis here is that all of the coefficients in the regression are zero, so this is a test of how significantly they differ from zero when taken together. This test requires the assumption that the error terms are normally distributed.

Testing the fit of the individual coefficients

Having estimated the parameters for the regression, it is also possible to test whether they are statistically different from zero or not on an individual basis – in other words, whether it would make any significant difference if each variable was omitted from the regression. To do this, the variance of the error terms, σ^2, must be estimated. The estimate is referred to here as s^2. Just summing the squared error terms and dividing by the number of observations would give a biased value of this variance, in the same way that using the formula for the population variance to calculate the sample variance gives a biased answer. The solution here is to divide by the number of observations, T, less the number of explanatory variables including any intercept term, N:

$$s^2 = \frac{SSE}{T-N}. \tag{11.23}$$

The square root of this value, s, is the standard error of the regression. The scalar s^2 is multiplied by $(X'X)^{-1}$ to give:

$$S_b = s^2(X'X)^{-1}. \tag{11.24}$$

where S is the sample covariance matrix for the vector of estimators, b. This is an $N \times N$ matrix, and the square root of the nth diagonal element is s_{b_n}, the standard error of the estimator b_n.

Having both a value for each estimator and a standard error for that value means that the significance of that value can be tested. To do this again requires the assumption that the error terms are normally distributed. Since the standard error is a sample rather than a population measure, the test used is a t-test, and the test statistic is:

$$\frac{b_n - \beta_n}{s_{b_n}} \sim t_{T-N}. \tag{11.25}$$

The null hypothesis is usually that β_n is zero, so the test is of the level of significance by which the coefficient differs from this value. A confidence level of 90% is the minimum level at which a coefficient is usually regarded as significant.

11.3.2 The method of maximum likelihood

Fitting a model to data using the method of maximum likelihood is very similar to the process used when fitting a distribution using the same technique. The main difference that the additional complexity of the model may well mean that iterative techniques are needed to find the parameters that maximise the likelihood function, L.

However, the method of maximum likelihood also allows the calculation of alternative statistics that can be used to test the goodness of fit of a particular model. *Tests of Alternative Dist's.*

The likelihood ratio test

The likelihood ratio test is used when comparing nested models. Two models are nested if a second model contains all of the independent variables of a first model plus one or more additional variables. The null hypothesis for the likelihood ratio test is that the additional variables give no significant improvement in the explanatory power of the model. The likelihood ratio test statistic, LR, is given below:

$$LR = -2\ln(L_1/L_2) \sim \chi^2_{N_2-N_1}, \tag{11.26}$$

where L_1 and L_2 are the values of the likelihood functions for the first and second models, whilst N_1 and N_2 are the numbers of independent variables in each, including the constant.

A feature of the likelihood ratio test which can often be a drawback is that it is suitable for comparing only nested models rather than alternative specifications. Information criteria avoid this drawback; however, unlike the likelihood ratio test, they offer only rankings of models with no way of describing the statistical significance of any difference between the models.

The Akaike information criterion

The Akaike information criterion (AIC) for a particular model is calculated from the likelihood function as follows:

$$\text{AIC} = 2N - 2\ln L, \tag{11.27}$$

where N is the number of independent variables in the model, including the constant. A lower value of the AIC indicates a better model

The Bayesian information criterion

The form of the Bayesian information criterion (BIC) is similar to that of the AIC. However, the BIC also takes into account the number of observations, T, and it is this that makes it 'Bayesian'. The formula for the BIC is:

$$\text{BIC} = N \ln T - 2\ln L. \tag{11.28}$$

As with the AIC, a lower value indicates a better model. However, the BIC has a more severe penalty for the addition of independent variables than the AIC, unless the number of observations is small. This means that BIC tends to lead to the less complex models being chosen than does the AIC.

11.3.3 Principal component analysis

PCA has already been discussed as a way of producing correlated random variables from a sample covariance matrix and a vector of sample means. It should also be recognised that this approach provides a fit of a dataset to a number of independent parameters with the relative importance of these parameters being seen by the size of its eigenvalue. This is particularly helpful if the purpose of fitting the data is to be produce stochastic projections, even more so if there is a desire to reduce the number of variables actually projected.

However, PCA is not easily able to attach any intuitive meaning to the factors it produces. This means that if a model is being fitted in order to investigate the influence of various factors, PCA is rarely helpful.

11.3.4 Singular value decomposition

Another form of least squares optimisation is singular value decomposition (SVD). This can be used to find a function that best fits a set of data when there are no independent variables on which a regression can be based.

The principle behind SVD is that a matrix X with M rows and N columns but a column rank of only R can be expressed as the sum of R orthogonal matrices. In this context, orthogonality means that none of the matrices can be expressed as a linear combination of any of the others. Each of these matrices can itself be expressed as the product of two vectors. The fact that X has a column rank of R rather than N implies that $N - R$ of the columns can be expressed as linear combinations of the other columns. Matrix X can therefore be broken down as follows:

$$X = L_1 U_1 V_1' + L_2 U_2 V_2' + \ldots L_R U_R V_R', \qquad (11.29)$$

or, writing out the matrices and vectors more completely:

$$
\begin{pmatrix}
X_{1,1} & X_{1,2} & \cdots & X_{1,N} \\
X_{2,1} & X_{2,2} & \cdots & X_{2,N} \\
\vdots & \vdots & \ddots & \vdots \\
X_{M,1} & X_{M,2} & \cdots & X_{M,N}
\end{pmatrix}
= L_1
\begin{pmatrix}
U_{1,1} \\
U_{2,1} \\
\vdots \\
U_{M,1}
\end{pmatrix}
\begin{pmatrix} V_{1,1} & V_{2,1} & \cdots & V_{N,1} \end{pmatrix}
$$

$$
+ L_2
\begin{pmatrix}
U_{1,2} \\
U_{2,2} \\
\vdots \\
U_{M,2}
\end{pmatrix}
\begin{pmatrix} V_{1,2} & V_{2,2} & \cdots & V_{N,2} \end{pmatrix}
$$

$$+\ldots$$

$$
+ L_R
\begin{pmatrix}
U_{1,R} \\
U_{2,R} \\
\vdots \\
U_{M,R}
\end{pmatrix}
\begin{pmatrix} V_{1,R} & V_{2,R} & \cdots & V_{N,R} \end{pmatrix}.
$$

$$(11.30)$$

Here, the vectors U_r, where $r = 1, 2, \ldots, R$, are orthogonal as are the vectors V_r. However, these vectors can be combined into matrices U and V, whilst the scalars L_r can be combined into a single diagonal $R \times R$ matrix:

$$
\begin{pmatrix}
X_{1,1} & X_{1,2} & \cdots & X_{1,N} \\
X_{2,1} & X_{2,2} & \cdots & X_{2,N} \\
\vdots & \vdots & \ddots & \vdots \\
X_{M,1} & X_{M,2} & \cdots & X_{M,N}
\end{pmatrix}
$$

$$
=
\begin{pmatrix}
U_{1,1} & U_{1,2} & \cdots & U_{1,R} \\
U_{2,1} & U_{2,2} & \cdots & U_{2,R} \\
\vdots & \vdots & \ddots & \vdots \\
U_{M,1} & U_{M,2} & \cdots & U_{M,R}
\end{pmatrix}
$$

$$
\times
\begin{pmatrix}
L_1 & 0 & \cdots & 0 \\
0 & L_2 & \cdots & 0 \\
\vdots & \vdots & \ddots & \vdots \\
0 & 0 & \cdots & L_R
\end{pmatrix}
\begin{pmatrix}
V_{1,1} & V_{2,1} & \cdots & V_{N,1} \\
V_{1,2} & V_{2,2} & \cdots & V_{N,2} \\
\vdots & \vdots & \ddots & \vdots \\
V_{1,R} & V_{2,R} & \cdots & V_{N,R}
\end{pmatrix},
\qquad (11.31)
$$

or, in more compact form:

$$X = ULV'. \qquad (11.32)$$

The scalars L_r are actually the square roots of the eigenvectors of XX' (and, for that matter, $X'X$), and are known as the singular values. These scalars, along with the matrices U and V can be found using an approach similar to the power method covered in PCA earlier, but applied to the original dataset rather than to the calculated covariance matrix. The approach involves calculating the vectors corresponding to decreasing values of L_r, with the largest being denoted L_1. The starting point is to take $V_1(0)$, an arbitrary vector of unit length with N elements. The simplest such vector is one whose elements are each equal to $1/\sqrt{N}$. This is pre-multiplied by the matrix of data, X to give $U_1^*(0)$, a vector of length M:

$$U_1^*(1) = XV_1(0). \qquad (11.33)$$

The vector U_1^* is then normalised, so it too has a unit length. This is done by dividing each element of U_1^* by the scalar $\sqrt{U_1^*(1)'U_1^*(1)}$:

$$U_1(1) = \frac{1}{\sqrt{U_1^*(1)'U_1^*(1)}} U_1^*(1). \qquad (11.34)$$

This new vector is then pre-multiplied by X' to give $V_1^*(1)$:

$$V_1^*(1) = X'U_1(1). \qquad (11.35)$$

This vector is then similarly scaled to give $V_1(1)$:

$$V_1(1) = \frac{1}{\sqrt{V_1^*(1)'V_1^*(1)}} V_1^*(1). \qquad (11.36)$$

This process continues for K iterations, at which point the proportional difference between $U_1(K)$ and $U_1(K-1)$ and between $V_1(K)$ and $V_1(K-1)$ is deemed to be sufficiently small. The resulting vectors $U_1 = U_1(K)$ and $V_1 = V_1(K)$ are the first left and first right singular vectors of the decomposition. The first singular value, L_1, is calculated as:

$$L_1 \approx \sqrt{V_1^*(K)'V_1^*(K)} \approx \sqrt{U_1^*(K)'U_1^*(K)}. \qquad (11.37)$$

Having found the first singular values and first left and right singular vectors, the process can be repeated to find the next $R-1$ vectors and values, U_r, U_r and L_r, where $2 \leq r \leq R$. However, the data matrix to which the method is applied changes each time. In particular:

$$X_r = X_{r-1} - L_{r-1} U_{r-1} V_{r-1}'. \qquad (11.38)$$

Whilst it is possible to carry out this process to identify all of the singular vectors and values, this technique is also often used to described the variation in a series of data using a small number of factors. In this way, it is similar to PCA. However, as mentioned above, SVD does not require a covariance matrix to be calculated, being performed on raw data.

11.4 Smoothing data

11.4.1 Splines

Sometimes, the main reason for fitting a model is to remove 'noise' from a dataset so that an underlying pattern can be seen. In this case, the data might just be fitted to a polynomial using time as the only independent variable. However, it is often difficult to achieve a good fit using a single function. An alternative approach is to use splines. A spline is a function that uses a number of different polynomial functions to fit a series of data.

Simple splines

Consider a series of T data points to which a smooth function is to be fitted. Rather than fitting a single curve, a number of separate curves, each following on from the previous curve, could instead be used. If each curve is a polynomial with at most M degrees – so includes terms raised to the Mth power – then the spline overall has a degree that is less than or equal to M.

The start and end of each curve is known as a knot. This means that with $N-1$ inner knots – where two curves meet – there are N curves. If these knots are equally spaced – that is, each curves covers the same number of data points – then the spline is said to be uniform.

To give the appearance of a single, smooth line from start to finish, it is important not only that start-point of each polynomial has the same co-ordinates as the end-point of the previous one, but that there is no sudden change in gradient. This can be achieved by ensuring that:

• the gradient of each polynomial is equal when they meet at a knot; and
• the rate of change of gradient for each is also equal at this point.

This means that both the first and second derivatives are continuous at each knot – there is no step-change in either. A commonly used spline which fits these criteria is the natural cubic spline. This fits a series of cubic functions to a dataset, ensuring that the two criteria outlined above hold.

There are a number of ways in which a cubic spline can be fitted, but a key decision that must be made is whether the spline is meant to interpolate between the points or to smooth across a dataset. Interpolation implies that N curves are fitted to T data points where $T = N+1$; smoothing, on the other hand, implies that $T = kN+1$ where k is some integer greater than one. For example, if there are sixteen data points and five separate curves are fitted, $T = 16$, $N = 5$ and $k = 3$. The constant k is also known as the knot spacing.

Even if the cubic spline is being used for smoothing, interpolation provides a good starting point. This involves using only the observations at the knots. Consider a situation where N curves are being fitted to $N+1$ data points. Let these data points have the co-ordinates x_n and y_n, where $n = 1, 2, \ldots, N+1$. Define each piece of the spline as:

$$f_n(x) = a_n + b_n(x - x_n) + c_n(x - x_n)^2 + d_n(x - x_n)^3, \qquad (11.39)$$

where $n = 1, 2, \ldots, N$. The fitting process is then as follows:

• for $n = 1, 2, \ldots, N+1$ set each $a_n = y_n$;
• for $n = 2, 3, \ldots, N+1$, set $\Delta x_n = x_n - x_{n-1}$;
• for $n = 2, 3, \ldots, N$, set $\alpha_n = 3(a_{n+1} - a_n)/\Delta x_{n+1} - 3(a_n - a_{n-1})/\Delta x_n$;
• set $\beta_1 = 1$, $\gamma_1 = 0$ and $\delta_1 = 0$;
• for $n = 2, 3, \ldots, N$, set $\beta_n = 2(x_{n+1} - x_{n-1}) - \gamma_{n-1}\Delta x_n$, $\gamma_n = \Delta x_{n+1}/\beta_n$ and $\delta_n = (\alpha_n - \delta_{n-1}\Delta x_n)/\beta_n$;
• set $\beta_{N+1} = 1$ and $c_{N+1} = 0$; and

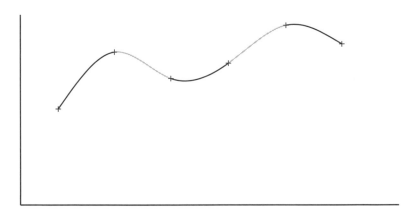

Figure 11.1 Interpolating natural cubic spline

- for $n = N, N-1, \ldots, 1$, set $c_n = \delta_n - \gamma_n c_{n+1}$, $b_n = (a_{n+1} - a_n)/\Delta x_{n+1} - \Delta x_{n+1}(c_{n+1} + 2c_n)/3$ and $d_n = (c_{n+1} - c_n)/3\Delta x_{n+1}$.

An example of a natural cubic spline fitted in this way is shown in Figure 11.1.

If the cubic spline is instead being used for smoothing, then the knot spacing determines the degree of smoothing. To be precise, a greater spacing gives a greater degree of smoothing. Choosing an appropriate knot spacing and applying the above process to the knots can be used to give a starting set of parameters for a smoothing spline. However, a smoother curve can then be found by allowing each a_n to take a value other than y_n. The other parameters can be determined in the same way as above, with the values of a_n being chosen to minimise $\sum_{t=1}^{T}[y_t - f_n(x_t)]^2$. If $k = 1$, then the result is an interpolating spline which passes through all of the points; however, for $k > 1$ a smoother spline is found, as shown in Figure 11.2. The expression to be minimised could also be altered to reflect heteroskedasticity or other data anomalies.

Basis splines

Basis splines, known as b-splines for short, are special types of splines with the following properties:

- each has $M+1$ polynomial curves, each of degree M;
- each has M inner knots;
- at each knot, derivatives up to order $M-1$ are continuous;
- each is positive on an area covered by $M+2$ knots and zero elsewhere;

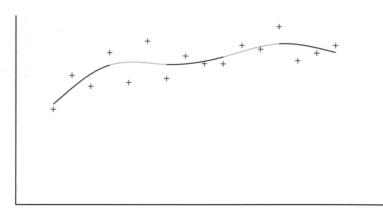

Figure 11.2 Smoothing natural cubic spline

- except at the boundaries, each overlaps with $2M$ polynomial curves of its neighbours; and
- for any value of x, there are $M+1$ non-zero b-splines.

If the knots are equally spaced, then each spline is identical. This means that another step is needed to fit these splines to a dataset. This step is to weight the individual splines. What this means is that each point is represented by the weighted sum of $M+1$ points from $N+1$ separate b-splines. So, for example, if the b-splines are quadratic, then each b-spline will be made up of three quadratic functions joining two knots, and each point on the fitted curve will be represented by the sum of three points, each from a separate b-spline, each spline having a different weight. Mathematically, this means that each smoothed data point $\hat{X}_t$, where $t = 1, 2, \ldots, T$, can be expressed in terms of a number of b-splines, $B_n(t)$, where $n = 1, 2, \ldots, N$, weighted by a value of A_n for each b-spline:

$$\hat{X}_t = \sum_{n=1}^{N} A_n B_n(t). \tag{11.40}$$

Note that for each t, there will be only $M+1$ non-zero values of $B_n(t)$.

The three sections of a quadratic b-spline are shown in the left-hand side of Figure 11.3. On the right-hand side, a group of unweighted b-splines are shown, with the dashed line above showing their sum.

Penalised splines

A potential issue with splines is they can lead to a model being over-fitted, removing useful information rather than just noise. One way to control for this

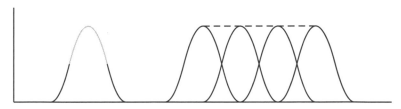

Figure 11.3 Unweighted basis splines

is to use penalised splines, or p-splines for short. The penalty is based on the level of variation in A_n and, implicitly, on the number of b-splines, N:

$$P = \sum_{n=3}^{N} (\Delta^2 A_n)^2. \tag{11.41}$$

where $\Delta^2 A_n = \Delta A_n - \Delta A_{n-1}$ and $\Delta A_n = A_n - A_{n-1}$. The closer the weighted b-splines are to the dataset, the greater the variation in A_n, so the greater the penalty. Also, the greater the number of b-splines, the greater the penalty.

The penalty is then incorporated into the measurement of likelihood to give a penalised likelihood function, PL:

$$PL = L - \frac{1}{2} \lambda \sum_{n=3}^{N} (\Delta^2 A_n)^2, \tag{11.42}$$

where L is the likelihood function and λ is a roughness parameter which balances fit and smoothness. In particular, when $\lambda = 0$ there is no penalty for roughness, whereas when $\lambda = \infty$ the result is a linear regression. As can be seen in Figure 11.4, the p-spline approach gives a smoother line than the b-spline

11.4.2 Kernel smoothing

Another approach to smoothing a dataset is to describe the values of data points in terms of surrounding observations. Such an approach is known as kernel smoothing. This approach also allows missing observations to be estimated.

All kernel functions are symmetrical. This means that the influence of observations above a particular point have the same weight as observations the same distance below that point. In mathematical terms, if a kernel function is defined as $k(u)$ for some input u, this means that $k(u) = k(-u)$. Also, the

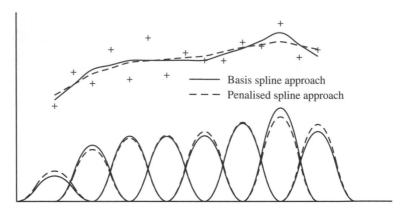

Figure 11.4 Weighted basis and penalised splines (sum on separate vertical scale)

total area under a kernel function must also sum to one, or:

$$\int_{-\infty}^{\infty} k(u)\mathrm{d}u = 1. \tag{11.43}$$

This means that a kernel function can be thought of as a type of probability density function. Various kernal density functions are shown in Figure 11.5.

There are a number of kernel functions available to use. Of the most useful, all but one are left- and right-bounded. This is helpful, because it means that sensible smoothing results can be obtained close to the upper and lower ends of dataset. Three common bounded kernel functions are the uniform kernel:

$$_U k(u) = \frac{1}{2} I(|u| \leq 1), \tag{11.44}$$

the triangular kernel:

$$_T k(u) = (1 - |u|) I(|u| \leq 1), \tag{11.45}$$

and the Epanechnikov kernel:

$$_E k(u) = \frac{3}{4} (1 - u^2) I(|u| \leq 1), \tag{11.46}$$

where $I(|u| \leq 1)$ is an indicator function which is one if the absolute value of u is less than or equal to one, and zero otherwise. The most used unbounded kernel function is the Gaussian or normal function which has the following form:

$$_N k(u) = \frac{1}{\sqrt{2\pi}} e^{-\frac{1}{2}u^2}. \tag{11.47}$$

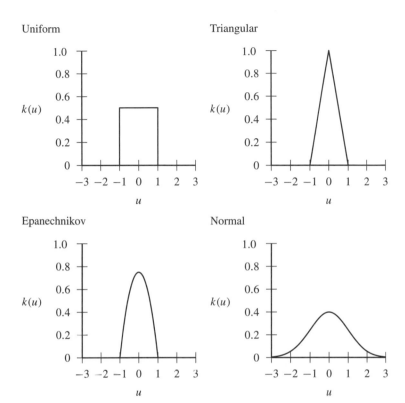

Figure 11.5 Various kernel functions

The ranges of these kernels are fixed, a fact that is clearest for the bounded kernels. However, it is generally desirable to be able to alter the range of observations used in the kernel smoothing process. This is done by including a bandwidth parameter, λ, in the smoothing formula used to give $\hat{X}_t$, the smoothed value of the raw observation X_t, as shown below in Equation (11.48):

$$k_\lambda(u) = \frac{1}{\lambda} k \left(\frac{u}{\lambda} \right). \tag{11.48}$$

This can then be used to give a smoothed value, $\hat{X}_t$, as follows:

$$\hat{X}_t = \frac{\sum_{s=1}^{T} k_\lambda (t-s) X_s}{\sum_{s=1}^{T} k_\lambda (t-s)}. \tag{11.49}$$

The denominator ensures that the kernel weights sum to one – whilst the area under the curve for any kernel is equal to one, the weights when applied to discrete data may not be, so an adjustment is required.

Example 11.5 The following table gives the central mortality rates for UK centenarian males from 1990 to 2005. What smoothed mortality rates are found if an Epanechnikov kernel with a bandwidth of three years is used?

Year	Central mortality rate ($m_{100,t}$)
1990	0.5230
1991	0.5009
1992	0.5085
1993	0.5626
1994	0.4659
1995	0.4841
1996	0.5646
1997	0.4912
1998	0.5137
1999	0.5249
2000	0.5484
2001	0.4739
2002	0.5221
2003	0.5634
2004	0.4842
2005	0.5293

With a bandwidth of three years, the first year for which a smoothed rate can be calculated is 1992. Combining the structure for the Epanechnikov kernel with the general structure for a kernel smoothing function gives:

$$\hat{m}_{100,1992} = \frac{\sum_{t=1990}^{1994} \frac{1}{3} \times \frac{3}{4}\left[1 - \left(\frac{1992-t}{3}\right)^2\right] m_{100,t}}{\sum_{t=1990}^{1994} \frac{1}{3} \times \frac{3}{4}\left[1 - \left(\frac{1992-t}{3}\right)^2\right]},$$

This gives a smoothed value for $\hat{m}_{100,1992}$ of 0.5151. Continuing this process gives the following values for $\hat{m}_{100,t}$:

Year	Central mortality rate ($m_{100,t}$)	Smoothed central mortality rate ($\hat{m}_{100,t}$)
1990	0.5230	–
1991	0.5009	–
1992	0.5085	0.5151
1993	0.5626	0.5081
1994	0.4659	0.5123
1995	0.4841	0.5106
1996	0.5646	0.5081
1997	0.4912	0.5169
1998	0.5137	0.5233
1999	0.5249	0.5156
2000	0.5484	0.5173
2001	0.4739	0.5220
2002	0.5221	0.5189
2003	0.5634	0.5182
2004	0.4842	–
2005	0.5293	–

The raw and smoothed data are shown graphically below:

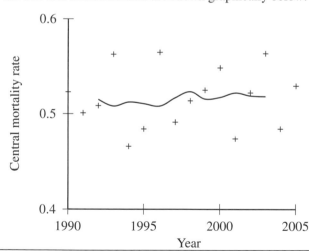

11.5 Using models to classify data

So far, the focus has been on explaining the value of an observation using the characteristics of an individual or firm. However, observations are sometimes in the form of categories rather than values, for example whether an individual is alive or dead, or whether a firm is solvent or insolvent. In this case, different types of models need to be used to analyse the data.

11.5.1 Generalised linear models

A generalised linear model (GLM) is a type of model used to link a linear regression model, such as that described in least squares regression, and a dependent variable that can take only a limited range of values. Rather than being calculated using a least squares approach, the method of maximum likelihood is more likely to be employed to fit such models.

The most common use for a GLM is when the dependent variable can take only a limited number of values, and in the simplest case there are only two options. For example, a firm can either default on its debt or not default; an individual can die or survive; an insurance policyholder can either claim or not claim. If trying to decide which underlying factors might have an impact on the option chosen, it is first necessary to give the options values of zero and one and to define them in terms of some latent variable. So, if Z_i is the event that is of interest (credit default, death, insurance claim and so on) for company or individual i, the relationship between Z_i and a latent variable Y_i is:

$$Z_i = \begin{cases} 0 & \text{if } Y_i \leq 0 \\ 1 & \text{if } Y_i > 0. \end{cases} \tag{11.50}$$

The vector Y contains values of Y_i for each i. This is then described in terms of a matrix of independent variables, X, and the vector of coefficients, β. It is possible to extend this to allow for more than two categories. In this case:

$$Z_i = \begin{cases} 0 & \text{if } Y_i \leq \alpha_1 \\ 1 & \text{if } \alpha_1 < Y_i \leq \alpha_2 \\ 2 & \text{if } \alpha_2 < Y_i \leq \alpha_3 \\ \vdots & \quad \vdots \\ N-1 & \text{if } \alpha_{N-1} < Y_i \leq \alpha_N \\ N & \text{if } Y_i > \alpha_N, \end{cases} \tag{11.51}$$

where $-\infty < \alpha_1 < \alpha_2 < \ldots < \alpha_N < \infty$.

However, as mentioned above, some sort of link function is needed to convert the latent variable into a probability. Two common link functions are:

- probit; and
- logit; and

The probit model

The probit model uses the cumulative distribution function for the standard normal distribution, $\Phi(x)$. If there are two potential outcomes, then the probit model is formulated as follows:

$$\Pr(Z_i = 1 | X_i) = \Phi(X_i'\beta), \tag{11.52}$$

where X_i is the vector of independent variables for company or individual i. Since Φ returns the cumulative normal distribution function – which is bounded by zero and one – when given any value between $-\infty$ and ∞, it is a useful function for using unbounded independent variables to explain an observation such as a probability that falls between zero and one.

It is possible to extend the probit model to allow for more than two choices, the result being an ordered probit model.

The logit model

The logit model uses the same approach, but rather than using the cumulative normal distribution it uses the logistic function to ensure that Z_i falls between zero and one. For two potential outcomes, this has the following form:

$$\Pr(Z_i = 1 | X_i) = \frac{e^{X_i'\beta}}{1 + e^{X_i'\beta}}. \tag{11.53}$$

The logistic function is symmetrical and bell-shaped, like the normal distribution, but the tails are heavier.

As with the probit model, it is possible to extend the logit model to allow for more than two choices, the result being an ordered logit model.

11.5.2 Survival models

Probit and logit models – in common with many types of model – tend to consider the rate of occurrence in each calendar year, or for each year of age. For example, when using a probit model to describe the drivers of mortality, the model could be applied separately for each year of age using data over a number of years. When record keeping was more limited and only an individual's age was known, then there were few alternatives to such an approach. However, dates of birth and death are now recorded and accessible, meaning that survival models can also be applied.

Survival models were developed for use in medical statistics, and the most obvious uses are still in relation to human mortality. However, there is no reason why such models could not be used to model lapses, times until bankruptcy or other time-dependent variables.

In relation to mortality, a survival model looks at $_t p_x$, the probability that an individual aged x will survive for a further period t before dying. Importantly, if an underlying continuous-time mortality function is defined, then the exact dates of entry into a sample and subsequent death can be allowed for.

The survival probability for an individual can be defined in terms of the force of mortality, μ_x – the instantaneous probability of death for an individual aged x, quoted here as a rate per annum – as follows:

$$_t p_x = \int_0^t \mu_{x+s} ds. \tag{11.54}$$

This leaves two items to be decided:

- the form of μ_x; and
- the drivers of μ_x.

There are a number of forms that μ_{x+s}, but a simple model might be $\mu_x = e^{\alpha + \beta x}$, also known as the Gompertz model (Gompertz, 1825). The next stage is to determine values for α and β. Ideally, these would be calculated separately for each individual n of N. For example:

$$\alpha_n = a_0 + \sum_{m=1}^{M} I_{m,n} a_m, \tag{11.55}$$

and

$$\beta_n = b_0 + \sum_{m=1}^{M} I_{m,n} b_m, \tag{11.56}$$

where a_0 and b_0 are the 'baseline' levels of risk, a_m and b_m are the additions required for risk factor m and $I_{m,n}$ is an indicator function which is equal to one if the risk factor m is present for individual n and zero otherwise. For example, a_1 and b_1 might be the additional loadings required if an individual was male, a_2 and b_2 the additional loadings for smokers and so on.

The next stage is to combine the survival probabilities into a likelihood function, and to adjust the values of the parameters to maximise the joint likelihood of the observations.

Unless a population is monitored until all lives have died, the information on mortality rates will be incomplete to the extent that data will be right-censored.

Furthermore, unless individuals are included from the minimum age for the model, data will be left-truncated. These two features should be taken into account when a model is fitted.

Whilst this approach has advantages over GLMs, in that the exact period of survival can be modelled without the need to divide information into year-long chunks, there are some shortcomings. In particular, logit and probit models can allow for complex relationships between risk factors and ages, whilst the survivor model approach required any age-related relationship to be parametric. Even parametric relationships that are more complex than linear ones can be difficult to allow for. It is worth considering the use of a GLM to determine the approximate shapes of any relationships that the factors have with age before deciding on the form of a survival model.

6 11.5.3 Discriminant analysis

Discriminant analysis is an approach that takes the quantitative characteristics of a number of groups, G, and weights them in such a way that the results differ as much as possible between the groups. Its most well-known application was for the Altman's Z-score (Altman, 1968). There are a number of ways of performing discriminant analysis, but most approaches – including those discussed here – require the assumption that the independent variables are normally distributed, either within each group (as in Fisher's linear discriminant) or in aggregate (as in linear discriminant analysis).

Discriminant analysis can be carried out for any number of groups; however, the most relevant in financial risk involve considering only two. In this regard, it is helpful to start with the original technique described by Sir Ronald Fisher (Fisher, 1936).

Fisher's linear discriminant

Fisher's linear discriminant was originally demonstrated as a way to distinguish between different species of flowers using various measurements of specimens' sepals and petals. However a more familiar financial example might the use of discriminant analysis to distinguish between two groups of firms, one that becomes insolvent and one that does not. These firms form the training set used to parameterise the model. Each firm will have exposure to M of risk factors, relating to levels of earnings, leverage and so on. For firm n of N, the financial measures are given by $X_{1,n}, X_{2,n}, \ldots, X_{M,n}$. The discriminant function for that firm is:

$$d_n = \beta_1 X_{1,n} + \beta_2 X_{2,n} + \ldots + \beta_M X_{M,n}, \tag{11.57}$$

where $\beta_1, \beta_2, \ldots, \beta_M$ are the coefficients that are the same for all firms. In particular, the values of the coefficients are chosen such that the difference between the values of d_n is as great as possible between the groups of solvent and insolvent firms, but as small as possible within each group. Histograms of poorly discriminated and well-discriminated data are shown in Figures 11.6 and 11.7.

Using this approach, the distance between two groups is $(\bar{d}_1 - \bar{d}_2)^2$, where $\bar{d}_1$ and $\bar{d}_2$ are the average values of d_n for each of the two groups. The term $\bar{d}_g$ is often referred to as the 'centroid' of group g. A vector $\bar{X}_1$ can be defined

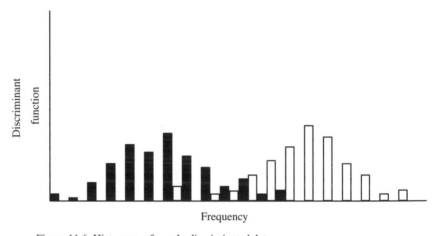

Figure 11.6 Histogram of poorly discriminated data

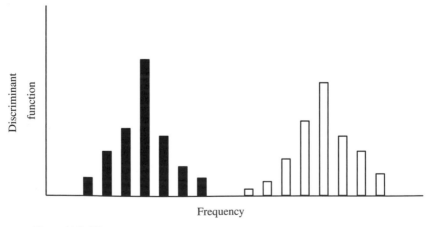

Figure 11.7 Histogram of well-discriminated data

as the average values of $X_{1,n}, X_{2,n}, \ldots, X_{M,n}$ for the first group and $\bar{X}_2$ can be defined as the corresponding vector for the second group. If the vector of coefficients, $\beta_1, \ldots, \beta_M$, is also defined as β, then it is clear that $\bar{d}_1 = \beta'\bar{X}_1$, $\bar{d}_2 = \beta'\bar{X}_2$, and so $(\bar{d}_1 - \bar{d}_2)^2 = (\beta'\bar{X}_1 - \beta'\bar{X}_2)^2$.

The variability within each of the groups requires the calculation of a covariance matrix between $X_{1,n}, X_{2,n}, \ldots, X_{M,n}$ for the first group, Σ_1, and for the second group, Σ_2. The total variability within the groups can then be calculated as $\beta'\Sigma_1\beta + \beta'\Sigma_2\beta$. This means that to both maximise the variability between groups whilst minimising it within groups, the following function needs to be maximised:

$$D_F = \frac{(\beta'\bar{X}_1 - \beta'\bar{X}_2)^2}{\beta}' \Sigma_1\beta + \beta'\Sigma_2\beta. \tag{11.58}$$

The numerator here gives a measure of the difference between the two centroids, which must be as large as possible, whilst the denominator gives a measure of the difference between the discriminant functions within each group, which should be as small as possible.

Since the data only make up a sample of the total population, Σ_1 and Σ_2 must be estimated from the data as S_1 for the first group and S_2 for the second. If the estimator of β that provides the best separation under Fisher's approach is b_F, then b_F can be estimated as:

$$b_F = (S_1 + S_2)^{-1}(\bar{X}_1 - \bar{X}_2). \tag{11.59}$$

This vector can also be used to determine the threshold score between the two groups, d_c:

$$d_c = b_F'(\bar{X}_1 - \bar{X}_2)/2. \tag{11.60}$$

This means that if the data described above are regarded as training data, then when another firm is examined the information on its financial ratios can be used to decide whether it is likely to become insolvent or not, based on whether the calculated value of d_n for a new firm n is above or below d_c.

However, sometimes even the best discrimination cannot perfectly distinguish between different groups. In this case, there exists a 'zone of ignorance' or 'zone of uncertainty' within which the group to which a firm belongs is not clear. This is shown in Figure 11.8. The zone of ignorance can be determined by inspection of the training set of data. For example, assume $\bar{d}_1$ lies below d_c, whilst $\bar{d}_2$ lies above it. If there are firms from group 1 whose discriminant values lie above d_c and firms from group 2 whose values are below it, then the zone of ignorance could be classed as the range between the lowest discriminant value for a group 2 firm up to the highest value for a group 1 firm.

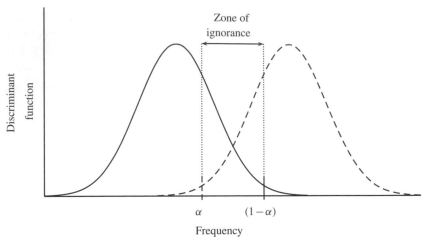

Figure 11.8 Zone of ignorance ($\alpha = 0.01$)

Furthermore, if there were sufficient observations, a more accurate confidence interval could be constructed.

The zone of ignorance can also be defined in terms of confidence intervals if it is defined in terms of the statistical distribution assumed, in particular if normality is assumed. Let $\bar{d}_1$ again lie below d_c, whilst $\bar{d}_2$ lies above it. Then calculate the standard deviations of the values of d_n for each of the two groups, $s_{\bar{d}_1}$ and $s_{\bar{d}_2}$ respectively. For a confidence interval of α, the zone of ignorance could be defined as follows:

$$\bar{d}_2 - s_{\bar{d}_2}\Phi^{-1}(1-\alpha) \text{ to } \bar{d}_1 + s_{\bar{d}_1}\Phi^{-1}(1-\alpha) \text{ if } \bar{d}_1 + s_{\bar{d}_1}\Phi^{-1}(1-\alpha) >$$
$$\bar{d}_2 - s_{\bar{d}_2}\Phi^{-1}(1-\alpha)$$
$$0 \text{ if } \bar{d}_1 + s_{\bar{d}_1}\Phi^{-1}(1-\alpha) \leq$$
$$\bar{d}_2 - s_{\bar{d}_2}\Phi^{-1}(1-\alpha).$$

$$(11.61)$$

Example 11.6 A group of policyholders has been classified into 'low net worth' (LNW) and 'high net worth' using Fisher's linear discriminant. If the LNW group discriminant functions have a mean of 5.2 and a standard deviation of 1.1, whilst the HNW group discriminant functions have a mean of 8.4 and a standard deviation of 0.6, where is the zone of ignorance using a one-tailed confidence interval of 1%?

In this dataset, $\bar{d}_1$ and $\bar{d}_2$ are equal to 5.2 and 8.4 respectively, whilst s_{d_1} and s_{d_2} are 1.1% and 0.6%. The upper-tail critical value for the normal distribution with a confidence interval of 1% is 2.326. The lower limit of the zone of ignorance is therefore:

$$8.4 - (0.6 \times 2.326) = 7.004,$$

whilst the upper limit is:

$$5.2 + (1.1 \times 2.326) = 7.759.$$

The zone of ignorance is therefore 7.004 to 7.759, and any individual whose discriminant function falls in this range cannot be classified within the confidence interval given above.

Linear discriminant analysis

One of the advantages of Fisher's linear discriminant is that it is relatively light on the assumptions required. In particular, an assumption of normally distributed observations is needed only to measure the probability of misclassification. However, if some further assumptions are made then a simpler approach can be used. This approach is linear discriminant analysis (LDA).

The main simplifying assumption is that the independent variables for the two groups have the same covariance matrix, so $\mathbf{6}_1 = \mathbf{6}_2 = \mathbf{6}$. This means that the function to be maximised becomes:

$$D_{LDA} = \frac{(\beta'\bar{X}_1 - \beta'\bar{X}_2)^2}{\beta'\Sigma\beta}. \tag{11.62}$$

If Σ is estimated from the data as S, then the estimator of β that provides the best separation under the LDA approach is b_{LDA}, which can be estimated as:

$$b_{LDA} = S^{-1}(\bar{X}_1 - \bar{X}_2). \tag{11.63}$$

The calculation of the threshold score, d_c, and the zone of ignorance is the same as for Fisher's linear discriminant.

Multiple discriminant analysis

It is possible to extend this approach to more than two classes. In this case rather than considering the distance of the centroids from each other, the distance of the centroids from some central point is used. If the average value of d_n

for all n is $\bar{d}$, then the distance of the centroid for each group g from this point is $\bar{d} - \bar{d}_g$. Considering the independent variables, a vector $\bar{X}$ can be defined as the average values of $X_{1,n}, X_{2,n}, \ldots, X_{M,n}$ for all firms, whilst a vector $\bar{X}_g$ can be defined as the average values of $X_{1,n}, X_{2,n}, \ldots, X_{M,n}$ for group g. The covariance of the group averages of these observations can be defined as:

$$\Sigma_G = \frac{1}{G} \sum_{g=1}^{G} (\bar{X} - \bar{X}_g)(\bar{X} - \bar{X}_g)'. \qquad (11.64)$$

This means that a new function needs to be maximised to give maximum separation between groups whilst minimising separation within groups:

$$D_{MDA} = \frac{\beta' \Sigma_G \beta}{\beta' \Sigma \beta}. \qquad (11.65)$$

11.5.4 The k-nearest neighbour approach

One of the main purposes of discriminant analysis is to find a way of scoring new observations to determine the group to which they belong. However, another approach is to use a non-parametric approach, and to consider which observations lie 'nearby'. This is the k-nearest neighbour (kNN) approach. It involves considering the characteristics of a number of individuals or firms that fall into one of two groups. These firms or groups form the training set used to parameterise the model. As before, these could easily be solvent and insolvent firms. When a new firm is considered, its distance from a number (k) of neighbours is assessed using some approach, and the proportion of these neighbours that have subsequently become insolvent gives an indication of the likelihood that this firm will also fail. The kNN approach is shown graphically in Figure 11.9.

The most appropriate measure of distance when M characteristics are being considered is the Mahalanobis distance, discussed in the context of testing for multivariate normality. The Mahalanobis distance between a new firm Y and one of the existing firms X_n measured using m characteristics of those firms, where $m = 1, 2, \ldots, M$, is:

$$D_{X_n} = \sqrt{(Y - X_n)' S^{-1} (Y - X_n)}. \qquad (11.66)$$

In this expression, Y and X_n are column vectors of length M containing the values of the M characteristics such as leverage, earnings cover and so on.

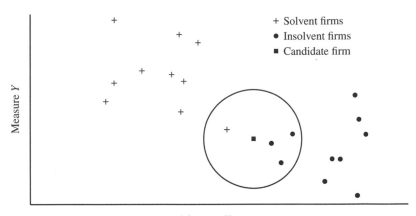

Figure 11.9 k-nearest neighbour approach

The matrix S contains estimates of the covariances between the two firms for these characteristics, calculated using historical data.

The Mahalanobis distance from firm Y must be calculated for all N firms, X_n, to see which the k nearest neighbours are. The score is then calculated based on the combination of the group to which X_n belongs and the distance of X_n from Y. Say, for example, an insolvent firm is given a score of one and a solvent firm is given a score of zero, k is taken to be 6 and firms X_1 to X_6 have the smallest Mahalanobis distances. In this case, the score for firm X is:

$$kNN_Y = \frac{\sum_{n=1}^{6} I(X_n)/D_{X_n}}{\sum_{n=1}^{6} 1/D_{X_n}}, \qquad (11.67)$$

where $I(X_n)$ is an indicator function which is one if X_n is insolvent and zero otherwise.

In the same way that there are a number of ways of calculating the distances between firms, there are also a number of ways of determining the optimal value of k. One intuitively appealing approach is to calculate the score for all firms whose outcome is already known using a range of values of k. For each firm, kNN_{X_i} is calculated using Equation (11.67) but excluding the X_n for $n = i$. For each i, the statistic $[kNN_{X_i} - I(X_n)]^2$ is calculated. These are summed over all $i = 1, 2, \ldots, N$, with the total being recorded for each value of k. The value of k used – the number of nearest neighbours – is the one that minimises $\sum_{i=1}^{N}[kNN_{X_i} - I(X_n)]^2$. However, this process can involve calculating a huge number of distances, so if this process is being used for example to assess a commercial bank's borrowers, it can quickly become unwieldy.

11.5.5 Support vector machines

Another approach to classifying data is to find the best way of separating two groups of data using a line (for two variables), plane (for three variables) or hyperplane (for more than three variables). The functions used to separate data in this way are known as support vector networks (SVMs).

Linear SVMs

A linear SVM uses a straight line – or its higher-dimensional alternative – to best separate two groups according to two or more measures. Consider again the two groups of solvent and insolvent firms, and two variables such as leverage and earnings cover.

In Figure (11.10), the two groups can clearly be divided by a single line. However, more than one line can divide the points into two discrete groups. Which is the best dividing line? One approach is to use tangents to each dataset. If pairs of parallel tangents are considered, then the *best* separating line can be defined as the line midway between the most separated parallel tangents.

This criterion can extended into higher dimensions, and expressed in mathematical terms. Consider a column vector X_h giving the co-ordinates of a point on a hyperplane. For a firm n, these co-ordinates could be the values of M financial ratios, each one corresponding to a dimension. A hyperplane can be defined as:

$$\beta' X_h + \beta_0 = 0, \tag{11.68}$$

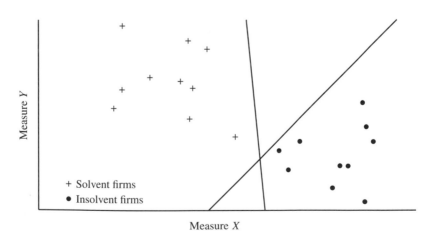

Figure 11.10 Linearly separable data – various separations

where β is the vector of M parameters and β_0 is a constant. The value of the expression $\beta' X_n + \beta_0$ can be evaluated for any vector of observations, X_n, for firm n. These firms constitute the training set used to parameterise the model. If $\beta' X_n + \beta_0 > 0$, for all firms in one group and $\beta' X_n + \beta_0 < 0$ for the other for a vector of parameters β, then Equation (11.68) can be said to be a separating hyperplane. To simplify this, a function $J(X_n)$ can be defined such that $J(X_n) = 1$ if firm n belongs to the first group, whilst $J(X_n) = -1$ if it belongs to the second group. This means that the separating hyperplane can be redefined as one where $J(X_n)(\beta' X_n + \beta_0) > 0$ for all n.

If the two groups are separable, then the degree of separation can be improved by finding a positive parameter C for which $\beta' X_n + \beta_0 \geq C$ for all firms in one group and $\beta' X_n + \beta_0 \leq -C$ for the other group. If the function $J(X_n)$ is used, this criterion can be redefined as $J(X_n)(\beta' X_n + \beta_0) \geq C$. The largest value of C for which this is true gives the best separating hyperplane, as described above. The parameters that provide this are the ones that minimise $\|\beta\|$, the norm of the vector β given by $\sqrt{\beta'\beta}$, subject to $J(X_n)(\beta' X_n + \beta_0) \geq C$. This is shown in Figure 11.11.

Once the parameters for a best separating hyperplane have been established, the vector of observations from a new firm can be input into the expression $\beta' X_i + \beta_0$. If the result is positive, then the firm can be included in the first group, whilst if is negative, then it can be included in the second.

Linear SVMs can still be used if the data are not linearly separable, but the constraints must be changed. In particular, rather than separation being given

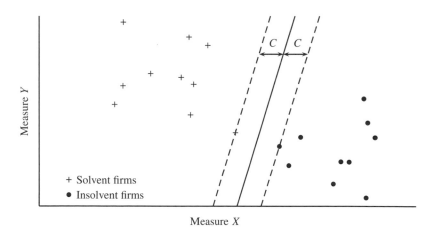

Figure 11.11 Best separating line

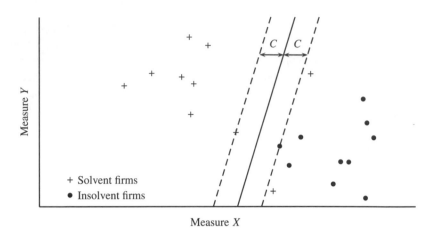

Figure 11.12 Best separating line – data not linearly separable

by the parameters that mean $J(X_n)(\beta' X_n + \beta_0) > 0$, an element of fuzziness is introduced, making the parameters satisfying $J(X_n)(\beta' X_n + \beta_0) > 1 - F_n$, where $F_n > 0$ is the degree of fuzziness for firm n. A penalty for this fuzziness can be introduced as $\Sigma_{n=1}^{N} f(F_n)$, where $f(F_n)$ is some function of the fuzziness measure F_n. The penalty function will typically be simply a scalar multiple of F_n. This means that the parameters are then the ones that minimise $\|\beta\| + \Sigma_{n=1}^{N} f(F_n)$ subject to $J(X_n)(\beta' X_n + \beta_0) > 1 - F_n$. The best separating hyperplane in this case remains $\beta' X_h + \beta_0 = 0$. This is shown in Figure 11.12.

The fuzziness parameters can also be used to define a zone of ignorance, such that any observation for a new firm within $\max(F_i)$ of the best separating hyperplane can be said to be unclassifiable. If many non-zero values of F_n are needed, then the distribution of these values can be used to determine a confidence interval for the zone of ignorance.

Non-linear SVMs

An alternative approach to using fuzziness to divide data that cannot be separated linearly is to use a non-linear SVM. In graphical terms, this means that rather than a straight line, a curve is used to separate data, as shown in Figure 11.13. This curve could be a kernel or a polynomial function. In mathematical terms, it means that the M elements of a vector of points on the hyperplane, X_h, which is denoted $X_{h,1}, X_{h,2}, \ldots, X_{h,M}$, are replaced by some function of each, given by $f_1(X_{h,1}), f_2(X_{h,2}), \ldots, f_M(X_{h,M})$.

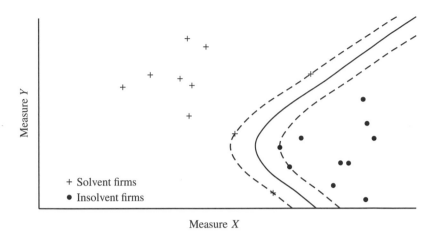

Figure 11.13 Best separating line – nonlinear support vector machine

Whilst it is possible in most cases to derive some form of separating hyperplane that always correctly classifies a set of training observations, there is a risk that the data will be over-fitted. The result can be that, whilst the training data are perfectly separated, new data will be misclassified.

11.6 Uncertainty

When fitting data to a model or a distribution, it is important to recognise that the fit might be incorrect. This can lead to greater certainty being ascribed to particular potential outcomes than is actually the case, resulting in suboptimal decisions being made. This is especially true if a model is being used to generate stochastic simulations.

There are three main sources of uncertainty:

- stochastic uncertainty;
- parameter uncertainty; and
- model uncertainty.

11.6.1 Stochastic uncertainty

Stochastic uncertainty occurs because only a finite number of observations are available. This aspect of uncertainty refers to the randomness in the observations themselves. As the number of observations increase, then the certainty in any model and its parameters also increases. However, stochastic uncertainty also exists in any outcomes predicted by a model. This uncertainty is reflected in the construction of stochastic models, discussed later.

11.6.2 Parameter uncertainty

Parameter uncertainty or risk refers to the use of inappropriate or inaccurate parameters or assumptions within models. As a result, incorrect or suboptimal decisions may be made. This uncertainty arises because the number of observations is finite. As a result, the parameters fitted to any model are not known with complete certainty. If projections are carried out under the assumption that stochastic volatility exists around unchanging parameters, then the range of projections will be too narrow.

There are a number of ways that parameter uncertainty can be allowed for. If least squares regression has been used and a covariance matrix for the parameters is available, then a multivariate normal distribution can be used to simulate the parameters which themselves are used in stochastic simulations. However, such covariances will not always be available – or relevant – when least squares regression has not been used to fit a model.

One approach to determining the confidence intervals for the parameters is to use the following process:

- fit a model to the data using the T data points available;
- simulate T data points using the model;
- re-fit the model to the simulated data points;
- record the parameter values; and
- repeat the process a large number of times, starting with the original data set each time.

This process gives a joint distribution for the parameters. This means that rather than using a single set of parameters to carry out the simulations, the simulated parameters can instead be used.

11.6.3 Model uncertainty

Model uncertainty or risk arises from the use of an inappropriate or inaccurate model when assessing or managing risks. However, the choice of model is not straightforward. When choosing a model, one of three assumptions must be made:

- that the true model or class of models is known;
- that the model used is an approximation to a known, more complex reality; or

- that the model used is an approximation to an unknown, more complex reality.

The third of these assumptions is the most common in financial modelling. This can lead to the wrong models being used for a number of reasons:

- the inappropriate projection of past trends into the future;
- the inappropriate selection of an underlying distribution; and
- the inappropriate number of parameters being chosen.

The inappropriate projection of past trends into the future has a number of causes. Errors in historical data can invalidate the fit of any model, but even if data are correct, they may be incomplete. A key example is in relation to insurance claims, which might be artificially low if there is no allowance for claims that have been incurred but not reported. It is also important to allow for any heterogeneity within the data. If a trend has more than one underlying driver, then these drivers should be identified and projected separately, allowing for dependencies between them. A good example is the improvement in mortality rates, which hides underlying trends for mortality improvements relating to various causes of death.

Even if the data are complete and correct, there is a risk that the distribution used to model them is inappropriate. The result can be that insufficient weight is given to the tails of a distribution, or that skew is not correctly allowed for. This is often due to there being insufficient observations to correctly determine the appropriate shape of the distribution. To determine the importance of distribution selection, it can be helpful to fit a range of distributions to a dataset.

Finally, the number of parameters chosen might be inappropriate. The broad principle used when choosing how many parameters to use is called the principle of parsimony. This states that where there is a choice between different fitted models, the optimal selection is the model with the fewest parameters. This reduces the degree of parameter estimation and should lead to more stable projections. However, a model with a small number of parameters may be over-simplified and reliant upon too many implicit assumptions, any of which could be inaccurate.

The conflict between goodness of fit and simplicity can be measured using a number of statistics that penalise the fit of a model for each increase in the number of parameters used. The adjusted R^2, Akaike information criterion and Bayesian information criterion all take the number of parameters into account to a greater or lesser extent. However, it is sensible to assess the parameters and the results using more than one model. If either change significantly from model to model, then this might be a cause for concern.

∮ 11.7 Credibility

When models are used to derive expected values for variables, the process used is often combined with an estimate of the variables directly from the data. For example, it might be possible to explain the underlying mortality rate for a group of annuitants in terms of a number of underlying risk factors, a process known as risk rating. However, the mortality experience of that group could also be used to explain the underlying mortality rate through a process known as experience rating. However, credibility analysis is by no means restricted to mortality risk, and the more general way in which experience and risk rating estimates are combined is covered here.

The combination of these estimates is carried out through the use of credibility. The broad approach is to derive a credibility-weighted estimate, $\hat{X}$, from the estimate calculated from historical experience, $\bar{X}$, and the estimate calculated from other sources, μ. The credibility given to the historical data is measured by Z, and all of these factors are linked as follows:

$$\hat{X} = Z\bar{X} + (1 - Z)\mu, \tag{11.69}$$

where $0 \leq Z \leq 1$. The greater the trust in the observed experience, the closer Z is to one; the more reliance that needs to be placed on other sources of information, the closer Z is to zero.

Three broad approaches to calculating credibility estimates are covered here:

- classical;
- Bühlmann; and
- Bayesian.

11.7.1 Classical credibility

The classical approach to credibility involves assessing the number of observations needed for a set of data to be fully credible, in other words for $Z = 1$. If there are fewer observations than this, then there is only partial credibility, with $0 < Z < 1$. Only if there are no historical data and only external sources of information are available – as might be the case with a new policyholder buying an insurance policy – will $Z = 0$.

Full credibility cannot exist, except at a given level of confidence for a given distance from the expected value. This means that full credibility exists only to the extent that the value calculated from historical data lies within a proportion p of the true value with a confidence level of $1 - \alpha$.

If an item such as the claim frequency or mortality rate is being considered, then the information needed depends on the approach being used. If deaths or claims are assumed to follow a binomial distribution, and the number of lives or policies is large enough for a normal approximation to be used to determine the level of full credibility, then both the number of lives or policies and the number of deaths or claims is needed. For example, let the claim rate estimated from observations, $\bar{X}$, be calculated as $\bar{X} = X/N$. The total number of claims is X, where $X = \sum_{n=1}^{N} \bar{X}_n$. In this expression, $\bar{X}_n$ is the average claim rate for policy n derived from historical data. The number of policies is given by N. The number of policies needed for full credibility is the smallest integer value of N satisfying the inequality:

$$\underbrace{\sqrt{N\bar{X}(1-\bar{X})}\Phi^{-1}\left[1-\left(\frac{\alpha}{2}\right)\right]}_{CI \quad \leq error} \leq pN\bar{X}. \tag{11.70}$$

The left-hand side of this expression gives the size of the confidence interval, calculated as the number of standard deviations for a given level of confidence, $\sqrt{N\bar{X}(1-\bar{X})}$ being the standard deviation. The term $\alpha/2$ is used because this is based on a two-tailed test, considering whether the number of claims is different from that expected. The right-hand side of Equation (11.70) gives the number of claims regarded as an acceptable margin of error, calculated as a percentage of the total number of claims, $N\bar{X} = X$. Abbreviating $\Phi^{-1}(1 - (\alpha/2))$ to Φ^{-1}, this expression can be rearranged and given in terms of N as:

$$N \geq \left(\frac{\Phi^{-1}}{p}\right)^2 \frac{1-\bar{X}}{\bar{X}}. \tag{11.71}$$

The lowest value of N for which this is true is N_F, the full credibility size of the population.

Example 11.7 From an initial population of 2,500 people, there are 175 deaths. Using a normal approximation to the binomial distribution, is this population large enough to give full credibility with a tolerance of 5% at the 90% level of significance? What is the smallest population that would give full credibility with a tolerance of 5% at the 90% level of significance?

The rate of mortality here is $125 \div 2,500 = 0.07$. The expected number of deaths is therefore $2,500 \times 0.07 = 175$ with a variance of $2,500 \times 0.07 \times (1-0.07) = 162.75$.

The $\Phi(-1.64) = 0.05$ and $\Phi(1.64) = 0.95$, so the 90% confidence interval for the expected rate of mortality is $175 \pm (1.64 \times \sqrt{118.75})$, or from
154 to 196 deaths. This is a range of 12% either side of the expected
number of deaths, so the population is not large enough to give full
credibility.

For full credibility, the confidence interval must be less than
the level of tolerance, so the population must be greater than
$(1.64/0.05)^2 \times (1 - 0.07)/0.07$. The lowest population size for which this
is true is 14,294.

If the number of deaths or claims is small in relation to the population or
number of policies, then a Poisson distribution can be assumed. This means
that if the number of lives or policies is large enough for a normal approximation to be used, then only the number of deaths or claims is needed to establish
whether full credibility has been achieved. Since under the Poisson distribution
the variance and the mean are identical, being $X = N\bar{X}$, Equation (11.70) can
be rewritten as:

$$\Phi^{-1}\sqrt{N\bar{X}} \le pN\bar{X}. \tag{11.72}$$

Reinstating X for $N\bar{X}$ and rearranging gives:

$$X \ge \left(\frac{\Phi^{-1}}{p}\right)^2. \tag{11.73}$$

The lowest value of X for which this is true is X_F, the full credibility number of claims or deaths. This expression can be used to construct a table of
the number of claims, deaths or more generally events required to give full
probability for given levels of confidence, α, and tolerance, p, as shown in
Table 11.1.

Of course, there will often be fewer than X_F events, so it will not be possible to estimate an item with full credibility. Consider, for example, the above
sample of N policyholders, and define the average number of claims per policyholder as $\bar{X}_n$. The expected rate of claims for all policyholders, μ, can be
estimated as $\bar{X} = \sum_{n=1}^{N} \bar{X}_n/N$. The variance of μ is equal to the variance of
$\sum_{n=1}^{N} \bar{X}_n/N$. This is equal to $1/N^2$ multiplied by the sum of the individual
variances. If the underlying Poisson mean for each policyholder is μ, then the
variance is $N\mu(1/N^2) = \mu/N$.

The Poisson mean, μ, can also be estimated from the total number of claims,
X, as $\mu = X/N$. Rearranging this to give $N = X/\mu$ and substituting this into
the above expression for the variance means that the variance of the claim rate
is also equal to μ^2/X, and the standard deviation is equal to $\mu/\sqrt{X}$.

Table 11.1. *Numbers of events required for full credibility*

		\multicolumn{7}{c}{p}					
		0.250	0.200	0.100	0.050	0.025	0.010	0.001
	0.250	22	34	133	530	2,118	13,234	1,323,304
	0.200	27	42	165	657	2,628	16,424	1,642,375
	0.100	44	68	271	1,083	4,329	27,056	2,705,544
α	0.050	62	97	385	1,537	6,147	38,415	3,841,459
	0.025	81	126	503	2,010	8,039	50,239	5,023,887
	0.010	107	166	664	2,654	10,616	66,349	6,634,897
	0.001	174	271	1,083	4,332	17,325	108,276	10,827,567

If the expected claim rate does not change, this means that the confidence interval for the claim rate is proportional only to $1/\sqrt{X}$. This is important when partial credibility is being considered, as one way of arriving at an estimate for the measure of partial credibility is to weight an estimate such that the variability is the same as if there were sufficient events for full credibility. This can be considered in terms of a confidence interval. The standard deviation of the estimated claim rate when there are X_F events, and therefore full credibility, is $\mu/\sqrt{X_F}$. If there is partial credibility since there are only X_P events where $X_P < X_F$, then the measure of credibility that would give the same standard deviation would be the value of Z for which $Z\mu/\sqrt{X_P} = \mu/\sqrt{X_F}$. In other words:

$$Z = \sqrt{\frac{N_P}{N_F}}. \tag{11.74}$$

As would be expected, the level of credibility increases with the number of events. However, the smooth increase turns into a horizontal line once full credibility has been reached. As can be seen, the level of full credibility is somewhat arbitrary. Various levels of classical credibility are shown in Figure 11.14. Other credibility measures aim to use less subjective approaches.

11.7.2 Bühlmann credibility

The basic formula for the Bühlmann estimate of credibility given N observations is:

$$Z = \frac{N}{N+K}, \tag{11.75}$$

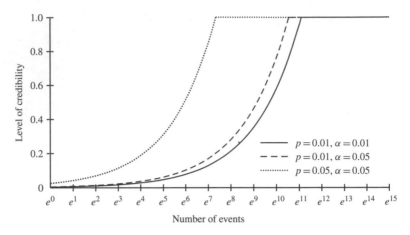

Figure 11.14 Classical credibility

where:

$$K = \frac{EPV}{VHM} \cdot \quad (11.76)$$

within group

among group

The term VHM is the variance of hypothetical means. Each hypothetical mean represents the average value for a particular combination or risk characteristics. For example, there might be several distinct groups of policyholders. The average claim rate for each group could be taken as the hypothetical mean, and the distribution of these averages would constitute the variance of hypothetical means. So if there were M types of policyholder, with claim rates of X_m and N_m policyholders in each type where $m = 1, 2, \ldots, M$, then the VHM would be calculated as:

$$VHM = \sum_{m=1}^{M} \frac{N_m}{N} X_m^2 - \left[\sum_{m=1}^{M} \frac{N_m}{N} X_m \right]^2, \quad (11.77)$$

where $N = \sum_{m=1}^{M} N_m$. The term EPV is the expected process variance. This captures the total uncertainty within each group. In aggregation, this is again weighted by the size of each group, so continuing the above example gives:

$$EPV = \sum_{m=1}^{M} \frac{N_m}{N} X_m (1 - X_m). \quad (11.78)$$

When added together, the EPV and the VHM give the total variance.

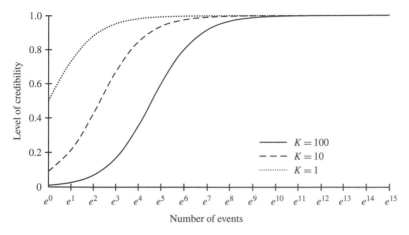

Figure 11.15 Bühlmann credibility

The Bühlmann credibility estimate also increases with the number of observations, but unlike the classical credibility estimate never reaches one as shown in Figure 11.15.

11.7.3 Bayesian credibility

Bayesian credibility derives from Bayes' theorem. This links the prior probabilities of two events and the conditional probability of one event on the other to give the posterior probability of the second event given that the first has occurred. In particular:

$$\Pr(X|Y) = \frac{\Pr(Y|X)\Pr(X)}{\Pr(Y)}. \tag{11.79}$$

So, for example, if:

- the probability that an individual smokes – the prior probability of X, $\Pr(X)$ – is 10%;
- the probability that an individual has life insurance – the prior probability of Y, $\Pr(Y)$ – is 20%;
- the probability that a smoker chosen at random has life insurance – the conditional probability of Y given X, $\Pr(Y|X)$ – is 30%; then
- the probability that an individual with life insurance is also a smoker – the posterior probability of X given Y $\Pr(X|Y)$ – is 30% × 10% ÷ 20% = 15%.

This can also be demonstrated by converting the probabilities into numbers, and seeing how 100 individuals would be categorised according to the above

Table 11.2. *Bayesian probabilities in terms of absolute numbers*

		Smoker		
		Yes	No	Total
	Yes	3	17	20
Insured	No	7	73	80
	Total	10	90	100

probabilities, as shown in Table 11.2. The first two probabilities, $\Pr(X)$ and $\Pr(X)$ give the row and column totals when multiplied by the total population of 100. Multiplying the third probability, $\Pr(Y|X)$, by the total of the first column gives a value for the top left cell, after which all other cells can be populated. However, simply dividing the same cell by the total of the first row gives $\Pr(X|Y)$.

The Bayesian approach can be applied to credibility. The item that credibility analysis is being used to derive is essentially the expected value of a quantity, such as a mortality or a claim rate, given a set of historical observations.

Whilst the algebra can get quite involved in some cases, there are instances where simple solutions can be found. These are where the prior and posterior distributions are conjugate.

Conjugate distributions

Distributions are conjugate if the posterior distribution is from the same family as the prior distribution. Two important examples are the beta-binomial and the gamma-Poisson cases.

Consider first the beta-binomial example applied to a portfolio of bonds. Assume that the number of bonds in this group is N, the total number of defaults is X and that defaults occur according to a binomial distribution.

The observed average rate of default is $\bar{X} = X/N$. However, the probability of default estimated from external factors is μ, which is assumed to have a beta distribution with parameters β_1 and β_2. This means that $E(\mu) = \beta_1/(\beta_1 + \beta_2)$. The probability of there being X defaults from a portfolio of N bonds can therefore be found by using $E(\mu) = \beta_1/(\beta_1 + \beta_2)$ in the calculation of the binomial probability formula. The result is that the posterior distribution of the expected probability of default also has a beta distribution with parameters

$X + \beta_1$ and $N - X + \beta_2$. This means that the expected number of defaults is:

$$E(\mu|X) = \frac{X + \beta_1}{N + \beta_1 + \beta_2}. \tag{11.80}$$

If $\bar{X}$ is substituted for X/N, and $E(\mu)$ is substituted for $\beta_1/(\beta_1 + \beta_2)$, this can be rewritten as:

$$E(\mu|X) = \frac{N}{N + \beta_1 + \beta_2}\bar{X} + \left(1 - \frac{N}{N + \beta_1 + \beta_2}\right)E(\mu)$$
$$= Z\bar{X} + (1 - Z)E(\mu), \tag{11.81}$$

where the credibility factor $Z = N/(N + \beta_1 + \beta_2)$. Note that since the statistical distribution of the external information is now important, $E(\mu)$ is now used in place of μ.

The Bayesian credibility estimate can be shown to be the same as the Bühlmann estimate. The EPV for a binomially distributed variable with an expected rate of μ is $E[\mu(1 - \mu)]$, whilst the VHM is equal to the variance of μ. Writing these in terms of β_1 and β_2 and substituting the results into the expression for the Bühlmann parameter K gives:

$$K = \frac{EPV}{VHM} = \frac{E[\mu(1 - \mu)]}{Var(\mu)} = \frac{E(\mu) - [E(\mu)]^2 - Var(\mu)}{Var(\mu)}$$
$$= \frac{[\beta_1/(\beta_1 + \beta_2)] - [\beta_1/(\beta_1 + \beta_2)]^2}{\beta_1\beta_2/(\beta_1 + \beta_2 + 1)(\beta_1 + \beta_2)^2} - 1$$
$$= \beta_1 + \beta_2. \tag{11.82}$$

Since the Bühlmann credibility formula is $Z = N/(N + K)$, substituting for K gives $Z = N/(N + \beta_1 + \beta_2)$, the same result as for the Bayesian approach.

Another useful combination of distributions is the gamma–Poisson conjugate pair. Assume that a variable such as the per-policy rate of insurance claims has a Poisson distribution with a mean of λ, and that λ itself has a gamma distribution with parameters β and γ. Assume also that the total number of claims observed from N policies is X. This time, the posterior distribution of λ given this information on claims – $E(\lambda|X)$ – is another gamma distribution whose parameters are now $X + \beta$ and $N + \gamma$. This means that the expected number of claims is:

$$E(\lambda|X) = \frac{X + \beta}{N + \gamma}. \tag{11.83}$$

Once again, some substitutions can convert this into a more recognisable credibility formula. If $\bar{X}$ is substituted for X/N, and $E(\lambda)$ is substituted for

β/γ, this can be rewritten as:

$$E(\lambda|X) = \frac{N}{N+\gamma}\bar{X} + \left(1 - \frac{N}{N+\gamma}\right)E(\lambda)$$
$$= Z\bar{X} + (1-Z)E(\mu), \tag{11.84}$$

where the credibility factor $Z = N/(N+\gamma)$. This too is equal to the Bühlmann credibility estimate, as can be shown by substituting $E(\lambda)$, the expected value of the Poisson mean (and variance), for the EPV and $Var(\lambda)$ for the VHM. For a gamma distribution with parameters β and γ, $E(\lambda) = \beta/\gamma$, whilst $Var(\lambda) = \beta/\gamma^2$. Substituting these into the expression for the Bühlmann parameter K gives:

$$K = \frac{EPV}{VHM} = \frac{E(\lambda)}{Var(\lambda)}$$
$$= \frac{\beta/\gamma}{\beta/\gamma^2} = \gamma. \tag{11.85}$$

Since the Bühlmann credibility formula is $Z = N/(N+K)$, substituting for K gives $Z = N/(N+\gamma)$, the same result as for the Bayesian approach.

11.8 Model validation

In all aspects of modelling, it is important to test the results of a model to ensure that they give reasonable results. Ideally, a model should be fitted to one subset of data and then tested on another independent sample of comparable size – if a model is tested using the data with which it was parameterised, then very little is proved about the model's effectiveness.

backtest

There are two types of testing that can be used, one in relation to time series and the other in relation to cross-sectional data.

11.8.1 Time series models

The testing process used in relation to time series models is known as back-testing. This involves fitting a model to data for one period, then seeing how well the model performs in a subsequent period. For example, a time series model might be used to try and predict equity returns based on a series of macro-economic variables for the period 1990–1999. Values for the same macro-economic variables could then be input into the model for the period 2000–2009, and the predicted equity returns compared with those observed for the same period.

This type of back-testing is particularly popular for testing trading strategies. In particular, it is intended to show that any anomalies do indeed offer an opportunity to make profitable trades, and are not just examples of temporary mis-pricing.

11.8.2 Cross-sectional models

For cross-sectional models where the dependent variable is a value, a similar approach to the back-testing method can be used. In this case, the data can be split into two groups rather than two time periods.

If the model is intended to classify firms or individuals into different groups, then a training set is used to provide the model parameters. The model can then be fitted to an independent data set to see how accurately it distinguishes between the various categories given the observations.

When testing cross-sectional models in this way it is important to ensure that there are no time effects – such as the impact of inflation on the data – that might result in the model appearing to be more accurate than is really the case.

11.9 Further reading

Many of the techniques discussed here are covered in more detail in Greene (2003) and similar books. However, more detail on particular areas is available in other texts.

Regressions are covered in Greene (2003) as well as countless books on econometrics such as Johnston and Dinardo (1997). In addition, Frees (2010) explores regression in an exclusively actuarial and financial context. Rebonato (1998) gives a good description of principal components analysis, whilst its practical application is also well described by Wilmott (2000).

The analysis of cross-sectional data is described more fully by Wooldridge (2002), and whilst Greene (2003) describes some smoothing techniques, penalised splines are best described by Eilers and Marx (2010), whilst smoothing across two rather than one dimension is discussed by Durbán et al. (2002).

Much of the recent work on data classification has been carried out in the context of credit modelling. As such, de Servigny and Renault (2004) – which describes a range of models in that context – gives a good overview. GLMs, with particular reference to insurance data, are discussed by de Jong and Heller (2008), whilst survival models and other approaches to dealing with life-contingent risks are covered in detail by Dickson et al. (2009).

Credibility is dealt with in detail by Bühlmann and Gisler (2005).

12

Extreme value theory

12.1 Introduction

In the above analysis, there is an implicit assumption that the distributions will be fitted to an entire dataset. However, when managing risk it is often the extreme scenarios that will be of most interest. There are two broad approaches to modelling such extreme events: the generalised extreme value distribution and the generalised Pareto distribution.

| 12.2 The generalised extreme value distribution

So far, the analysis has concentrated on distributions that relate to the full range of data available, or to the tail of a sample of data. However, another approach is to consider the distribution of the highest value for each of a number of tranches of data. This is the area of generalised extreme value theory. The starting point here is to consider the maximum observations from each of a sample of independent, identically distributed random variables, X_M. As the size of a sample increases, the distribution of the maximum observation $H(x)$ converges to the generalised extreme value (GEV) distribution. The cumulative distribution function is shown below in Equation (12.1):

$$H(x) = \Pr(X_M \leq x) = \begin{cases} e^{-\left(1+\gamma\frac{x-\alpha}{\beta}\right)^{-\frac{1}{\gamma}}} & \text{if } \gamma \neq 0; \\ \\ e^{-e^{-\left(\frac{x-\alpha}{\beta}\right)}} & \text{if } \gamma = 0. \end{cases} \qquad (12.1)$$

In this formulation, α and β are the location and scale parameters, analogous to the mean and standard deviation for the whole distribution. As with the mean and standard deviation, α can take any value, whilst β must be positive.

The value for which the expression is evaluated, x, must be greater or equal to α.

The parameter γ is the shape parameter of the distribution. With the GEV distribution, this parameter determines the range of distributions to which the extreme values belong. It does this by giving a particular distribution that has the same shape as the tail of a number of other distributions:

If $\alpha > 0$, then the distribution is a Fréchet-type GEV distribution. The Fréchet-type GEV distribution has a tail that follows a power law. This means that the extreme values could in fact have come from Student's t-distribution, the Pareto distribution or the Lévy distribution. Which of these distributions the full dataset might follow is irrelevant: the behaviour of observations in the tail – which is the important thing – will be the same.

If $\alpha = 0$, then the distribution is a Gumbel-type GEV distribution. Here, the tail will be exponential, as with the normal and gamma distributions and their close relatives.

If $\alpha < 0$, then the distribution is a Weibull-type GEV distribution. This has a tail that falls off so quickly that there is actually a finite right endpoint to the distribution, as with the beta, uniform and triangular distributions. Given that EVT is used when there is concern about extreme observations, this suggests that Weibull-type GEV distribution is of little interest in this respect.

A 'standard' GEV distribution can be created by setting $\alpha = 0$ and $\beta = 1$, as shown in Equation (12.2). The cumulative distributions for Fréchet-, Gumbel- and Weibull-types of this standard distribution are shown in Figure 12.1.

$$H(x) = \begin{cases} e^{-(1+\gamma x)^{-\frac{1}{\gamma}}} & \text{if } \gamma \neq 0; \\ e^{e^{-x}} & \text{if } \gamma = 0. \end{cases} \quad (12.2)$$

It is straightforward to differentiate the GEV distribution function to give the density function, as shown for the standard distribution in Equation (12.3). This is helpful as it allows us to see more clearly the shape of the tails for different values of γ. Density functions are shown in Figure 12.2

$$h(x) = \begin{cases} (1+\gamma x)^{-\left(1+\frac{1}{\gamma}\right)} e^{-(1+\gamma x)^{-\frac{1}{\gamma}}} & \text{if } \gamma \neq 0; \\ e^{-(x+e^{-x})} & \text{if } \gamma = 0. \end{cases} \quad (12.3)$$

A confusing point to note is that the Weibull distribution does not necessarily have a tail that corresponds to a Weibull-type GEV distribution. This

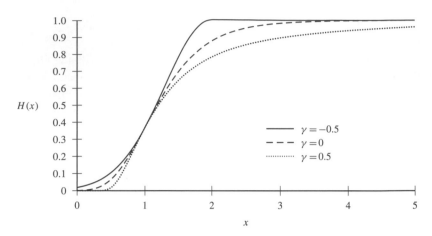

Figure 12.1 Various GEV distribution functions

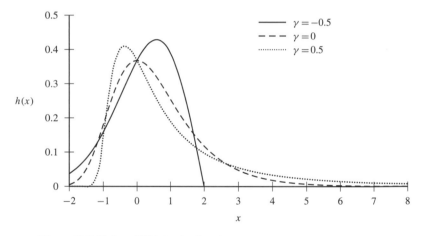

Figure 12.2 Various GEV density functions

is because there are a number of different versions of the Weibull distribution, only some of which have a finite end point; others – including the one
described in this book – have exponential tails.

To fit the GEV distribution, the raw data must be divided into equally sized
blocks. Then, extreme values are taken from each of the blocks. There are
two types of information that might be taken, and thus modelled. The first is
simply the highest observation in each block of data. This is known as the
return level approach, and the result is a distribution of the highest observation
per block size. So if each block contained a thousand observations, the result of

the analysis would be the distribution of the highest observation per thousand. The second approach is to set a level above which an observation could be regarded as extreme. Then, the number of observations in each block would be counted and modelled using a GEV distribution. In this case, if each block contained a thousand observations, the result would be the distribution of the rate of extreme observations per thousand. This is known as the return period approach.

The size of the blocks is crucial, and there is a compromise to be made. If a large number of blocks is used, then this means that there are fewer observations in each block. If the return level approach is used, this translates to less information about extreme values – a rate per hundred observations does not give as much information about what is 'extreme' as a rate per thousand. However, the large number of blocks means a large number of 'extreme' observations, so the variance of the parameter estimates is lower. If, on the other hand, fewer and larger blocks of data are used, then the information in each group about what is extreme is greater under the return level approach. However, with fewer blocks the variance of the parameter estimates is higher.

This can be seen in Figure 12.3. The first column of numbers shows the return level approach calculated using a block size of five. The result is the distribution of one in five events. The third column divides the data into only two blocks. The result is information on the distribution of more extreme one-in-ten events, but the distribution is based on only two observations rather than four. The choice of block size appears to be less important for the return period approach, since the total number of extreme events is five in both column 2 and column 4. However, since the result is divided into the number of observation per blocks, a similar issue arises when the parameters for the GEV distribution are being calculated.

A major drawback of the GEV approach is that by using only the largest value or values in each block of data, it ignores a lot of potentially useful information. For example, if the return level approach is used and there are a thousand observations per block, then 99.9% of the information is discarded. For this reason, the generalised Pareto distribution is more commonly used.

12.3 The generalised Pareto distribution

The generalised Pareto distribution has been described already, but it is actually an important limiting distribution. In particular, consider $X - u$, the distribution of a random variable, X, in excess of a fixed hurdle, u, given that X is greater than u. If the observations are independent and identically distributed,

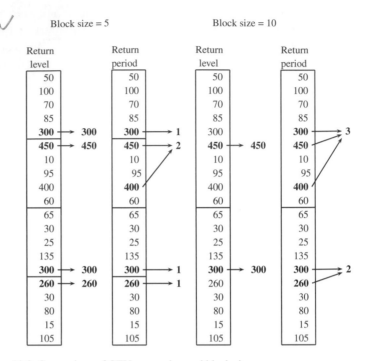

Figure 12.3 Comparison of GEV approaches and block sizes

then, as the threshold u increases, the conditional loss distribution – whatever the underlying distribution of the data – converges to a generalised Pareto distribution. The conditional cumulative distribution function, $G(x)$, is shown in Equation (12.4):

$$G(x) = \Pr(X - u \le x \mid X > u) = \frac{F(x+u) - F(u)}{1 - F(u)}$$

$$= \begin{cases} 1 - \left(1 + \dfrac{x}{\beta\gamma}\right)^{-\gamma} & \text{if } \gamma \neq 0; \\[2ex] 1 - e^{-\frac{x}{\beta}} & \text{if } \gamma = 0. \end{cases}$$

$$(12.4)$$

As discussed earlier, γ and β are the shape and scale parameters and whilst β must be positive, γ can take any value. If $\gamma = 0$, the formula reduces to the exponential distribution; if $\gamma > 0$, the result is the Pareto distribution, which

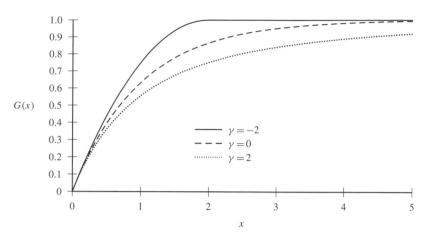

Figure 12.4 Various generalised Pareto distribution functions

follows a power law; and if $\gamma < 0$, x not only has a lower bound of zero, but also an upper bound of $-\beta\gamma$.

As with the GEV distribution, a standardised version of the generalised Pareto distribution can be defined by setting $\beta = 1$. This gives the conditional distribution shown in Equation (12.5). Different cumulative distribution functions are shown in Figure 12.4.

$$G(x) = \begin{cases} 1 - \left(1 + \dfrac{x}{\gamma}\right)^{-\gamma} & \text{if } \gamma \neq 0; \\ \\ 1 - e^{-x} & \text{if } \gamma = 0. \end{cases} \tag{12.5}$$

The generalised Pareto distribution function can be differentiated to give the density function, which gives a clearer idea of the shapes of the distribution. It is defined in Equation (12.6), with the result being shown for different values of γ in Figure 12.5.

$$g(x) = \begin{cases} \left(1 + \dfrac{x}{\gamma}\right)^{-(1+\gamma)} & \text{if } \gamma \neq 0; \\ \\ e^{-x} & \text{if } \gamma = 0. \end{cases} \tag{12.6}$$

The key with all distributions where only the tail is being considered is to choose the correct threshold. If it is too high, then there will be insufficient data to parameterise the distribution; however, if it is too low, then is it not just

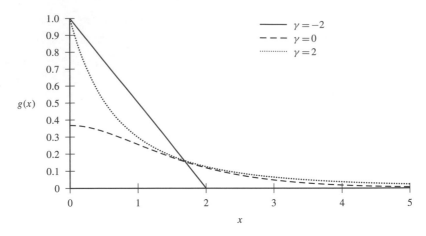

Figure 12.5 Various generalised Pareto density functions

the tail that is being considered. This is particularly important for the generalised Pareto distribution, which is only the limit of the conditional distribution
described if u is infinite, so is only a good approximation if u is sufficiently
high. In some cases, the value of the threshold will be clear from the context of
the work being done, but if it is not, then a suitable compromise between these
competing considerations will be needed.

One approach to choosing the threshold, u, is to consider the distribution of
the empirical mean excess function, $e(u)$, as u increases. This is defined as:

$$e(u) = \frac{\sum_{n=1}^{N}(X_n - u)I(X_n > u)}{\sum_{n=1}^{N} I(X_n > u)}, \tag{12.7}$$

where $I(X_n > u)$ is an indicator function that is equal to one if $X_n > u$ and zero
otherwise. The way in which $e(u)$ changes as u increases gives an indication
of whether the data being modelled is actually from the tail of the distribution
or not. Consider a distribution such as the normal distribution. As observations
move from the centre of the distribution to the right, the gradient of the distribution starts to decrease. However, after a time, it begins to flatten out, as
observations move from the body to the tail. This means that if $e(u)$ is plotted
against u, the value of $e(u)$ will initially fall sharply before levelling off as the
tail is approached. The value of $e(u)$ plotted against u for the normal distribution is shown in Figure 12.6, and it can be seen that the function becomes
increasingly linear in the tail of the distribution. However, this would simply
suggest that u should be as high as possible – in reality, another consideration is that real data are finite. Using too high a value of u will give values of

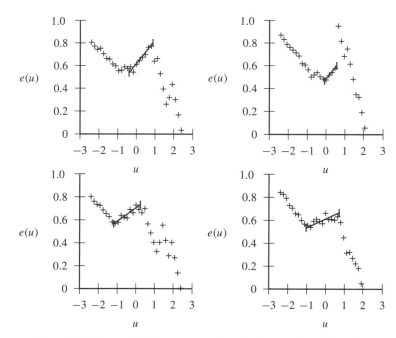

Figure 12.6 The empirical mean excess loss function – points of linearity

$e(u)$ that are no longer linear relative to u due to the sparse nature of the data. This means that when considering a body of data the appropriate value of u is not only one where $e(u)$ has *become* a linear function of u but also where it *remains* so. However, in practice it can be difficult to determine the value of u for which this is the case.

12.4 Further reading

The theoretical framework underlying extreme value theory is interesting, but involved. An alternative explanation of the principles can be found in Dowd (2005). Further details, including derivations of the distributions discussed here, can be found in McNeil *et al.* (2005), whilst de Haan and Ferreira (2006) give a comprehensive overview of this subject.

13

Modelling time series

13.1 Introduction

Many risks that are measured develop over time. As such, it is important that the ways in which these risks develop are correctly modelled. This means that a good understanding of time series analysis is needed.

13.2 Deterministic modelling

There are two broad types of model: deterministic and stochastic. At its most basic, deterministic modelling involves agreeing a single assumption for each variable for projection. The single assumption might even be limited to the data history, for example the average of the previous monthly observations over the last twenty years.

With deterministic approaches, prudence can be added only through margins in the assumptions used, or through changing the assumptions. A first stage might be to consider changing each underlying assumption in turn and noting the effect. This is known as sensitivity analysis. It is helpful in that it gives an idea of the sensitivity of a set of results to changes in each underlying factor, thus allowing significant exposures to particular risks to be recognised. However, variables rarely change individually in the real world. An approach that considers changes in all assumptions is therefore needed.

This leads us to scenario analysis. This is an extension of the deterministic approach where a small number of scenarios are evaluated using different pre-specified assumptions. The scenarios used might be based on previous situations, but it is important that they are not restricted to past experience – a range of possible futures is considered. This is the key advantage to scenario testing: a range of 'what if' scenarios can be tested, whether or not they have occurred in the past. However, this does not mean that all possible scenarios can be covered – the scenarios will always have been limited by what is thought to be plausible by the modeller. Another important limitation of scenario analysis is that it gives no indication of how likely a scenario is to occur. This is important when risk treatments are being considered, the cost will be considered in the context of the potential impact of the risk but also its likelihood.

The scenarios themselves might be given in quite general terms, such as 'high domestic inflation, high unemployment'. These scenarios need to be converted into assumptions for the variables of interest. It is important that each scenario is internally consistent and that the underlying assumptions reflect both the overall scenario and each other. Once any responses to risk have been taken, it is then important to carry out the scenario analysis again to ensure that the risk responses have had the desired effect. It is important that the effect of the scenario on the risk response is taken into account. It is also important to review both the types of scenarios and their assumptions on a regular basis.

It is possible to consider only extremely bad scenarios. Such an approach might be described as stress-testing. This has the advantage of focussing the mind on what might go wrong, but there are upside as well as downside risks. A strategy that minimises losses in the event of adverse scenarios might not be a good strategy if no profits are made in the good times. Positive strategies, and even middle-of-the-road outcomes, need to be considered when a strategy is being assessed.

Although stochastic modelling – described below – is increasingly popular, there is still a role for deterministic modelling. For example, regulators can find it useful to compare the effect of a range of consistent scenarios on a number of firms. Deterministic modelling is also more appropriate when there is insufficient information to build a complex stochastic model, as will generally be the case with new risks. Extreme events are also, by definition, so rare that the probabilities obtained from a stochastic model might not be reliable. Deterministic modelling can, on the other hand, allow consistent extreme scenarios to be considered without a need to assess their likelihood.

13.3 Stochastic modelling

Stochastic modelling is a far broader category than deterministic modelling. In a way, it seems similar to scenario testing but it differs from it in a key respect. In stochastic modelling, each run is drawn randomly from a distribution, rather than being predetermined. The broad relationships are defined, but the actual outcomes are down to chance.

13.3.1 Bootstrapping

In stochastic modelling, the first distinction is between bootstrapping, or re-sampling, and forward-looking approaches. For bootstrapping, all that is needed is a set of historical data for the variables being modelled. For example, historical monthly data for the last twenty years could again be used. However, rather than simply using this as a single 'run' of data, modelling is carried out by randomly selecting a slice of data, in this case the data from a particular month. This forms the first observation. This observation is then 'replaced' and another month is randomly chosen. This means that a relatively small data set can be used to generate a large number of random observations.

The main advantage of bootstrapping is that the underlying characteristics of the data and linkages between data series are captured without having to resort to parametrisation. However, any inter-temporal links in the data, such as serial correlation, are lost, and there is an implicit assumption that the future will be like the past. This assumption is not necessarily valid. Bootstrapping is also difficult if there is limited history for a particular asset class.

[margin note: markov]
[margin note: ° future = same as past]

13.3.2 Forward-looking approaches

A forward-looking approach, on the other hand, determines the future distribution explicitly. Whilst this might be with the benefit of past data, the approach does not stick slavishly to the results of such observations. Forward-looking approaches also require another decision to be made, and that is whether to use a factor- or a data-based approach. The former looks at the factors that determine the observations. These factors are modelled and their relationship used to derive the results for the observation in question. The data-based approach starts from the premise that understanding the drivers of a dataset does not improve the understanding of the observations, and modelling the data directly gives superior results when compared with a factor-based approach (or comparable results with less effort). Whereas a factor-based approach may result in a correlation pattern emerging for related datasets, these linkages must be explicitly modelled with a data-based approach.

[margin note: driver vs statistical]

The factors underlying a model can be found through regression analysis. For example, if one group of variables (say returns on individual shares) were thought to depend on a small number of factors (say short-term interest rates, long-term interest rates and price inflation), linear regressions could be run for each share to find an appropriate model:

$$Y_{n,t} = \beta_{0,n} + \beta_{1,n}X_{1,t} + \beta_{2,n}X_{2,t} + \beta_{3,n}X_{3,t} + \epsilon_{n,t}, \qquad (13.1)$$

where Y_{n_t} is the value of variable n at time t (say company n's share price), $\beta_{0,n}$ is a constant for that variable, $X_{1,t}$, $X_{2,t}$ and $X_{3,t}$ are the values of the three underlying factors at time t (say short- and long-term interest rates together with price inflation), $\beta_{1,n}$, $\beta_{2,n}$ and $\beta_{3,n}$ are the weights of these factors for variable n, and $\epsilon_{n,t}$ is an error term representing the difference between the true value and its estimate.

Once this model has been fitted, the underlying factors are projected, and their values used to imply the values of the variables based on the factors.

Factor-based models can be structured in several layers as cascade models. For example:

- price inflation can be modelled as a random walk;
- short-term interest rates can be modelled as a random variable changing in response to price inflation;
- long-term interest rates can be modelled as a random variable changing partly in response to short-term interest rates;
- equity dividends can be modelled as a function of short-term interest rates and price inflation; and
- equity returns can be modelled as a function of short- and long-term interest rates and equity dividends.

The factor-based approach can lend itself to modelling inter-temporal relationships between variables, particularly if the linkages between the factors are not necessarily contemporaneous. Whilst this can also be done in data-based models, more preparation of the data is needed.

Even if a data-based approach is used, there may be some aspects of a model for which a factor-based approach remains appropriate. Derivatives, particularly options, where the relationship between the price of the instrument and that of the underlying is complex but defined, provide a prime example.

13.3.3 Random numbers

When carrying out stochastic simulations, random numbers are needed to provide the range of outcomes. However, the numbers provided in most computer programs are not truly random, but 'pseudo-random'. This means that whilst they might appear to follow no discernible pattern, there is an underlying mathematical process at work.

There are a number of properties that pseudo-random numbers produced for the purposes of simulation should have, that is they should:

- be replicable;
- have a long period;

- be uniformly distributed over a large number of dimensions; and
- exhibit no serial correlation.

It is important that a series of random numbers used for simulations can be replicated when required. Having such a series means that it is easy to check the results from a simulation since the results can be exactly reproduced. It also makes it easier to see the effects of any changes to the model.

If pseudo-numbers are replicable, then there is an increased risk that the series of random numbers will begin to repeat itself eventually. In order that this repetition does not invalidate any simulations, it is important that the period before which a series repeats itself is sufficiently long.

It is also important that the distribution of pseudo-random numbers is apparently random, not just in a single dimension, but also if the numbers are projected into more than one dimension – for example, if a single column of numbers is divided into two, three or more supposedly independent series.

Furthermore, it is not enough that the distribution of the numbers is random – there should also be no clear link between any pseudo-random number and the number previously generated. In other words, serial correlation should be absent.

One popular pseudo-random number generator (PRNG) is the Mersenne twister. This is based on the digits extracted from a very large Mersenne prime number (a number with calculated as $2^N - 1$). The outputs appear to be random using a wide range of tests, it has a very long period before repetition of $2^19937 - 1$ iterations (more than $43 \times 10^{6,000}$) and can generate a large series of pseudo-random numbers very quickly.

13.3.4 Market consistency

For any forward-looking approach, it is interesting to consider the extent to which the projections are market-consistent. At the most basic level, this might involve comparing expected values from a model with those seen in the market; however, it is also possible to derive implied volatility expectations and even implied correlations from option prices. This is not to say that these market-consistent figures are perfect. In particular, the impact of demand and supply can mean that market prices do not necessarily reflect sensible estimates of future values. This can be a result of persistent market features such as a liquidity premium for less-liquid asset classes, or it can occur during market stresses when forced sales depress prices of some assets. However, it is difficult to identify the extent to which market prices are different from economic values, so care should be taken if values that are not market consistent are used.

13.4 Time series processes

Whether a factor- or data-based approach is used, the way in which the series develops over time is a crucial part of the stochastic modelling process. There are a number of ways in which series can be modelled, from very simple processes to very complex ones. To begin, however, it is helpful to consider the concept of stationary and non-stationary processes.

13.4.1 Stationarity

Stationarity is an important concept in time series analysis, as it determines the extent to which past data can be used to make predictions about the future. A strictly stationary process is one where, if you take any two sets of data points from a single set of observations, the joint distribution of those two sets will not depend on which sets you choose. This can be put in mathematical terms by considering a set of observations, X_t, where $t = 1 \ldots T$. Two subsets of observations can be taken from the data, $X_r \ldots X_s$ and $X_{r+k} \ldots X_{s+k}$ where $r, s \geq 1$ and $r + k, s + k \leq T$. The set of data $X_r \ldots X_s$ has a joint distribution function $F(X_r, \ldots, X_s)$. If this distribution function is equal to $F(X_{r+k}, \ldots, X_{s+k})$ for all k, then the series is strictly stationary. If a process is strictly stationary, then its characteristics do not change over time, including the relationships between observations in different periods.

Strict stationarity is very restrictive. Many other series still have properties that make them easy to analyse without necessarily following the rigid rules of strict stationarity. Many of these will be weakly stationary of order n, where n is a positive integer. This means that a time series is stationary up to the nth moment, but not necessarily beyond. The most used form of weak stationarity is second-order, or covariance stationarity. Taking the above series, a covariance stationary process is one where each subset of observations has the same defined mean, and the same defined covariance with observations for a given lag. In mathematical terms, this means that $E(X_t)$ is fixed for all t, and that $E(X_t X_{t+k})$ is also fixed, depending only on k.

Some series are not stationary due to the presence of a fixed time trend. If the removal of this trend from the data results in a stationary series, then the series is said to be trend stationary. Similarly, whilst a series of observations might not be stationary, a series made up of the differences of the observations might be. Such a series is said to be difference stationary. Trend and difference stationary processes are described in more detail below.

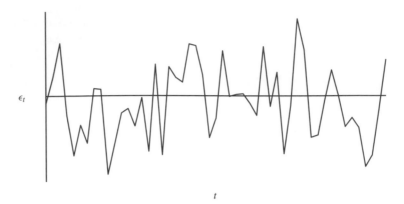

Figure 13.1 Strict white noise process

$$\mathcal{E}_i \sim N(0, \sigma^2) \, \forall \, i$$

13.4.2 White noise processes

The building block for many time series is known as a basic white noise process, ϵ_t, covering observations made at time t where $t = 1 \ldots T$. This is a stochastic process that oscillates around zero (the expected value of ϵ_t is zero) with a fixed variance (the expected value of ϵ_t^2 is equal to σ^2, which is fixed), and where no observation is correlated with any previous observation (the covariance of ϵ_s and ϵ_t is zero). This makes a white noise process at least covariance stationary. If the process is made up of independent, identically distributed random variables with a fixed, finite variance, then it is strictly stationary and known as a strict white noise process. Such as process is shown in Figure 13.1.

In itself, this process is not particularly representative of anything useful; however, it is the building block for many other processes, and it forms the series or errors or residuals in subsequent models.

A common assumption for the distribution of ϵ_t is that it follows a normal distribution, and this is important for many financial models.

13.4.3 Fixed values and trends

A number of data series might be assumed to oscillate around a value other than zero. Similarly, other series have trends, implying that they oscillate around a steadily changing value. For example, some asset prices might be assumed to increase linearly with time (although this would will be a gross simplification in most cases). The formulae for these situations are given in Equation (13.2) and Equation (13.3) respectively, with graphical representations shown in

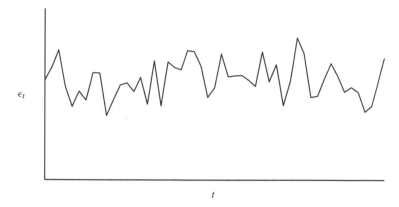

Figure 13.2 Stationary process

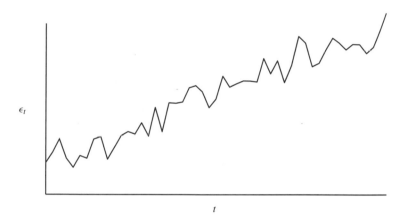

Figure 13.3 Trend-stationary process

Figure 13.2 and Figure 13.3:

$$X_t = \alpha_0 + \epsilon_t, \tag{13.2}$$

$$X_t = \alpha_0 + \alpha_1 t + \epsilon_t, \tag{13.3}$$

where X_t is the observation of the variable at time t. Equation (13.2) is a stationary process, whilst Equation (13.3) is trend stationary.

If the error term is assumed to have a normal distribution, then it might not be appropriate for X_t to represent the raw data being modelled. In particular, if the data being modelled are from a variable such as an asset price, which can take only positive values, then a common approach is for X_t to be the natural logarithm of the variable in question. This approach is used in many financial models.

13.4.4 Inter-temporal links

The simple processes above assume that there are no links between the values seen in one period and those observed in prior periods. However, processes with such links do exist, and can appear similar to the series described above.

Autoregressive processes

Consider the single-period autoregressive – or AR(1) – process in Equation (13.4):

$$X_t = \alpha_0 + \alpha_1 X_{t-1} + \epsilon_t. \tag{13.4}$$

This will give a similar pattern to Equation (13.2), except that rather than an oscillation around a fixed value, there will be a tendency for X_t to move towards – or away from – its previous value. The smaller the value of α_1, the more strongly the series is drawn to a fixed value. This tendency towards a fixed value is mean reversion. An important point to note is that for this series to tend towards a fixed value – in other words, to be at least covariance stationary – it is necessary that $|\alpha_1| < 1$. In this case, the variance of the series is $\sigma^2/(1 - \alpha_1^2)$ and the fixed value towards which the series tends is its mean, μ, which is equal to $\alpha_0/(1 - \alpha_1)$. These results are true regardless of the distribution of ϵ_t.

A potential issue with this formulation is that the series can easily return negative values. A slight modification can be added to reduce the chance of this happening, if it is important that only positive values are returned. In particular, the volatility can be modified so that it is proportional to the square root of the previous value of the series:

$$X_t = \alpha_0 + \alpha_1 X_{t-1} + \sqrt{X_{t-1}}\epsilon_t. \tag{13.5}$$

This means that as the value of the series falls, so does the volatility. This series is not *guaranteed* to remain positive unless the time scale becomes infinitesimally small, in which case negative values cannot occur if $X_0 \geq \sigma^2/2$

This more basic AR(1) process can be generalised to a p-period or AR(p) process, as shown in Equation (13.6):

$$X_t = \alpha_0 + \alpha_1 X_{t-1} + \alpha_2 X_{t-2} + \ldots + \alpha_p X_{t-p} + \epsilon_t. \tag{13.6}$$

The conditions for stationarity are more complicated here. First, a polynomial equation must be constructed from the parameters of the autoregressive function:

$$f(z) = 1 - \alpha_0 - \alpha_1 z - \alpha_2 z^2 - \ldots - \alpha_p z^p = 0. \qquad (13.7)$$

For the original equation to be at least a covariance stationary series, the roots of Equation (13.7) must 'lie outside the unit circle', meaning that the length of the p-dimensional vector must exceed one. This is easiest to appreciate using a smaller number of dimensions. For example, consider the situation where $p = 2$. This polynomial will have two roots, z_1 and z_2. If, when plotted on a two-dimensional chart, the co-ordinate described by the two roots lies outside a circle with a radius of one centred on the origin, then the roots lie outside the unit circle. If $p = 3$, then the three-dimensional co-ordinate must lie outside a unit sphere; alternatively, however, this criterion can be recast as being that each *pair* of co-ordinates – say z_1 and z_2 – must lie outside a two-dimensional circle with a radius of $\sqrt{1 - z_3^2}$. This is equivalent to looking at the relevant slice of the sphere. Numerically, the criterion can be fulfilled by simply squaring each root of the equation, summing the squares and square-rooting the sum. If the result is greater than one, then the roots lie outside the unit circle.

This is shown graphically in Figure 13.4. The triangle, with roots 0.8, 0.3 and 0.3 has a length of $0.8^2 + 0.3^2 + 0.3^2 = 0.82$, so lies within the unit sphere (if the sum is less than one, then the square root of the sum will be as well); however, the square, with roots 0.8, 0.2 and 0.9 has a length of $0.8^2 + 0.2^2 + 0.9^2 = 1.49$, so lies outside. Since the third co-ordinate is the same for both shapes, the first two co-ordinates can be compared with a circle of radius $\sqrt{1 - 0.9^2} = 0.6$. As expected, the triangle lies inside the circle, whilst the square is outside.

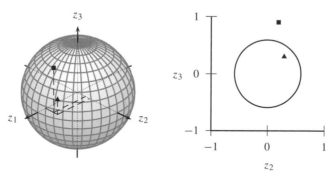

Figure 13.4 Points inside and outside a unit sphere

For an AR(1) model, if $|\alpha_1|$ is not between 1 and -1, then the series becomes unstable. However, there is an important situation where this instability forms a widely used process: if $\alpha_1 = 1$, the result is a random walk. If a constant, α_0, is also present, then the result is a random walk with drift, the drift being α_0 per period. This is shown in Equation (13.8):

$$X_t = \alpha_0 + X_{t-1} + \epsilon_t. \qquad (13.8)$$

Since $\alpha = 1$, this is not a stationary process; however, if it is transformed by defining $\Delta X_t = X_t - X_{t-1}$, then the resulting process is at least covariance stationary, as shown in Equation (13.9):

$$\Delta X_t = \alpha_0 + \epsilon_t. \qquad (13.9)$$

Integrated processes

If differencing is required once to arrive at a stationary series, as is the case here, the series is said to be difference stationary. More specifically, it can be referred to as an integrated process of order one, or I(1). An I(2) process is characterised as $\Delta X_t - \Delta X_{t-1}$, or $\Delta^2 X_t$, and this process can be generalised through the repetition of the differencing process d times to give an I(d) process, $\Delta^d X_t$.

A trend-stationary process such as Equation (13.3) and a difference-stationary process such as Equation (13.8) can be difficult to distinguish visually, as the two types of time series can look similar. This is clear from Figures 13.3 and 13.5. One way to test which process a time series follows is to use a Dickey–Fuller test (Dickey and Fuller, 1979, 1981). This involves

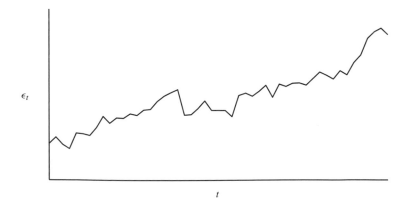

Figure 13.5 Difference-stationary process

regressing Δx_t on the lagged dependent variable and a time trend:

$$\Delta X_t = \alpha_0 + \alpha_1 t + \alpha_2 X_{t-1} + \epsilon_t. \tag{13.10}$$

The constant, α_0, and the time trend, t, are only included if they appear to be significant in basic regressions. Regressions are discussed in more detail in the section on fitting models. The Dickey–Fuller test involves testing whether α_2 is significantly different from zero by comparing the test statistic – α_2 divided by its standard error – with the critical values calculated by Dickey and Fuller. Special tables are needed for the critical values because if α_2 is close to one, the standard error reported in this test will be biased downwards. This means that a traditional t-test would fail to reject the null hypothesis.

Moving average processes

So far, the error term in the equation has continued to be simply a white-noise stochastic process. However, the changes from period to period can also be linked. Using Equation (13.2) as the starting point, a single-period moving average – or MA(1) is given as Equation (13.11):

$$X_t = \epsilon_t + \beta \epsilon_{t-1}. \tag{13.11}$$

This can be generalised to a q-period or MA(q) process:

$$X_t = \epsilon_t + \beta_1 \epsilon_{t-1} + \beta_2 \epsilon_{t-2} + \ldots + \beta_q \epsilon_{t-q}. \tag{13.12}$$

The assumption that the residuals in one period are not correlated with those in a prior period is an important part of many analyses. In particular, when considering financial time series the presence of serial correlation would imply that future asset prices depend at least in part on past returns. This in turn would allow the possibility of arbitrage – risk-free profits – which is prohibited in many economic models. Tests have therefore been developed to detect serial correlation in the residuals, in particular the Durbin–Watson test (Durbin and Watson, 1950, 1951). The Durbin–Watson test statistic, d, is calculated from the error terms as:

$$d = \frac{\sum_{t=2}^{T} (\epsilon_t - \epsilon_{t-1})^2}{\sum_{t=1}^{T} \epsilon_t^2}. \tag{13.13}$$

The null hypothesis is that $d = 2$ and that no serial correlation is present. If d is significantly less than 2, then there is significant positive serial correlation – in other words, for an MA(1) model, β is positive and successive observations are positively correlated; if d is significantly greater than 2, then there is significant negative serial correlation – β is negative and successive observations are negatively correlated.

In practice, two critical values for the Durbin–Watson test statistic are given, d_L and d_U, for each level of significance. If $d < d_L$, then there is evidence of significant positive serial correlation at that level; if $d > d_U$, then there is no evidence of significant positive serial correlation; and if $d_L < d < d_U$, then the test is inconclusive. For negative serial correlation, the same test is carried out with $4 - d$ replacing d.

If the test is being carried out with an ARMA model (described below), then the test statistic needs to be modified, and Durbin's (1970) h-statistic is used instead. This is calculated as:

$$h = \left(1 - \frac{d}{2}\right)\sqrt{\frac{T}{1 - Ts_{\alpha_1}^2}}, \qquad (13.14)$$

where $s_{\alpha_1}^2$ is the squared standard error for the coefficient of X_{t-1}. The distribution of the statistic h tends to towards a standard normal distribution as T tends to infinity.

It is possible to express an AR series in MA terms and *vice versa*. For example, consider again a simple AR(1) process:

$$X_t = \alpha_0 + \alpha_1 X_{t-1} + \epsilon_t. \qquad (13.15)$$

It is also possible to express the lagged term in the same form:

$$X_{t-1} = \alpha_0 + \alpha_1 X_{t-2} + \epsilon_{t-1}. \qquad (13.16)$$

Substituting this back into the first equation gives:

$$X_t = \alpha_0 + \alpha_0 \alpha_1 + \alpha_1^2 X_{t-2} + \epsilon_t + \alpha_1 \epsilon_{t-1}. \qquad (13.17)$$

This process can be continued indefinitely. Ultimately, providing the absolute value of α_1 is less than one, the coefficient on the lagged X_t term tends to zero and the constant term tends to $\alpha_0/(1 - \alpha_1)$. This means that an AR(1) process can also be described as the following infinite moving average process:

$$X_t = \frac{\alpha_0}{1 - \alpha_1} + \epsilon_t + \alpha_1 \epsilon_{t-1} + \alpha_1^2 \epsilon_{t-2} + \dots. \qquad (13.18)$$

A similar process can be used to convert a moving average process into an autoregressive one. The first stage is to take an MA(1) process:

$$X_t = \epsilon_t + \beta\epsilon_{t-1}. \tag{13.19}$$

Rearranging this to give an expression in terms of ϵ_t gives:

$$\epsilon_t = X_t - \beta\epsilon_{t-1}. \tag{13.20}$$

This expression can itself be lagged:

$$\epsilon_{t-1} = X_{t-1} - \beta\epsilon_{t-2}. \tag{13.21}$$

Substituting this back into the original MA(1) equation gives:

$$X_t = \epsilon_t + \beta X_{t-1} - \beta^2\epsilon_{t-2}. \tag{13.22}$$

This process can be carried on indefinitely, meaning that an MA(1) process can also be described as the following infinite autoregressive process:

$$X_t = \epsilon_t + \beta X_{t-1} - \beta^2 X_{t-2} + \beta^3 X_{t-3} + \dots. \tag{13.23}$$

ARIMA processes

Similar expressions can be derived for more general AR(p) and MA(q) processes. These two processes can also be found in a single expression, described as an autoregressive moving average (ARMA) process. A further layer – integration – may also be added to give an ARIMA (p,d,q) process. This is a process where an I (d) series can be modelled by an ARMA(p,q) series, as shown in Equation (13.24):

$$\Delta^d X_t = \alpha_0 + \alpha_1\Delta^d X_{t-1} + \alpha_2\Delta^d X_{t-2} + \dots + \alpha_p\Delta^d X_{t-p} +$$
$$\epsilon_t + \beta_1\epsilon_{t-1} + \beta_2\epsilon_{t-2} + \dots + \beta_q\epsilon_{t-q}, \tag{13.24}$$

or, more compactly:

$$\Delta^d X_t = \alpha_0 + \sum_{i=1}^{p}\Delta^d \alpha_i X_{t-i} + \epsilon_t + \sum_{j=1}^{q}\beta_j\epsilon_{t-j}. \tag{13.25}$$

Fitting ARIMA models

ARIMA models can be fitted – after suspected integration has been removed – by looking at the patterns of serial correlation in data. One approach to investigating these serial correlations is to use a correlogram. This compares the level of serial correlation at different lags in a dataset with the correlations implied by different ARMA models. The horizontal axis gives the lag at which the serial correlation is calculated, h, whilst the serial correlation for lag h is estimated as r_h:

$$r_h = \frac{\sum_{t=h+1}^{T}(X_t - \bar{X})(X_{t-h} - \bar{X})}{\sum_{t=1}^{T}(X_t - \bar{X})^2}. \tag{13.26}$$

If such a process is expressed as an MA(q) process using the techniques described above, then the implied correlation between successive observations for a lag of h, ρ_h, is given by the following expression:

$$\rho_h = \frac{\sum_{i=0}^{\infty} \beta_i \beta_{i+|h|}}{\sum_{i=0}^{\infty} \beta_i^2}. \tag{13.27}$$

In most cases, $\beta_0 = 1$, as is implicit in the moving average processes described above. Sample corellograms are shown in Figures 13.6 and 13.7, together with the raw data.

If there is any doubt as to whether the data are integrated and as to the order of integration, then it is worth constructing a separate correlogram for each degree of integration, d, since d is likely to be either zero, one or two.

Checking the fit of ARIMA models

Whilst this approach can give an indication of the level of serial correlation, the best way to test between candidate models is to compare statistics such as the AIC or BIC. If the models are nested, then the likelihood ratio test can also be used.

A more objective way of checking model fit is to consider the residuals from a fitted model. Consider, for example, an AR(1) model. If the estimates of α_0 and α_1 are $\hat{\alpha}_0$ and $\hat{\alpha}_1$, then the calculated residual at time t, $\hat{\epsilon}_t$, is given by:

$$\hat{\epsilon}_t = X_t - \hat{\alpha}_0 - \hat{\alpha}_1 X_{t-1}. \tag{13.28}$$

The $\hat{\epsilon}_0$ poses a problem, since it requires a value for X_{-1}. One solution is to set $\hat{\epsilon}_0 = 0$ and $X_{-1} = \bar{X}$. Similar approaches can be used for other ARIMA models with a greater lagging period. Once the calculated residuals have been calculated, then they can be tested, the test being whether they form a white noise process. A correlogram is again a useful tool here.

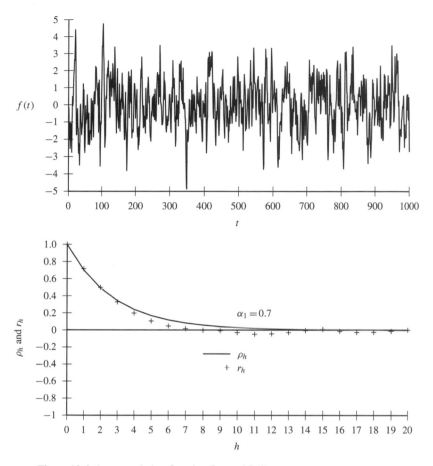

Figure 13.6 Autocorrelation function for an AR(1) process

Prediction with ARIMA processes

Consider an ARMA(1,1) model:

$$X_t = \alpha_0 + \alpha_1 X_{t-1} + \epsilon_t + \beta_1 \epsilon_{t-1}. \qquad (13.29)$$

Looking forward one period, this equation can be used to derive a value for X_{t+1}:

$$X_{t+1} = \alpha_0 + \alpha_1 X_t + \epsilon_{t+1} + \beta_1 \epsilon_t. \qquad (13.30)$$

If the values at time t have been observed, then taking expectations on both sides of this equation gives the expected value of X_{t+1}, $E(X_{t+1})$:

$$E(X_{t+1}) = \alpha_0 + \alpha_1 X_t + \beta_1 \epsilon_t, \qquad (13.31)$$

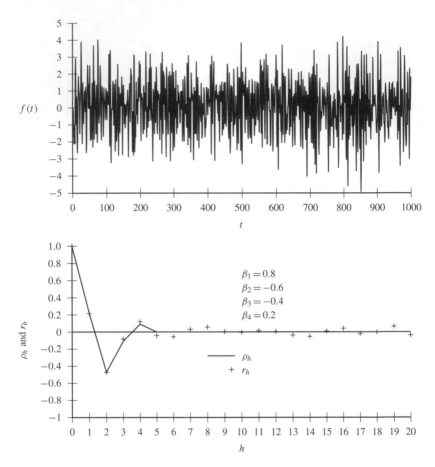

Figure 13.7 Autocorrelation function for an MA(4) process

since $E(\epsilon_{t+1}) = 0$. Looking ahead two periods, X_{t+2} is expressed as:

$$X_{t+2} = \alpha_0 + \alpha_1 X_{t+1} + \epsilon_{t+2} + \beta_1 \epsilon_{t+1}. \qquad (13.32)$$

Substituting the Equation (13.30) into Equation (13.32) gives:

$$X_{t+2} = \alpha_0 + \alpha_1(\alpha_0 + \alpha_1 X_t + \epsilon_{t+1} + \beta_1 \epsilon_t) + \epsilon_{t+2} + \beta_1 \epsilon_{t+1}. \qquad (13.33)$$

Taking expectations and simplifying gives:

$$E(X_{t+2}) = \alpha_0(1 + \alpha_1) + \alpha_1^2 X_t + \alpha_1 \beta_1 \epsilon_t. \qquad (13.34)$$

This can be generalised to give:

$$E(X_{t+h}) = \alpha_0(1 + \alpha_1 + \ldots + \alpha_1^{h-1}) + \alpha_1^h X_t + \alpha_1^{h-1} \beta_1 \epsilon_t$$

$$= \alpha_0 \sum_{i=0}^{h-1} \alpha_1^i + \alpha_1^h X_t + \alpha_1^{h-1} \beta_1 \epsilon_t. \tag{13.35}$$

13.4.5 Seasonality

Another important feature of some time series is seasonality. This means that there is regular seasonal variation in a statistical series, such that values are generally higher than an underlying trend at some points in the period and below it in others. The period in question can be a day, week, month or year. There are a number of ways that seasonality can be dealt with. One is through the use of an ARIMA model, since seasonality is essentially an autoregressive process. However, it is also possible to use seasonal dummy variables.

A dummy variable is a variable that takes the value of one if a certain condition holds and zero otherwise. For example, if annual seasonality were thought to exist in quarterly time series data that otherwise followed a simple trend, then the following model could be used:

$$X_t = \alpha_0 + \alpha_1 d_1 + \alpha_2 d_2 + \alpha_3 d_3 + \alpha_4 t + \epsilon_t. \tag{13.36}$$

In this equation, d_1 is a dummy variable taking a value of one if X_t is an observation from the first quarter and zero otherwise, d_2 is equal to one only if X_t relates to the second quarter and d_3 is one only if X_t relates to the third quarter. There are only three dummy variables since otherwise there would be an infinite number of parametrisations for this equation. In general terms, the number of dummy variables must be one less than the number of 'seasons'.

13.4.6 Structural breaks

However, there are potential complications. First, constant values might not be constant indefinitely, and trends may change. These changes are known as structural breaks. Two types of break are possible. The first is a step-change or jump in the value the series; the second is an alteration in the rate of change of the series.

An example of the step-change break is the jump diffusion model described by Merton (1976). A similar effect can be added to Equation (13.9), the random walk with drift, by adding a discrete random term that is usually zero – in other words, a Poisson variable. If the average size of the jump when it does occur is

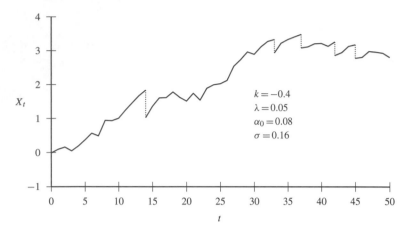

Figure 13.8 Jump diffusion model

k and $P_t(\lambda)$ is a Poisson random variable with mean Poisson mean λ, then the revised model is:

$$\Delta X_t = (\alpha_0 - \lambda k) + \epsilon_t + k P_t(\lambda). \tag{13.37}$$

The term λk is deducted from the drift term so that the overall average rate of drift stays at α_0. An example of the cumulative returns generated by such a model is given in Figure 13.8, with the dotted lines showing where Poisson-determined jumps occur. Here, the error term, ϵ_t, is assumed to be normally distributed with a fixed variance of σ^2. The model is specified to simulate random crashes of 40% occurring on average once every twenty years, following a random Poisson process. The long-term rate of increase in stock prices is assumed to be 8% with a volatility of 16% per annum.

The second type of structural break is more subtle, and is characterised by a change in the rate of change of a variable. In Equation (13.3), this could be a change in the time trend; in Equation (13.4), the mean to which a series reverts; and in Equation (13.9), a change in the rate of drift. The latter is shown graphically in Figure 13.9. Here, a series where the error term has a standard deviation is compared with an equivalent that has no volatility. From this it can be seen how hard changes in trend can be hard to spot. Structural breaks can, however, be identified using a test such as the Chow test (Chow, 1960). A Chow test involves splitting a set of observations into two subsets, one before and one after a supposed structural break. First a model is fitted to the full set of observations, and the residuals from this model are squared and summed to give SSR. Then the model is fitted to the first subset of observations to calculate the sum of squared residuals for this subset, SSR_1, after which the

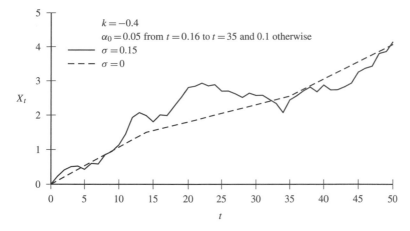

Figure 13.9 Changes in the trend rate of growth

same process is carried out for the second subset to give another sum of squared residuals, SSR_2. The test statistic, CT, is then calculated as:

$$CT = \frac{(SSR - (SSR_1 + SSR_2))/k}{(SSR_1 + SSR_2)/(N_1 + N_2 - 2k)}, \tag{13.38}$$

where N_1 and N_2 are the number of observations in the first and second subset respectively, and k is the number of parameters in the model, including any constant terms. This has an F-distribution with k and $N_1 + N_2 - 2k$ degrees of freedom. The null hypothesis under the Chow test is that the parameters for the two subsets are not significantly different from the parameters for the full dataset.

13.4.7 Heteroskedasticity

An important assumption for the white-noise process is that the variance does not change over time. However, this is arguably not true for many real-life time series, where broad patterns of stability are interspersed with periods of relatively high volatility. Data which do not have a constant level of volatility are said to be heteroskedastic.

ARCH models

One way of modelling this feature is to use an autoregressive conditional heteroskedasticity (ARCH) model. The first stage is to redefine X_t and ϵ_t as shown in Equation (13.39):

$$X_t = \epsilon_t = Z_t \sigma_t, \qquad \text{2 is strict} \tag{13.39}$$
$$\text{white noise}$$

where Z_t is a random variable following a strict white noise process with a mean of zero and a unit standard deviation at time t, and σ_t is the standard deviation also at time t, which unlike the fixed value of σ used earlier will now change over time. The simplest form is a single-period ARCH(1) model. Here, the standard deviation in time t is still linked to the long-term variance, σ^2, but also to the size of previous errors as shown in Equation (13.40):

$$\sigma_t^2 = \alpha_0 + \alpha_1 X_{t-1}^2. \quad = \alpha_o + \alpha_1 Z_{t-1}^2 \, \delta_{t-1}^2 \, (13.40)$$

In this equation, $\alpha_0 > 0$ and $\alpha_1 \geq 0$. This is at least a covariance stationary process – that is, it has a finite variance – if $\alpha_1 < 1$. In this case, the process has a variance of $\alpha_0/(1-\alpha_1)$. These results are true regardless of the distribution of X_t. However, the conditions for strict stationarity do depend on the distribution of the error terms. For example, an ARCH(1) process where Z_t follows a standard normal distribution is strictly stationary if $\alpha_1 < 2e^\eta \approx 3.562$, where η is the Euler–Mascheroni constant (which is equal to around 0.557). However, such a series is still not covariance stationary if $\alpha_1 \geq 1$. This leads to the interesting situation where a strictly stationary process with a finite variance is also weakly stationary, whilst a strictly stationary process with an infinite variance is not. Figure 13.10 shows three ARCH(1) processes whose error terms are normally distributed. The first two are strictly (but not covariance) stationary, whilst the third is not. The vertical axis is a defined as follows:

$$f(X_t) = \begin{cases} \ln(\ln(X_t)) & \text{if } X_t > 1; \\ -\ln(\ln(-X_t)) & \text{if } X_t < -1; \\ 0 & \text{otherwise.} \end{cases} \quad (13.41)$$

It is interesting to consider the higher moments of an ARCH(1) process. A strictly stationary ARCH(1) process has finite moments of order $2m$ if $E(Z_t^{2m}) < \infty$ and $\alpha_1 < E(Z_t^{2m})^{-1/m}$. In this case, the excess kurtosis, κ, can be calculated as:

$$\kappa = \frac{E(Z_t^4)(1-\alpha_1^2)}{1-\alpha_1^2 E(Z_t^4)} - 3. \quad (13.42)$$

Further lagged error terms can be added to an ARCH(1) process to create an ARCH(p) model:

$$\sigma_t^2 = \alpha_0 + \alpha_1 X_{t-1}^2 + \alpha_2 X_{t-2}^2 + \ldots + \alpha_p X_{t-p}^2. \quad (13.43)$$

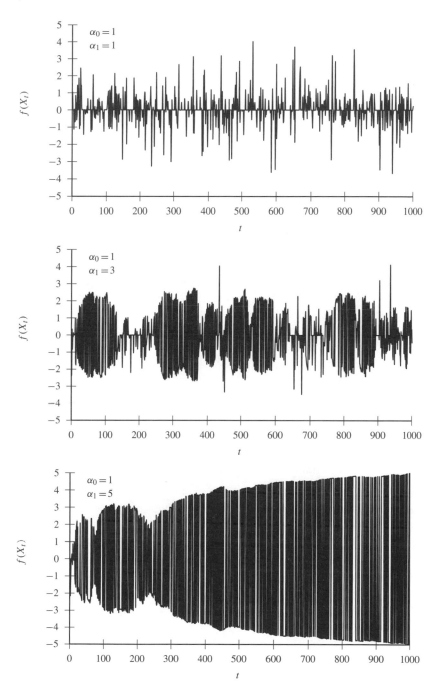

Figure 13.10 Various ARCH(1) processes

Here, $\alpha_0 > 0$ and $\alpha_1, \alpha_2, \ldots, \alpha_p \geq 0$. As for an AR($p$) model, this series is covariance stationary only if the roots of the polynomial constructed from $\alpha_1, \alpha_2, \ldots, \alpha_p$ lie outside the unit circle.

GARCH models

The logical extension of an ARCH model is known as a generalised autoregressive conditional heteroskedastic (GARCH) model. Starting again with the simplest form, a GARCH(1,1) process is defined as:

$$\sigma_t^2 = \alpha_0 + \alpha_1 X_{t-1}^2 + \beta_1 \sigma_{t-1}^2. \tag{13.44}$$

Since $X_t = Z_t \sigma_t$, this can also be written:

$$\sigma_t^2 = \alpha_0 + \left(\alpha_1 Z_{t-1}^2 + \beta_1\right)\sigma_{t-1}^2. \tag{13.45}$$

A GARCH(1,1) series will be covariance stationary if $\alpha_1 + \beta_1 < 1$, and the variance in this case will be $\alpha_0/(1 - \alpha_1 - \beta_1)$. If $E((\alpha_1 Z_t^2 + \beta_1)^2) < 1$, then the excess kurtosis of this series can be calculated as:

$$\kappa = \frac{E(Z_t^4)(1 - (\alpha_1 + \beta_1)^2)}{1 - (\alpha_1 + \beta_1)^2 - (E(Z_t^4) - 1)\alpha_1^2} - 3. \tag{13.46}$$

In practice, a GARCH(1,1) model as shown in Equation (13.44) will capture many of the volatility features of a time series – and it is much easier to analyse than higher-order alternatives. However, it is worth considering the form of a GARCH(p,q) process:

$$\sigma_t^2 = \alpha_0 + \alpha_1 X_{t-1}^2 + \alpha_2 X_{t-2}^2 + \ldots + \alpha_p X_{t-p}^2 +$$
$$\beta_1 \sigma_{t-1}^2 + \beta_2 \sigma_{t-2}^2 + \ldots + \beta_p \sigma_{t-p}^2, \tag{13.47}$$

or, more compactly:

$$\sigma_t^2 = \alpha_0 + \sum_{i=1}^{p} \alpha_i X_{t-i}^2 + \sum_{j=1}^{q} \beta_j \sigma_{t-j}^2. \tag{13.48}$$

In this model, $\alpha_0 > 0$, $\alpha_1, \alpha_2, \ldots, \alpha_p \geq 0$ and $\beta_1, \beta_2, \ldots, \beta_p \geq 0$. This model is covariance stationary if $\sum_{i=1}^{p} \alpha_i + \sum_{(j=1)}^{q} \beta_j < 1$.

If the term V_t is defined as $\sigma_t^2(Z_t^2 - 1) = X_t^2 - \sigma_t^2$, then substituting for σ_t^2 in Equation (13.48) gives:

$$X_t^2 = \alpha_0 + \sum_{i=1}^{p} \alpha_i X_{t-i}^2 + \sum_{j=1}^{q} \beta_j (X_{t-j}^2 - V_{t-j}) + V_t. \tag{13.49}$$

This can be further rearranged to give:

$$X_t^2 = \alpha_0 + \sum_{i=1}^{\max(p,q)} (\alpha_i + \beta_i) X_{t-i}^2 - \sum_{j=1}^{q} \beta_j V_{t-j} + V_t. \tag{13.50}$$

This process, given in terms of X_t^2 rather than σ_t^2, is known as a squared GARCH(p, q) process.

A special case of the GARCH(p,q) model occurs when $\sum_{i=1}^{p} \alpha_i + \sum_{j=1}^{q} \beta_j = 1$. This gives an integrated GARCH or IGARCH model. Starting with Equation (13.50), consider a simple GARCH(1,1) model:

$$X_t^2 = \alpha_0 + (\alpha_1 + \beta_1) X_{t-1}^2 - \beta_1 V_{t-1} + V_t. \tag{13.51}$$

For this to be an IGARCH(1,1) model, $\alpha_1 + \beta_1$ must equal one. This also means that $\beta_1 = 1 - \alpha_1$, giving:

$$X_t^2 = \alpha_0 + X_{t-i}^2 - (1 - \alpha_1) V_{t-1} + V_t. \tag{13.52}$$

Defining ΔX_t^2 as $X_t^2 - X_{t-1}^2$, this can be rewritten as:

$$\Delta X_t^2 = \alpha_0 - (1 - \alpha_1) V_{t-1} + V_t. \tag{13.53}$$

There are a number of other extensions to GARCH models that can be used, but one of the most useful is simply to incorporate GARCH errors into ARIMA models. This gives a flexible structure that can take into account many features of a wide range of time series.

Fitting ARCH and GARCH models

ARCH and GARCH models can be fitted using the maximum likelihood approach discussed earlier. The starting point in this case, however, is a conditional likelihood function. As before, this describes the joint probability that $X_t = x_t$ where $t = 1, 2, \ldots, T$, but this time with the likelihood being conditional on all previous observations of X_t back to the known starting point, $X_0 = x_0$. The conditional likelihood function is given by:

$$L = \prod_{t=1}^{T} f(x_t | x_{t-1}, x_{t-2}, \ldots, x_0,). \tag{13.54}$$

For an ARCH(1) model, L is calculated as:

$$L = \prod_{t=1}^{T} \frac{1}{\sigma_t} f\left(\frac{X_t}{\sigma_t}\right). \tag{13.55}$$

In this equation, $f(X_t/\sigma_t)$ is the density function used to model observation t. This must have a mean of zero and a standard deviation of one, and the standard normal distribution is an obvious candidate. As before, $\sigma_t^2 = \alpha_0 + \alpha_1 X_{t-1}^2$. The likelihood function can then be maximised using numerical approaches.

The same likelihood function exists for a GARCH(1,1) model. However, because in this model the volatility is defined by $\sigma_t^2 = \alpha_0 + \alpha_1 X_{t-1}^2 + \beta_1 \sigma_{t-1}^2$, a value is needed for σ_0^2. Unlike X_0, σ_0 is unobservable. As a consequence, a value must be chosen. This could be the sample variance for the whole dataset, or just for the first few observations.

Checking the fit of ARCH and GARCH models

The goodness of fit of a GARCH model can be checked by examining the residuals from that model. The process is similar to that used for ARIMA models, but since the residuals are expected to show changing variance it is important to try and remove this variation. Consider, for example, an AR(1) model whose variance has a GARCH(1,1) process. The residuals from the AR(1) process, $\hat{\epsilon}_t$, can be calculated as described above. The estimated volatility in time t for a GARCH(1,1) process, $\hat{\sigma}_t^2$, can then be calculated from Equation (13.44) as:

$$\hat{\sigma}_t^2 = \hat{\alpha}_0 + \hat{\alpha}_1 X_{t-1}^2 + \hat{\beta}_1 \sigma_{t-1}^2. \tag{13.56}$$

The estimates $\hat{\epsilon}_t$ and $\hat{\sigma}_t$ are then compared to give the standardised residual $\hat{Z}_t$:

$$\hat{Z}_t = \frac{\hat{\epsilon}_t}{\hat{\sigma}_t}. \tag{13.57}$$

These standardised residuals should form a white noise process, and should be tested to see whether this is the case. As with the ARMA model analysis, there is an issue that no estimates can be calculated for the early values of $\hat{\epsilon}_t$ and $\hat{\sigma}_t$. Possible solutions include setting both to zero and/or to the average values of each series.

Volatility forecasting with ARCH and GARCH processes

Once data have been fitted to an ARCH or GARCH model, such a model can then be used to forecast volatility. The process is almost identical to that used for forecasting the values of ARMA models. Consider a GARCH(1,1) model. As implied above, the best estimate of the variance in period $t+1$, $\hat{\sigma}_{t+1}^2$, can be expressed in terms of known information at time t as:

$$\hat{\sigma}_{t+1}^2 = \alpha_0 + \alpha_1 X_t^2 + \beta_1 \sigma_t^2. \tag{13.58}$$

Recall that $X_t = Z_t \sigma_t$, so $X_t^2 = Z_t^2 \sigma_t$. Since Z_t is normally distributed with a mean of zero, Z_t^2 has a χ_1^2 distribution. This means that $E(Z_t^2) = 1$, so:

$$E(X_{t+1}^2) = \hat{\sigma}_{t+1}^2. \tag{13.59}$$

Moving forward one period and using Equation (13.59), Equation (13.58) can be used to give the expected variance at time $t+2$, although this time the right-hand side of the equation includes some expected values:

$$E(X_{t+2}^2) = \hat{\sigma}_{t+2}^2 = \alpha_0 + \alpha_1 E(X_{t+1}^2) + \beta_1 \hat{\sigma}_{t+1}^2. \tag{13.60}$$

Substituting Equation (13.59) into the right-hand side of Equation (13.60) gives:

$$E(X_{t+2}^2) = \hat{\sigma}_{t+2}^2 = \alpha_0 + (\alpha_1 + \beta_1)\hat{\sigma}_{t+1}^2. \tag{13.61}$$

Then substituting Equation (13.58) back into this expression gives:

$$E(X_{t+2}^2) = \hat{\sigma}_{t+2}^2 = \alpha_0 + \alpha_0(\alpha_1 + \beta_1) + (\alpha_1 + \beta 1)(\alpha_1 X_t^2 + \beta_1 \sigma_t^2). \tag{13.62}$$

This expression can be generalised to predict volatility h periods in the future as:

$$
\begin{aligned}
E(X_{t+h}^2) = \hat{\sigma}_{t+h}^2 &= \alpha_0[1 + (\alpha_1 + \beta_1) + \ldots + (\alpha_1 + \beta_1)]^{h-1} \\
&\quad + (\alpha_1 + \beta 1)^{h-1}(\alpha_1 X_t^2 + \beta_1 \sigma_t^2) \\
&= \alpha_0 \sum_{i=0}^{h-1} (\alpha_1 + \beta_1)^i + (\alpha_1 + \beta 1)^{h-1}(\alpha_1 X_t^2 + \beta_1 \sigma_t^2).
\end{aligned}
$$

$$\tag{13.63}$$

5 13.5 Data frequency

One issue that crops up frequently in time series analysis is data scarcity. One possible way of dealing with this is to calculate statistics from higher frequency data and to scale the results to the appropriate time scale. Consider, for example, a range of N asset classes, each of whose returns are independently and identically distributed according to a normal distribution with means given by the column vector $\bar{X}$ and covariances given by the $N \times N$ matrix Σ. If means and covariances are required over a time scale T times as long, then the revised means are given as $T\bar{X}$ and the covariances by $T\Sigma$. For example, if annual statistics are required and monthly data are available, then multiplying all means and covariances by 12 will give annual statistics.

Furthermore, if the means are all taken to be zero and the short-term data is used to calculate a metric such as expected shortfall or Value at Risk (VaR), then these aggregated statistics can be calculated for longer time scales by multiplying by $\sqrt{T}$.

The assumption of zero means is reasonable if the periods talked about are short term – days rather than months – but the scaling approach is not ideal for other reasons. Firstly, the series being scaled might not be normally distributed. In particular, they may be from leptokurtic distributions. Serial correlation may also be present, or have changing volatility, all of which make scaling inaccurate.

In these cases, it is more appropriate to use the shorter-term data to parameterise a stochastic model based on the shorter time frame, and to calculate measures such as expected shortfall of VaR from multi-period simulations.

13.6 Discounting

Having constructed the time series, it is often necessary to calculate a present value of the projected cash flows. This requires discounting, and the choice of an appropriate discount rate. If the current time is defined as $t = 0$, then the present value, V_0, from the value at some future time t, V_t, using an interest rate per period, r, is calculated using the following relationship:

$$V_0 = \frac{V_t}{(1+r)^t}. \tag{13.64}$$

In this equation, the term r represents the rate of interest as a proportion of an initial amount invested. It is also possible to express the interest rate in terms of a discount on the final amount received:

$$V_0 = V_t(1-d)^t, \tag{13.65}$$

where d is the rate of discount. An important third approach is to use the force of interest, s. This can be thought of as representing the amount of interest being continually added to an initial investment rather than paid as a lump sum at the end of the period or discounted from the investment at the start. The force of interest relates the initial and terminal amounts as follows:

$$V_0 = V_t e^{-st}. \tag{13.66}$$

Discounted values of assets and liabilities can be compared to determine the existence of a surplus or a deficit. In stochastic simulations, each scenario can also be discounted to determine the number of times that particular level is breached. Discounting can also be used to determine whether or not a project should be undertaken on or not.

The choice of discount rate is not a trivial matter, and depends on the purpose for which a value is being discounted.

For example, if the item being discounted is an amount that a party is obliged to pay, then the logical starting point is the risk-free rate of interest. This is because the resulting amount could then, in theory, be invested in a risk-free security offering that rate of interest to arrive at the amount needed in the future to meet the liability. If the discounted value of liability on this basis is less than the market value of the assets, then there is a surplus of assets over liabilities; if the opposite is true, then there is a deficit.

However, this assumes that there exists a risk-free investment that exactly matches the required payment, thus giving a risk-free rate of interest to use. If such an investment does not exist, then the discount rate must be reduced to allow for this.

Furthermore, if the assets are not invested in risk-free investments, then the volatility in investment returns could result in assets falling below the level of the liabilities at some point in the future. This risk could be expressed as an additional liability or as a further reduction in the discount rate used. In these cases, an alternative approach is to agree an appropriately low probability of insolvency, to stochastically simulate the assets and liabilities, to determine the current value of assets that would be needed to ensure the agreed level of solvency, and to calculate the implied discount rate that then sets the present value of the liabilities equal to the market value of the assets.

Even if the risk free approach is used, it is not entirely clear what the risk-free rate of interest is. Whilst the obvious starting point is the yield on government bonds, the fact that these are so easily traded compared to all other securities means that their price includes a liquidity premium. This means that the yields are lower than genuinely risk-free yields would be.

Taking these factors into account means that discounting at a predetermined rate is not necessarily the best way to determine the value liabilities. An alternative approach is to project the liabilities and the assets held in respect of those liabilities using a stochastic model. Rather than discounting the liabilities, the expected return on each of the assets is incorporated into the model. Then, rather than determining whether sufficient assets are being held by comparing two numbers, the decision could be taken based on the proportion of scenarios for which the liabilities are paid before the assets run out. If this proportion is sufficiently high – say, 95% – then the institution owing the liabilities could be said to be solvent with that level of confidence. Having sufficient assets would therefore depend not only on the amount of assets but also on what those assets were. If required, the discount rate can be calculated as the rate which sets the present value of liabilities equal to the market value of assets when the value of assets is exactly sufficient to meet the liability cash flows with an agreed level of confidence.

Determining the value of an obligation is not the only reason for discounting a set of cash flows to obtain a present value. It might also be of interest to calculate what an obligation is actually worth. This means that the possibility that the cash flows making up a liability will not be received for some reason must be considered. In practice, this means making some allowance in the discount rate for credit risk. The addition for credit might be available from published information. For example, if the obligation is from a listed company, then credit default swaps might exist whose prices give an indication of the market's view of the likelihood of insolvency for a firm. If the obligation is from one of a large number of individuals, as might be the case with a portfolio of loans, then an average historical rate of default might be used.

However, again there are issues. The main one is that the probability of default will be linked to other risks, so it is often inappropriate to deal with this risk simply by adjusting the discount rate – again, a projection-type approach would be more appropriate, with the probability of default being treated as a random variable linked to other risks.

Another approach to determining an appropriate discount rate, typically applied to the choice of whether or not to pursue a project, is based on the capital asset pricing model (CAPM). The CAPM says that the expected return from an investment X, r_X, is a function of four things:

- the risk-free rate of return, r^*;
- the rate of return available from the universe of investment opportunities, r_U;
- the uncertainty of the return from the universe of investment opportunities, as measured by its estimated variance, σ_U^2; and
- the covariance of the return from investment X and the return on the universe of investment opportunities, $\sigma_X \sigma_U \rho_{X,U}$.

These are linked as follows:

$$r_X = r^* + \frac{\sigma_X \sigma_U \rho_{X,U}}{\sigma_U^2} \left(r_U - r^* \right). \tag{13.67}$$

Since the standard deviation of the investment in the numerator and denominator of the second term cancel, this can be rewritten as:

$$r_X = r^* + \frac{\sigma_X}{\sigma_U} \rho_{X,U} \left(r_U - r^* \right). \tag{13.68}$$

The term $\sigma_X \rho_{X,U}/\sigma_U$ is often referred to collectively as the beta of investment X, β_X. The line described by the relationship between r_X and β_X is known as the security market line. The above expression means that, relative to the risk-free rate of return:

- the greater the volatility of the investment relative to the universe of investment opportunities, the greater the expected return – this is a reward for uncertainty; and
- the greater correlation the investment has with the universe of investment opportunities, the greater the expected return – this is a reward for the lack of diversification.

For an organisation considering a particular project, 'expected return' can be read as 'required return' and 'the universe of investment opportunities' can be read as 'the existing portfolio'. In other words, the discount rate an organisation uses when considering a project should reflect the uncertainty of that project and how it relates to the existing portfolio of projects that an organisation has. The security market line also becomes the project market line, shown in Figure 13.11.

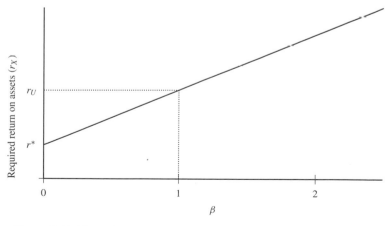

Figure 13.11 The project market line

There are some important caveats here. First, the accuracy of the required return depends on the stability of the volatilities of the market and the project returns as well as their correlation. These parameters are frequently unstable. A more practical issue is that when comparing the expected returns from the CAPM for a particular period with the returns that were actually achieved, it is difficult to find any meaningful relationship between the two.

13.7 Further reading

There are a number of books on different aspects of time series analysis. A comprehensive source of information is Hamilton (1994), but this is also quite an advanced text; a good starting point is the work of Box and Jenkins (1970), whose eponymous approach to fitting a time series is still widely used. Many econometric books such as Johnston and Dinardo (1997) also cover aspects of time series analysis.

Merton (1992) provides a detailed, though complex, guide to continuous-time finance. More up-to-date concepts are discussed by McNeil *et al.* (2005) with a less technical treatment given by Dowd (2005). Hull (2009) discusses time series analysis in the context of derivative valuation, as does Wilmott (2000).

14

Quantifying particular risks

14.1 Introduction

Many of the approaches described above are used directly to quantify particular types of risk. These applications are described in this chapter, together with some specific extensions that can also be used to determine levels of risk. Since different risks can affect different types of institutions in different ways, several approaches are sometimes needed to deal with a single risk. The links between various risks and the implications for quantification are also discussed.

When quantifying particular risks, it is important that these risks are modelled consistently with each other. In particular, it is important that assets and liabilities are modelled together, so that their evolution can be mapped. This is the basic principle of asset-liability modelling.

As part of this process, it is also important to consider the level of assets and liabilities throughout the projection period, not just at the ultimate time horizon. If the modelling suggests that action should be taken at points within the projection time horizon, then the projection should be re-run taking these actions into account. This is known as dynamic solvency testing or dynamic financial analysis. *management action*

14.2 Market and economic risk

14.2.1 Characteristics of financial time series

Before discussing the way in which market and economic risks can be modelled, it is worth considering some important characteristics of financial time series, particularly in relation to equity investments.

In spite of the assumptions in many models to the contrary, market returns are rarely independent and identically distributed. First, whilst there is little

311

obvious evidence of serial correlation between returns, there is some evidence that returns tend to follow trends over shorter periods and to correct for excessive optimism and pessimism over longer periods. However, the prospect of such serial correlation is enough to encourage trading to neutralise the possibility of arbitrage. In other words, serial correlation does not exist to the extent that it is possible to make money from it – the expected return for an investment for any period is essentially independent from the return in previous periods, and for short periods is close to zero.

Whilst there is no apparent serial correlation in a series of raw returns, there is strong serial correlation in a series of absolute or squared returns: groups of large or small returns in absolute terms tend to occur together. This implies volatility clustering. It is also clear that volatility does vary over time, hence the development of ARCH and GARCH models.

The distribution of market returns also appears to be leptokurtic, with the degree of leptokurtosis increasing as the time frame over which returns are measured falls. This is linked to the observation that extreme values tend to occur close together. In other words, very bad (and very good) series of returns tend to follow each other. This effect is also more pronounced over short time horizons.

Given that exposure to equities usually comes from an investment in a portfolio of stocks, it is also important to consider the characteristics of multivariate return series. The first are of interest is correlation – or, more accurately, co-movement. Correlations do exist between stocks, and also between asset classes and economic variables. However, these correlations are not stable. They are also not fully descriptive of the full range of interactions between the various elements. For example, whilst the correlation between two stocks might be relatively low when market movements are small, it might increase in volatile markets. This is in part a reflection of the fact that stock prices are driven by a number of factors. Some relate only to a particular firm, others to an industry, others still to an entire market. The different weights of these factors at any particular time will determine the extent to which two stocks move in the same way.

Whilst correlations do exist (or appear to exist) between contemporaneous returns, there is little evidence of cross-correlation – in other words, the change in the price of one stock at time t does not generally have an impact on any other stock at time $t + 1$. However, if absolute or squared returns are considered, then cross-correlation does appear to exist. This is a reflection of the fact that rising and falling levels of volatility can be systemic, affecting all stocks and even all asset classes.

Similarly, extreme returns often occur across stocks and across time series, meaning that not only are time series individually leptokurtic, but that they have *jointly* fat tails.

14.2.2 Modelling market and economic risks

When modelling market and economic risks, the full range of deterministic and stochastic approaches can be used, although some asset classes require special consideration. A good example is in relation to bootstrapping. This is not necessarily appropriate for modelling the returns on bonds without some sort of adjustment. The reason for this can be appreciated if a period of falling bond yields is considered. This will lead to strong returns for bonds. However, since yields will be lower at the end of the period than at the start, the potential for future reductions in bond yields – and thus increases in bond prices – will be lower. A suitable adjustment might be to base expected future returns on current bond yields and to use bootstrapping simply to model the deviations from the expected return.

The combination of impacts on the prices of individual securities and whole asset classes means that factor-based approaches are often used to model the returns on portfolios of stocks and combinations of asset classes. For example, a factor-based approach to modelling corporate bonds might start by recognising that the returns on this asset class can be explained by movements in the risk-free yield, movements in the credit spread, coupon payments and defaults. These can each be modelled and combined to give the return for the asset class. The interactions of these four factors and other financial variables would also need to be modelled. For example, defaults could be linked to equity market returns. With a factor-based approach, complex relationships between asset class returns arise because of the linkages between the underlying factors. These models can also be specified to include heteroskedasticity through the use of ARCH or GARCH processes.

Rather than trying to determine which factors drive various securities or asset classes, it is possible instead to use a data-based multivariate distribution. The most common approach is to assume that the changes in the natural logarithms of asset classes are linked by a multivariate normal distribution. If this approach were being used to describe the returns on a range of asset classes – say UK, US and Eurozone and Japanese equities and bonds – the following process could be used to generate stochastic returns:

- decide the scale of calculation, for example daily, weekly, monthly or annual data;
- decide the time frame from which the data should be taken, choosing an appropriate compromise between volume and relevance of data;
- choose the indices used to calculate the returns, ensuring that each is a total return index, allowing for the income received as well as changes in capital values;

late the returns for each asset class as the difference between the natural logarithm of index values;

- calculate the average return for each asset class over the period for which data are taken;
- calculate the variance of each asset class and the covariances between them;
- simulate series of multivariate normal distributions with the same characteristics using Cholesky decomposition.

One issue with this approach is that it can involve calculating a very large number of data series if the number of asset classes increases, or if individual securities are to be simulated. An alternative approach is to use a dimensional reduction technique instead. In this case, principal component analysis (PCA) could instead be used to determine the extent to which unspecified factors affect the returns. The following procedure could then be followed instead:

- decide the scale of calculation, for example daily, weekly, monthly or annual data;
- decide the time frame from which the data should be taken, choosing an appropriate compromise between volume and relevance of data;
- choose the indices used to calculate the returns, ensuring that each is a total return index, allowing for the income received as well as changes in capital values;
- calculate the returns for each asset class as the difference between the natural logarithm of index values;
- calculate the average return for each asset class over the period for which data are taken;
- deduct the average return from the return in each period for each asset class, leaving a matrix of deviations from average returns;
- use PCA to determine the main drivers of the return deviations;
- choose the number of principal components that explains a sufficiently high proportion of the past return deviations;
- project this number of independent, normal random variables with variances equal to the relevant eigenvalues;
- obtain the projected deviations from the expected returns for each asset class by weighting these series by the appropriate elements of the relevant eigenvectors; and
- add these returns to the expected returns from each asset class.

The PCA approach is particularly helpful if bonds of different durations are being modelled, since a small number of factors drives the majority of most bond returns. In particular, changes in the level and slope of the yield curve explain most of the change.

A drawback with the PCA approach is that multivariate normality is a data requirement rather than a computational nicety. Having already seen that the data series for many asset classes are not necessarily normal, it should be clear that this could pose problems. However, if all series are to be modelled instead, then it is also possible to use an approach other than a jointly normal projection of the natural logarithms of returns. This could mean using another multivariate distribution, or taking more tailored steps. In particular, copulas could be used to model the relationship between asset classes, whilst leptokurtic or skewed distributions can be used to model each asset class individually. The extent to which the process moves away from a simple multivariate normal approach depends to a large extent on the volume of data that is available – it is difficult to draw any firm conclusions on the shape of the tails of a univariate or multivariate basis if only a handful of observations are available on which to base any decision.

14.2.3 Expected returns

When carrying out stochastic or deterministic projections, assumptions for future returns are required. Whilst the returns experienced over the period of data analysis might be used, they typically reflect only recent experience rather than a realistic view of future returns. The forward-looking view can come from subjective, fundamental economic analysis, or through quantitative techniques.

Whatever approach is used, it is important that tax is allowed for, either in the expected returns or at the end of the simulation process. This is particularly important if different strategies are being compared – such a comparison should always be made on an after-tax basis.

Government bonds

For domestic government bonds that are regarded as risk-free, a reasonable estimate for the expected return can be obtained from the gross redemption yield on a government bond of around the same term as the projection period. This represents the return received on a bond if held until maturity, subject to being able to invest any income received at the same rate. Better estimates of expected bond returns are discussed in the section on interest rates. Note that the expected return is not based on the gross redemption yield on a bond of the same term as those held. The expected return on a risk-free bond for a given term is approximately equal to the yield on a bond of that term, and if a different return were available over the same period on a bond with a different term, then this would imply that an arbitrage opportunity existed.

Some adjustment might be made if the projection period and term of investments differ significantly to allow for a term premium. If yields on long-term bonds are consistently higher than those on short-term bonds, then this might be due to investors perceiving long-date assets as being riskier due to greater interest rate or credit risk, and requiring an additional premium to compensate. However, since long-dated bonds are *less* risky for some investors – in particular those with long-dated liabilities – term premiums cannot be taken for granted, and their presence will vary from market to market.

A better estimate of the expected return could be obtained by constructing a forward yield curve to take into account the term structure of the bonds held in a portfolio, but this level of accuracy is spurious in the context of stochastic projections.

For overseas government bonds that are also regarded as risk free, a good approximation for the expected return in domestic currency terms is, again, the gross redemption yield on a *domestic* government bond of around the same term as the projection period. If it were anything else, then this would again imply an arbitrage opportunity. The fact that domestic and overseas government bond yields differ can be be explained by an expected appreciation or depreciation in the overseas currency.

Corporate bonds

For risky bonds where there is a chance of default, the expected return must be altered accordingly. Such bonds will usually be corporate bonds, although many government bonds are not risk free, particularly if there is any doubt over the ability of that government to be able to honour its debts. A starting point is the credit spread, which represents the additional return offered to investors in respect of the credit risk being taken. There are a number of ways in which the credit spread can be measured, the three most common being:

- nominal spread;
- static spread; and
- option adjusted spread.

The nominal spread is simply the difference between the gross redemption yields of the credit security and the reference bond against which the credit security is being measured (often a treasury bond). This is attractive as a measure because it is a quick and easy measure to calculate. It ignores a number of important features and should be regarded as no more than a rule of thumb if the creditworthiness of a particular stock is being analysed; however, for the purposes of determining an additional level of return for an asset class as a whole, it offers a reasonable approximation.

A more accurate measure is the static spread. This is defined as the addition to the risk-free rate required to value cash flows at the market price of a bond. On a basic level, this appears to be similar to the nominal spread mentioned above; however, rather than just considering the yield on a particular bond, this approach considers the full risk-free term structure and the constant addition that needs to be added to the yield at each duration to discount the payments back to the dirty price.

Finally, the option adjusted spread is similar to the static spread, but rather than looking at a single risk-free yield curve, it allows for a large number of stochastically generated interest rates, such that the expected yield curve is consistent with that seen in the market. The reason this is of interest is that by using stochastic interest rates it is possible to value any options that are present in the credit security that might be exercised only when interest rates reach a particular level.

One of the interesting features of credit spreads is that they have historically been far higher than would have been needed to compensate an investor for the risk of defaults, at least judging by historical default rates. One suggested reason is that the spread partly reflects a risk premium. This is a premium in a similar vein to the equity risk premium, effectively a 'credit beta', designed to reward the investor for volatility relative to risk-free securities.

Another suggestion has been that the spread is partly a payment for the lower liquidity of corporate debt when compared to government securities. This means that the reward is given for the fact that it might not be possible to sell (or at least sell at an acceptable price) when funds are required. Similarly, it costs more to buy and sell corporate bonds than to buy and sell government bonds, and part of the higher yield on corporate debt may be a reflection of this. The size of the effect would depend on the frequency with which these securities were generally traded.

Another argument used is that although credit spreads have historically been higher than justified by defaults, this does not reflect how bad markets could get. In other words, the market could be pricing in the possibility of as yet unseen extreme events. Furthermore, the pay-off profile from bonds is highly skewed – the potential upside is limited (redemption yields cannot fall below zero), but the potential downside is significant (the issuer can default).

If, investors dislike losses more than they like gains, then this might mean that investors require additional compensation for this skewness.

Taxation can also play a part in the yield differences in some jurisdictions. Corporate bonds are sometimes treated less favourably than government bonds for some individuals. For example, capital gains on government bonds might be tax free but not so on corporate bonds. This effect might result in spreads being higher than otherwise might be the case.

There is also evidence that the correlation between credit spreads and interest rates is typically negative. For example, when an economy is growing quickly, when credit spreads might be lower, then the expectation might be that interest rates would be raised. For investors focussing on absolute returns, this negative correlation might be reflected in a lower required credit spread as it offers diversification for an investor seeking absolute returns.

As valid as these arguments are, they suggest that the additional risk should be reflected in the volatility side of any modelling, or in liquidity planning, rather than in an adjustment to the additional return. This risk premium should be based on the credit spread, historical default rates and, if relevant, taxation.

Historical risk premiums

Whilst it is possible to derive expected returns from market priced for bonds – albeit with adjustments for corporate bonds – the uncertainty surrounding the income of other investments means that such an approach is not possible for them. So for equities and property in particular, a different approach is needed. One approach is to consider the historical risk premium available for these asset classes. If this is being done on an annual basis, the return on a risk-free asset class should be deducted from the rate of return on the risky asset class each year for which historical data are available. The arithmetic average of these annual risk premiums should then be taken and this can be used as the annual expected return for any forward-looking analysis.

However, if past returns involve any changes as to the views on an asset class, then this will result in an historical premium that it is not reasonable to anticipate in the future. For example, if investors anticipated higher levels of risk at the start of the period than they did at the end, then prices would increase accordingly. This would raise the historical risk premium as prices would be higher at the end relative to their starting value, but the higher starting value would lower the expected future premium as a higher price would be paid for future earnings. This suggests that the historical risk premium should be altered for this re-rating.

The capital asset pricing model

One way of ensuring that risk premiums for different asset classes are consistent is to price them according to the capital asset pricing model (CAPM). This has already been discussed in the context of choosing a discount rate, but the model can also be used to calculate consistent expected returns across asset classes. To recap, the capital asset pricing model links the rate of return on an individual investment X, r_X, with the risk-free rate of return, r^*, and the return available from investing in the full universe of investment opportunities, r_U, as follows:

$$r_X = r^* + \beta_X \left(r_U - r^* \right), \tag{14.1}$$

where $\beta_X = \sigma_X \rho_{X,U} / \sigma_U$, σ_X and σ_U being the standard deviations of investment X and the universe of investment opportunities respectively. This approach still requires an estimate to be made of the overall risk premium for investing in risky securities. There is also an important caveat. Consider a UK institution considering investment in both UK and Japanese equities. When measured in sterling terms, Japanese equities will seem very volatile when compared with UK equities, suggesting that the former should have a high beta. To a Japanese investor, the opposite will be true, but it is inconsistent to have differences in risk premium that change depending on the currency of calculation. One way of dealing with this is to consider the volatility of each asset class in its domestic currency, whilst allowing for exchange rate risk in the calculation of correlations. This essentially means that the additional volatility seen in an asset class arising from exchange rate movements is not rewarded when the expected return is calculated.

14.2.4 Benchmarks

When considering market risk in particular, it is important that the risk is assessed relative to an appropriate benchmark. A good benchmark should be:

- unambiguous – components and constituents should be well defined;
- investable – it should be possible to buy the components of a benchmark and track it;
- measurable – it should be possible to quantify the value of the benchmark on a reasonably frequent basis;
- appropriate – it should be consistent with an investor's style and objectives;
- reflective of current investment opinion – it should contain components about which the investor has opinions (positive, negative and neutral); and
- specified in advance – it should be known by all participants before the period of assessment has begun.

There are also a number of specific criteria against which a benchmark can be measured in its appropriateness in a particular instance:

- the benchmark should contain a high proportion of the securities held in the portfolio;
- the turnover of the benchmark's constituents should be low;
- benchmark allocations should be investable position sizes;
- an investor's active position should be given relative to the benchmark;
- the variability of the portfolio relative to the benchmark should be lower than its volatility relative to the market portfolio;
- the correlation between $r_X - r_U$ and $r_B - r_U$ should be strongly positive;
- the correlation between $r_X - r_B$ and $r_B - r_U$ should be close to zero; and
- the style exposure of the benchmark and the portfolio should be similar.

In this list, r_U is the market return, r_B is the benchmark return and r_X is the portfolio return.

Most of the analysis on benchmarks assumes that the benchmark will be some sort of market index. However, assets are almost always held in respect of some liability, and these liabilities are the true underlying benchmark. If indices are used, then when a decision is made on the strategic level of risk to be taken this is reflected partly in the choice of indices and partly in the return targets given to those managing the assets. However, an alternative approach is to use the liabilities themselves as the benchmark. In this case, liabilities are usually converted to a series of nominal or inflation-linked cash flows which are then discounted at a risk-free rate of interest. The performance of the investments is then measured against the change in value of these cash flows.

The performance target relative to a benchmark is determined by the investors risk appetite. This is discussed in more detail later.

14.2.5 The Black–Scholes model

One important type of market risk requiring a special approach is the valuation of a financial option. The payment from an option is a function of X_t, the price of the underlying asset X at some future time t, also known as the spot price, and the exercise price of an option, E. There are two broad categories of option: a call option gives the buyer the right, but not the obligation, to buy the underlying asset at a fixed price, E, at some point in the future; a put option gives the buyer the right, but not the obligation, to sell the underlying asset at a fixed price, E, at some point in the future, the delivery date. In both cases, whoever sells or 'writes' the option is obliged to enter into the transaction. Whilst in many cases the expiry date and the delivery date will coincide, an option can

Table 14.1. *Option pay-offs*

Price (X_t)	Call option (buyer)	Call option (writer)	Put option (buyer)	Put option (writer)
4	0	0	6	−6
8	0	0	2	−2
12	2	−2	0	0
16	6	−6	0	0

expire – that is, cease to trade – before the underlying asset is delivered. This is particularly true for options on commodities.

Options can take many forms, common ones being American (exercisable at any time before expiry), Bermudan (exercisable on specific dates) or European (exercisable only at expiry). Whilst the European option is not the most common, it is the easiest to analyse. In particular, the Black–Scholes model assumes that the option being valued is European.

The buyer of a call option exercising that option at time t will receive a pay-off of $\max(X_t - E, 0)$; the buyer of a put option exercising that option at time t will receive a pay-off of $\max(E - X_t, 0)$. Table 14.1 shows the payoffs that would be due for different values of X_t at time t given an exercise price of ten units.

This series of pay-offs does not look particularly attractive for anyone writing put or call options. However, what this does not show is the initial cost of the option. This can be deducted from the present value of any payment received by the buyer of an option and from the present value of any payment made by the writer of an option. This means that if an option is not exercised, then the buyer of that option will have lost the price of that option, whilst the writer will have gained the same amount. If the pay offs are discounted to time $t = 0$ at a rate of r, then the table can be rewritten to include the cost of the call and put options, C and P respectively, as shown in Table 14.2 and Figure 14.1.

A more complete idea of option returns can be seen if a pay-off diagram is used. The horizontal axis shows the price of the underlying asset, with the exercise price marked, whilst the vertical axis shows the pay-off to the buyer and the writer of each option.

It is possible to use a stochastic approach to price an option, and for more complicated options with a range of exercise dates, this is the only way. However, for a European option, which has a fixed exercise date, the price of an option can be derived using the Black–Scholes formula. The price of a call option at time $t = 0$, C_0, where the current price of the underlying asset is X_0, the exercise price is E, with exercise taking place at a fixed time, T, and where

Table 14.2. *Option pay-offs (present values including premiums)*

Price (X_t)	Call option (buyer)	Call option (writer)	Put option (buyer)	Put option (writer)
4	$-C$	C	$\dfrac{6}{(1+r)^t} - P$	$P - \dfrac{6}{(1+r)^t}$
8	$-C$	C	$\dfrac{2}{(1+r)^t} - P$	$P - \dfrac{2}{(1+r)^t}$
12	$\dfrac{2}{(1+r)^t} - C$	$C - \dfrac{2}{(1+r)^t}$	$-P$	P
16	$\dfrac{6}{(1+r)^t} - C$	$C - \dfrac{6}{(1+r)^t}$	$-P$	P

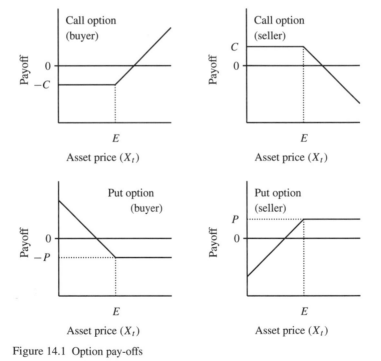

Figure 14.1 Option pay-offs

the continuously compounded risk-free rate of interest is r^* is:

$$C_0 = X_0 \Phi(d_1) - E e^{-r^* T} \Phi(d_2), \tag{14.2}$$

and the price of the corresponding put option, P_0 is:

$$P_0 = -X_0\Phi(-d_1) + Ee^{-r^*T}\Phi(-d_2). \tag{14.3}$$

In these equations:

$$d_1 = \frac{\ln(X_0/E) + (r^* + \sigma_X^2/2)T}{\sigma_X\sqrt{T}}, \tag{14.4}$$

$$d_2 = \frac{\ln(X_0/E) + (r^* - \sigma_X^2/2)T}{\sigma_X\sqrt{T}}, \tag{14.5}$$

$\Phi(d)$ is the cumulative normal distribution calculated at d and σ_X is the standard deviation of returns for the underlying asset. It is also possible to calculate the price of a put option using put–call parity. Consider a portfolio consisting of a share with a price at time $t = T$ of X_T and a put option with an exercise price of E. If X_T is above E, then the portfolio pays out X_T, since the option is worthless; however, if the X_T is below E, then the portfolio will still pay E since the put option allows the share to be sold at this price. Now consider a second portfolio consisting of a risk-free zero-coupon bond paying E at time $t = T$, and a call option with an exercise price of E. If X_T is above E, then the portfolio pays out X_T, this time since the call option will pay out $X_T - E$, which when added to the bond's payment of E gives a total of X_T; however, if the X_T is below E, then the portfolio will still pay E since although the call option is worthless, the bond will pay E. In other words, the pay-offs of the two portfolios are the same. If the risk-free rate of interest is r^*, then the value of the bond at time $t = 0$ is Ee^{-r^*T}, so the put and call options valued at time $t = 0$ can be related as follows:

$$C_0 + Ee^{-r^*T} - P_0 + X_0. \tag{14.6}$$

The two components of both Equation (14.2) and Equation (14.3) are essentially:

- the current value of the asset, X_0, multiplied by a factor between zero and one; and
- the present value of the exercise price, E, discounted continuously at the risk-free rate from time zero until time T, e^{-r^*T}, multiplied by another factor between zero and one.

The first factor for the call option represents the probability that the value of the asset exceeds the exercise price, and the probability that it falls below the exercise price for the put option; the second factor represents the probability that this does not happen.

If the asset is providing a regular flow of income, then these equations can be modified by deducting the continuous rate of income, r_D, from the risk-free rate in d_1 and d_2, and by replacing the term X_0 in the formulae for the call and put options with $X_0 e^{-r_D T}$. Both of these have the effect of discounting the income flow from the current value of the asset. The resulting expressions for call and put options are:

$$C_0 = X_0 e^{-r_D T} \Phi(d_1) - E e^{-r^* T} \Phi(d_2), \qquad (14.7)$$

and:

$$P_0 = -X_0 e^{-r_D T} \Phi(-d_1) + E e^{-r^* T} \Phi(-d_2). \qquad (14.8)$$

In these equations:

$$d_1 = \frac{\ln(X_0/E) + (r^* - r_D + \sigma_X^2/2)T}{\sigma_X \sqrt{T}}, \qquad (14.9)$$

$$d_2 = \frac{\ln(X_0/E) + (r^* - r_D - \sigma_X^2/2)T}{\sigma_X \sqrt{T}}. \qquad (14.10)$$

It is helpful to look at the way in which option prices move as the parameters change. As X_0 increases relative to the exercise price, E, the value of a call option increases and the value of a put option falls. This makes sense, as this relative movement makes it more likely that a call option will be exercised since X_T is more likely to exceed E; the same relative movement makes it less likely that a put option will be exercised.

An increase in r^* has the same effect, as it increases the rate of growth of the value of the asset. However, an increase in σ_X increases the value of both put and call options. This is essentially because higher volatility increases the likely range of values – higher and lower – that the asset will have. Similarly, as T increases, if there are no dividends, then the likelihood that either option will be exercised increases. However, if dividends are present, then an increase in T will have an indeterminate effect.

The Black–Scholes model has a number of assumptions that are not necessarily appropriate. For example, there is the assumption that the returns on an asset, defined as the difference between the logarithms of asset values, follow a normally distributed random walk with fixed values for all of the parameters. It also requires that a perfect hedge is always available and that there are no transaction costs. These conditions are often, if not always, absent in practice. However, the model offers a good 'first cut' estimate of the value of a financial option.

2 **14.3 Interest rate risk**

14.3.1 Interest rate definitions

A particular economic or financial variable that is of specific interest to pension schemes and life insurance companies is the interest rate, since it used to discount long-term liabilities. The rate may be a risk-free rate or some other metric as discussed in the section on discounting. However, the phrase 'interest rate' has a number of meanings, and it is important to define them more clearly before considering how to model them.

Spot rates

The most widely understood use of the interest rate is to describe the spot rate of interest. The t-year spot rate is essentially the gross redemption yield on a theoretical t-year zero-coupon bond, or the annualised rate of interest you would receive if you held a t-year zero-coupon bond until it matured. It is also usually expressed as a force of interest, s_t, particularly in the context of the money market. This means that rather than being given as the interest earned on an initial amount of assets, or the discount at which a final payment is made, it is given as a continuously compounded value, representing the rate at which interest is earned bit-by-bit over the whole period. However, the discretely compounded interest rate, r_t, is more likely to be used when valuing other assets and liabilities. Whilst the analysis below works in terms of continuously compounded rates, the discretely compounded rate can easily be obtained using the following relation:

$$\frac{1}{1+r_t} = e^{-s_t}. \tag{14.11}$$

The t-year spot rate is often taken to be simply the gross redemption yield on a t-year coupon-paying bond, but this can be a crude approximation if yields vary significantly by term. A more robust approach to calculating the spot rates for various terms – and thus creating a spot rate curve – is to use bootstrapping. Whilst this procedure has the same name as the approach used to simulate random variables, it is a very different approach. In this context, bootstrapping involves constructing a spot rate curve from the gross redemption yields on a series of bonds with a range of terms. This involves the following stages:

- calculate the 'dirty' price (in other words, the price allowing for any accrued interest) of the bond with the shortest term, t_1, given its gross redemption yield;

- calculate the continuously compounded rate of interest on the interest and principal payments that would give this price – this is the spot rate for the term t_1;
- calculate the dirty price of the bond with the next shortest term, t_2;
- calculate the value of interest receivable on this bond at time t_1 using the spot yield calculated for this term, and deduct the result from the price of the bond;
- calculate the continuously compounded rate of interest on the remaining interest and principal payments that would give the price net of the earlier payments – this is the spot rate for the term t_2;
- continue this process until all spot rates have been found.

Example 14.1 Consider five bonds with terms of one to five years, each paying annual interest payments or coupons of 5%, with the next coupons all being due one year from now. The gross redemption yields for these bonds are given below:

Term (t)	Gross redemption yield (r_t)
1	5.200 %
2	5.250 %
3	5.350 %
4	5.350 %
5	5.600 %

Calculate the prices of these bonds, and the continuously compounded spot rates of interest for each of the five maturities.

The first value needed is the price of the first bond. If a redemption payment of 100 is assumed, the 5% interest gives a coupon payment of 5 at the same time. This means that given a gross redemption yield of 5.20%, the price of the first bond is:

$$BP_1 = \frac{100+5}{1+0.0520} = 99.81.$$

Remembering that spot yields are given in force of interest terms, this means that the one-year spot rate, s_1, is given by:

$$99.81 = 105e^{-s_1}.$$

Rearranging this gives a value of 5.069% for s_1. Similarly the dirty price for the two-year bond is given by discounting all payments at the gross redemption yield:

$$BP_2 = \frac{5}{1+0.0520} + \frac{100+5}{(1+0.0520)^2} = 99.54.$$

According to the principle of no arbitrage, the value of the coupon in the first year can be found by discounting it at the one-year spot rate. This means that the two-year spot rate must satisfy the following equation:

$$99.54 = 5e^{-s_1} + 105e^{-s_2}.$$

Substituting the known value for s_1 and rearranging gives a value for s_2 of 5.116%. This process can be repeated to find s_3, s_4 and s_5, whose values are given in the table below:

Term (t)	Bond price BP_t	Spot interest rate (s_t)
1	99.81	5.069%
2	99.54	5.116%
3	99.05	5.219%
4	98.77	5.216%
5	97.44	5.479%

Forward rates

Spot rates give the rate of interest applicable from the $t = 0$ to some future point in time. However, if considering the evolution of interest rates over future periods, it is helpful to talk in terms of forward rates. The one-year forward rate of interest for a maturity of t, f_t, years is the rate of interest applying from period $t - 1$ to t. Spot and forward rates are linked in the following way:

$$e^{-s_T T} = e^{-\sum_{t=1}^{T} f_t}. \tag{14.12}$$

This means that in practice f_t can be calculated as $ts_t - (t - 1)s_{t-1}$ for single-period forward rates of interest.

Example 14.2 Using the information in Example 14.3.1, calculate the continuously compounded forward rates of interest for the maturities of one to five years.

The first forward rate, f_1, is simply equal to s_1. Using the relationship in Equation (14.12), f_2 can be calculated as $2 \times s_2$ less s_1, so:

$$f_2 = (2 \times 0.05116) - 0.05069 = 0.05163.$$

Similarly, f_3 can be calculated as $3 \times s_3$ less $2 \times s_2$, so:

$$f_3 = (3 \times 0.05219) - (2.05116) = 0.05425.$$

This process can be continued to find all values of f_t, resulting in the values given below:

Term (t)	Forward interest rate (f_t)
1	5.069%
2	5.163%
3	5.425%
4	5.207%
5	6.531%

It is also possible to have forward rates covering periods shorter than one year, and of particular interest is the instantaneous forward rate. For a maturity of t, this gives the rate of interest applying at that exact point in time. The relationship between spot and forward rates is shown in Figure 14.2.

✔ 14.3.2 Single-factor interest rate models

One approach to modelling future interest rates is to use a single-factor interest rate model. Such a model can be used to generate a series of future interest rates – forward rates of interest – that can be reconstructed into a complete yield curve for each series of simulations.

The Ho–Lee model

Consider for example the basic random walk with drift model discussed earlier. This can be written in terms of Δr_t, the difference between r_t and r_{t-1}, where r_t is the interest rate payable from time $t-1$ to time t:

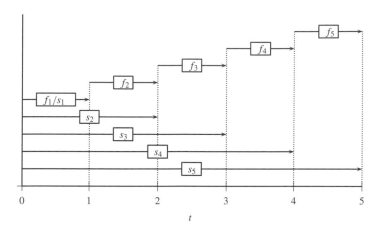

Figure 14.2 Spot and forward rates of interest

$$\Delta r_t = \alpha \Delta t + \epsilon_t, \qquad (14.13)$$

where Δt is the period of time between $t-1$ and t and ϵ_t is a normally distributed random variable with a variance of σ^2. This process can be used to generate the values of a whole series of bonds of different terms as part of a single simulation. However, it is more difficult to ensure that the prices of the bonds are consistent with their market prices. In order to do this, the constant α must be replaced by a time-varying term, α_t, which is used to calibrate the model from market data:

$$\Delta r_t = \alpha_t \Delta t + \epsilon_t, \qquad (14.14)$$

where Δt is the difference between t and $t+1$. The calibration is done by deriving forward rates of interest from the market prices of the bonds, and setting α_t equal to the forward rate for time t.

This model becomes a continuous-time interest rate process known as the Ho–Lee model (Ho and Lee, 1986) as Δt tends to zero.

The Vasicek and Hull–White models
One potential drawback of the Ho–Lee model is that it does not allow for the possibility that interest rates might revert to some predetermined value. To do so means building an autoregressive model. Consider for example the basic AR(1) series described earlier. This can be specified differently, explicitly stating the value to which the series reverts, again in terms of an interest rate r_t applying from time t to $t+1$, in other words:

$$\Delta r_t = (\alpha - \beta r_{t-1})\Delta t + \epsilon_t. \qquad (14.15)$$

In this equation, β – which must be greater than or equal to zero – gives the speed of mean reversion whilst α/β is the level to which the series reverts. For infinitesimally small values of Δt, this becomes a continuous-time interest rate process known as the Vasicek model (Vasicek, 1977).

However, unlike the Ho–Lee model, the Vasicek model has only a fixed value of α. This makes it difficult to fit the model accurately to market data. The Hull–White model (Hull and White, 1994a, 1994b) corrects for this by replacing α with α_t, a time-varying parameter as used in the Ho–Lee model. The Hull–White model is, like the other models described here, a continuous-time model. However, if converted to a discrete-time model, then the result can be expressed as follows:

$$\Delta r_t = (\alpha_t - \beta r_{t-1})\Delta t + \epsilon_t. \tag{14.16}$$

The Cox–Ingersoll–Ross model

The Hull–White extension of the Vasicek model gives a result that can be calibrated more easily to market data. However, an even more realistic fit can be obtained by allowing the volatility of ϵ_t to be time varying through the use of σ_t^2 rather than the fixed σ^2. The strength of mean reversion, β, can also be allowed to change over time to better reflect views about interest rate changes.

Another serious practical issue with the Vasicek model is that it can easily return negative values, and negative interest rates are not economically possible in normal circumstances. One solution is to modify the volatility so that it is not just time varying through the use of σ_t^2 in place of σ^2, but so that the volatility is also proportional to the square root of the previous value of the series. In discrete time, this gives the following model:

$$\Delta r_t = (\alpha_t - \beta_t r_{t-1})\Delta t + \sqrt{r_{t-1}}\epsilon_t. \tag{14.17}$$

Here, the expected interest rate in each period is given by α_t/β_t. As noted earlier, this series is not guaranteed to remain positive unless Δt becomes infinitesimally small, in which case negative values cannot occur if $\alpha_t \geq \sigma_t^2/2$. The continuous-time process that emerges as Δt tends to zero is known as the Cox–Ingersoll–Ross interest rate model.

The Black–Karasinski model

Another approach to avoiding negative interest rates is to apply the Vasicek model to the natural logarithm of interest rates, in the same way that the natural logarithm of asset values is often modelled:

$$\Delta \ln r_t = (\alpha_t - \beta_t \ln r_{t-1})\Delta t + \epsilon_t. \tag{14.18}$$

As with the Cox–Ingersoll–Ross model (Cox *et al.*, 1985), α_t, β_t and σ_t^2 are all time varying. If implemented in continuous time, the interest rate process that this gives rise to is the Black–Karasinski model (Black and Karasinski, 1991).

14.3.3 Multi-factor interest rate models

The one-factor approach to modelling interest rates has limitations. It can work well for simulating a single interest rate of a particular term, typically a short-term value such as the three-month or the one-year spot rate. It can also be used to simulate the movement of a different spot rate with a longer term. However, it is not so good for modelling different points on a yield curve. In particular, changes in long-term spot interest rates are determined by the cumulative projected forward rates of interest. This means that there is no capacity for long-term rates to change as a result of changes in *expected* future short-term spot rates that do not subsequently materialise.

This is important if more than one point on a yield curve is being modelled, something that is particularly relevant when considering long-term liabilities. Here, changes in the shape of the yield curve can have an important effect on the value of the liabilities even if the average level of interest rates along the yield curve remains unchanged.

The simplest group of models that deals with this issue is the two-factor family, which simultaneously models the spot rates of interest for two distinct maturities, usually the short-term and long-term rates.

The Brennan–Schwartz model

One of the earliest two-factor models is the Brennan–Schwartz model (Brennan and Schwartz, 1982). This is another continuous-time model, but can be written in discrete-time form as follows:

$$\Delta r_{1,t} = \left[\alpha_1 + \beta_1(r_{2,t-1} - r_{1,t-1})\right]\Delta t + r_{1,t-1}\epsilon_{1,t}, \tag{14.19}$$

and:

$$\Delta r_{2,t} = r_{2,t-1}(\alpha_2 + \beta_2 r_{1,t-1} + \gamma_2 r_{2,t-1})\Delta t + r_{2,t-1}\epsilon_{2,t}, \tag{14.20}$$

where the $r_{1,t}$ is the short-term rate of interest at time t and $r_{2,t}$ is the long-term rate of interest at time t. This model says the following about changes to interest rates:

- changes in short-term interest rates vary in proportion to the steepness of the yield curve (that is, the extent to which long-term rates exceed short-term rates;
- the volatility of short-term interest rates is proportional to the level of short-term rates;

- changes in long-term interest rates vary in proportion to the product of long- and short-term rates;
- changes in long-term interest rates vary in proportion to the square of the level of long-term rates; and
- the volatility of long-term interest rates is proportional to the level of long-term rates.

14.3.4 PCA–based approaches

Whilst the Brennan–Schwartz approach offers the prospect of more complex interest rate curves than can be obtained from single-factor modelling, this model is difficult to parameterise, and also exists for only two factors A more general approach can be obtained by starting to fit a single-factor model to each spot interest rate of interest, observing the correlations between these interest rates, and then using correlated normal random variables to project future spot rates. However, the term structure of interest rates means that it is particularly suited to dimensional reduction techniques such as principal component analysis (PCA). In particular, such approaches make it easier to model changes in the shape of the curve as well as just the level.

PCA can be applied to interest rates using the following process:

- decide the scale of calculation, for example daily, weekly, monthly or annual data;
- decide the time frame from which the data should be taken, choosing an appropriate compromise between volume and relevance of data;
- take the gross redemption yields for bonds of a range of maturities, with the yields for each maturity forming a distinct series;
- for each series, calculate the average interest rate over the period for which data are taken;
- deduct the average interest rates from the interest rate in each period within each series;
- use PCA to determine the main drivers of the deviations from the average interest rate;
- choose the number of principal components that explains a sufficiently high proportion of the past return deviations;
- project this number of independent, normal random variables with variances equal to the relevant eigenvalues;
- obtain the projected deviations from the expected interest rates for each maturity by weighting these series by the appropriate elements of the relevant eigenvectors; and
- add the expected interest rates, derived from bond prices for market consistency, to these simulated yields to give projections of future yield curves.

This process can be applied directly to gross redemption yields or to their natural logarithms, the latter approach avoiding the possibility of negative yields. However, it is more commonly applied directly to forward rates of interest.

It can also be applied to bond prices rather than interest rates. In this case, the variables analysed would be the deviations from the average return for each bond, with returns being measured as the difference between successive logarithms of prices. This is an important alternative approach – if bond prices are being modelled, then other asset prices can be modelled as well. As a result, interest rates can be modelled consistently with other financial variables. However, it is important to note that if bond returns are modelled, then the increase in bond volatility as term increases can result in the results being more influenced by changes at the long end of the yield curve.

Example 14.3 Determine the first five principal components for the UK forward rate curve using daily forward rates of interest from the end of 1999 to the end of 2009 as provided by the Bank of England. Verify the results by considering the correlations between the excess returns for the various forward rates.

Recall from Chapter 11 that each principal component is the combination of an eigenvalue which represents the volatility of a particular independent factor, and an eigenvector which contains the weights of this factor that contribute to the returns of each variable. Each eigenvector can be represented by a single line, showing the weights of each factor for each forward rate:

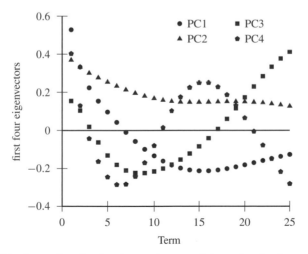

This shows that the dominant change of shape for the forward rate curve over this period is for rates at opposite ends of the curve to move in opposite directions – given a random number scaled by the first eigenvalue, the movement of the one-year forward rate in respect of the first principal component would be around 0.5 times this random number, whilst the twenty-five year forward rate would move by −0.1 times the same random number. The second most important move is a move in the same direction, as evidenced by the fact that all values of the second eigenvector are above the horizontal axis. The third principal component produces changes in short and long forward rates that are in the opposite direction to those in the middle of the curve, whilst the fourth principal component produces still more involved changes. The first four eigenvalues are shown below:

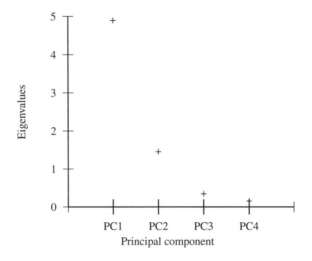

These show that the first principal component is by far the most dominant. The dominance of the first principal component – and its shape – can be verified by looking at the correlations between the various excess returns.

		Term				
		1	7	13	19	25
Term	1	1.0000	0.1649	−0.6919	−0.6715	−0.4710
	7	0.1649	1.0000	0.3870	0.2894	0.2137
	13	−0.6919	0.3870	1.0000	0.9364	0.5965
	19	−0.6715	0.2894	0.9364	1.0000	0.7904
	25	−0.4710	0.2137	0.5965	0.7904	1.0000

The correlations between the terms shown above correspond largely with the magnitudes and signs in the first eigenvector, supporting the results of the PCA.

✓ 14.3.5 Deriving price changes from interest rates

If bonds are being modelled rather than interest rates, it is still possible to calculate the interest rates implied by the changes in bond prices. One approach is to calculate the interest rate by calculating the duration and convexity of bonds whose returns are modelled, and then to calculate the change in interest rate implied by the simulated return on the bonds. Consider a bond issued by firm X paying a coupon of c_X times the face value of the bond at the end of each year, redeemable T years in the future. If the gross redemption yield of the bond is r_X, then the price of the bond per unit of notional value, BP_X, is:

$$BP_X = \sum_{t=1}^{T} \left[\frac{c_X}{(1+r_X)^t} \right] + \frac{1}{(1+r_X)^T}. \tag{14.21}$$

The modified duration[1] is a measure of the sensitivity of the price of a bond to a level change in the gross redemption yield. It is defined as:

$$BD_X = \frac{1}{(1+r_X)BP_X} \sum_{t=1}^{T} \left[\frac{tc_X}{(1+r_X)^t} \right] + \frac{T}{(1+r_X)^T}. \tag{14.22}$$

Using just the duration, the change in the price of a bond, ΔBP_X, for a change in the yield, Δr_X, is given by:

$$\Delta BP_X = -BP_X BD_X \Delta r_X. \tag{14.23}$$

However, this approximation of the relationship as linear is a very crude approximation to the real relationship, giving inaccurate values for the change in bond price for anything other than very small changes in yield. A better approximation can be obtained by including the convexity of the bond in the calculation of the change in price. Convexity – the rate of change of price for a given level change in the gross redemption yield – is defined as:

$$BC_X = \frac{1}{BP_X} \sum_{t=1}^{T} \left[\frac{t(t+1)c_X}{(1+r_X)^{t+2}} \right] + \frac{T(T+1)}{(1+r_X)^{T+2}}. \tag{14.24}$$

Including convexity in the calculation of the change in bond price, ΔBP_X, gives the following formula:

$$\Delta BP_X = -BP_X \left[BD_X \Delta r_X - \frac{1}{2} BC_X (\Delta r_X)^2 \right]. \tag{14.25}$$

---

[1] The Macauley duration, which gives the average term to payment of a series of cash flows, is equal to $(1+r_X)BD_X$.

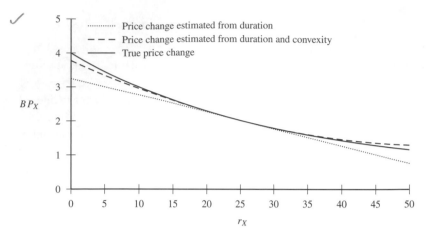

Figure 14.3 The relationship between price and yield

The true relationship between price and yield for a hypothetical twenty-year bond with a 5% per annum rate of interest initially priced at par (that is, with a yield of 5% per annum) is compared with the approximate relationships derived using modified duration alone and combined with convexity in Figure 14.3.

The equation for ΔBP_X can be rearranged to give the approximate change in price for a given change in gross redemption yield. Since the equation including convexity – the more accurate of the two – is a quadratic equation, there are two values for Δr_X that will satisfy this equation for a given value of ΔBP_X. However, since ΔBP_X is always a downward-sloping function of Δr_X, it is always the smaller root of the equation that is used:

$$\Delta r_X = \frac{BD_X - \sqrt{BD_X^2 + 2BC_X(\Delta BP_X/BP_X)}}{BC_X}. \tag{14.26}$$

A similar approach can also be used for modelling the impact of a change in the price of a corporate bond relative to a risk-free benchmark on the credit spread.

It is also possible to bypass the modelling of interest rates completely by calculating the value that liabilities would have had with historical rates of interest, examining the interaction of the liabilities with the assets, and then projecting the liabilities as though they were another asset class.

If the interest rate maturity structure is important – which it often is for the discounting of liabilities – then it might be desirable to use a specific interest rate model to project discount rates.

✓ 14.3.6 The Black model

The single-factor models in particular are also used to price interest rate derivatives. However, a model exists that can provide a closed-form valuation of options, allowing for the fact that changes in the spot price of the asset are unlikely to be lognormally distributed. This model is the Black model (Black, 1976), and it instead assumes that the *forward* price of the asset follows a lognormal random walk. The price of a call option under this approach is:

$$C = e^{-r^*T} \left[F_0 \Phi(d_1) - E \Phi(d_2) \right], \tag{14.27}$$

and the price of the corresponding put option, P is:

$$P = e^{-r^*T} \left[-F_0 \Phi(-d_1) + E \Phi(-d_2) \right], \tag{14.28}$$

where F_0 is the forward price at time zero of a contract on an underlying asset deliverable at time T, and all other parameters are as for the Black–Scholes model. If $F_0 = X_0 e^{r^*T}$, then this model in fact reduces to the Black–Scholes model.

3 14.4 Foreign exchange risk

Foreign exchange risk has already been mentioned in the context of market risk, but it is closer in nature to interest rate risk. This is because foreign exchange risk can be modelled in terms of the returns on cash deposits held in different currencies.

This means that the best way to model currency risk in a multi-asset context is to include short-term money market assets in any model being developed.

As noted before, the expected appreciation or depreciation in a currency is given by the difference in interest rates across different countries. More precisely, the discretely compounded spot rates in two currencies for a maturity of t, $r_{X,t}$ and $r_{Y,t}$ and the changes in exchange rate can be related as follows:

$$\frac{e_0}{e_T}(1 + r_{Y,T}) = 1 + r_{X,T}, \tag{14.29}$$

where e_T is the expected exchange rate at future time T expressed in terms of units of currency Y receivable per unit of currency X. This value is known when $t = 0$. The reasoning behind this equation is as follows. Investing a single unit of currency X at time $t = 0$ would yield a value of $1 + r_{X,t}$ units at time $t = T$ if invested in a risk-free asset with this maturity. However, an investor could instead take the single unit of currency X and exchange it for e_0 units of currency Y. Investing in this currency would yield a value at time $t = T$ of $e_0(1 + r_{Y,T})$. This could then be converted back to currency X at exchange rate e_T, the final amount in terms of currency X then being $(e_0/e_T)(1 + r_{Y,T})$. If the possibility of arbitrage is to be excluded – since it is possible to enter in to currency forward agreements – then these two end results must be equal.

A corollary of this analysis is that if modelling is carried out in a single currency, then currency risk is not rewarded by additional return, since it can easily be hedged away.

14.5 Credit risk

14.5.1 The nature of credit risk

Credit risk can manifest itself in large number of ways, each requiring a different method of assessment. The sources of credit risk for different types of financial organisation have been discussed in detail, but they can all be placed at some point on a scale. At one end, there are sources of credit risk such as counter-party risk relating to derivative contracts, or the risk of sponsor insolvency for pension scheme members. In these cases, the creditworthiness of the organisation in question is the main issue, although the links between this risk and others faced by an institution are important. At the other end is credit risk arising from investment in portfolios of credit derivatives or from the issue of loans and mortgages. In these cases, the interaction between the various credit exposures is as important as the assessment of each individual credit risk.

In relation to fixed-interest investments, credit risk is sometimes taken to include risks other than default, such as the risk that the credit spread will widen. Whilst this is an important risk, and one that is reflected in the price of these investments, it is essentially a market risk rather than a credit risk. Credit risk is defined here solely as the risk of default – in other words, the risk that monies owed are unable to be repaid.

This definition of credit risk does, though, have two components:

- the probability of default; and
- the magnitude of loss given that default has occurred.

The purpose of modelling credit risk is therefore twofold: to determine how likely a credit event is to occur; and to determine the extent of loss that will be incurred. In this way, it is similar to the analysis of insurance risks to the extent that both incidence and intensity need to be modelled. In fact, the modelling of the probability of default can be regarded as being particularly similar to the modelling of non-life insurance risks: high-quality credit risks are analogous to low-probability events such as catastrophe insurance, whereas low-quality credit risks resemble higher frequency lines such as motor insurance.

Most of the analysis is indeed concerned with determining the value of the first item. However, it is often possible to recover assets from defaulting firms and individuals, so ignoring the value of any recoveries can mean that the financial implications of default are overestimated.

A major issue with credit risk is that is relates to an institution or individual other than that holding the risk. This is an issue because of the fact that it is more difficult for the organisation holding the risk to get reliable information on the risk posed by the institution or individual creating it. For example, a bank that has lent money to a business is exposed to the risk of the business defaulting. The business clearly knows more about this risk than the bank, and in fact has an incentive to ensure that the bank knows more about the positive aspects and less about the negative aspects of the business.

It is possible to ask more questions in order to gain a clearer understanding of the level of credit risk faced, but as is discussed later it is important that the additional cost of acquiring this information does not outweigh the benefit of gaining the information. Some institutions, typically credit rating agencies, will always carry out this in-depth analysis. Others, such as banks, will typically rely on standardised questions. However, in many cases credit risk will be assessed using only publicly available data. This can be of a very high quality, particularly if there are disclosure rules in place such as those associated with Basel II. Most firms that have quoted shares also have to provide particular disclosures to comply with the rules of the stock exchanges on which they are listed, but the disclosure requirements for unlisted and particularly private companies are much less.

14.5.2 Qualitative credit models

The most common type of qualitative credit model is the type developed by a credit rating agency, which leads to a credit rating. These types of models are essentially risk management frameworks for the analysis of firms, so specific models are discussed in more detail in that context later on.

It is possible to try and build a model along the same lines as those used by the rating agencies. Such an approach could involve a range of factors assessing the firm, its industry and the broader economic environment. The assessment process would ultimately be subjective, but could include meetings with the firm under consideration, analysis of financial ratios and an assessment of various economic indicators.

Much of the analysis will centre on assessing the risk of default. However, this is not the only risk faced. In particular, if the investment is in a marketable security, there is the risk that the perceived creditworthiness will change. This is reflected in a change in the credit spread, and the risk of spread widening – and a subsequent fall in the value of a security – is also important to assess.

A key feature that affects both of these risks is the seniority of a debt. Not all debt has the same priority on the wind-up of a company. In particular, the more senior issues have an earlier call on any assets remaining. This means that analysing the seniority of an issue is crucial. Similarly, the presence of collateral is important. If a debt is secured on some collateral, which reverts to the lender in the event of a default, then this suggests a lower level of risk. However, there are different types of collateral. The more liquid this collateral is, the better the terms of a loan will be. For example, industrial machinery might be difficult to value accurately, and there may be only a limited second-hand market, so its attractiveness as collateral might be limited. Also, the proceeds from a particular asset are only as valuable as the firms ability to generate income from that asset.

If an internal qualitative model is built, the first decision needed is whether the model is intended to reflect the risk over the economic cycle (as with the rating agencies) or over a shorter time horizon. The choice of approach will be determined by the use to which the model may be put. For example, in calculating the pension protection fund (PPF) risk-based levy, a long-term approach might probably be more appropriate; however, a pension fund assessing the strength of the employer covenant might prefer to take account of the risk over the short term. This choice relates to both the calibration of the model and the inputs used for scoring.

Qualitative models have the advantage that factors beyond the quantitative ones can be allowed for. However, this can also lead to excessive subjectivity, which can cause a number of problems. For example, there might be a lack of consistency in the ratings given across different sectors, or even different analysts. Even if this consistency is achieved, the meaning of credit ratings might change over time as the economic environment changes. A related problem is that ratings might fail to distinguish correctly between the creditworthiness at a particular point in time and over the economic cycle.

However, even if a rating is intended to reflect risk over the economic cycle, it should still be modified in response to a fundamental change in the nature of a firm or the broader environment. The subjective nature of qualitative modelling can lead to a reluctance to change a credit rating rapidly. This is a behavioural bias known as anchoring, as it arises from a reluctance to move too far and too quickly from some existing anchor – such as an existing credit rating. The nature of the qualitative process can also limit the speed with which ratings are revised. However, the qualitative approach is still the most widely used by rating agencies, and the most frequently adopted by other organisations building their own models.

14.5.3 Quantitative credit models

Most credit models are quantitative in nature. This means that they take some financial variable of an entity and use them to give a score to that entity. The score might have a meaning such as a probability of default, or it might simply be a ranking of the relative creditworthiness of a range of entities.

It is also important to recognise that if a quantitative model uses bond returns, then the return profile for is highly skewed. This is because the best return that can be obtained from a bond held to maturity is where all coupons and redemption payments are received in full and on time. For corporate bonds, this will mean a marginally higher return than the 'expected' return. However, the worst return is where the bond defaults and no payments are received at all.

There are three broad types of quantitative credit model: credit scoring, structural and reduced form. The first type uses features of an entity to arrive at a score that represents the likelihood of its insolvency. Probit and logit models, Altman's Z-score, the k-nearest neighbour approach and support vector machines all fall into this category. Structural models, on the other hand, model the value of an entity rather than relying on accounting ratios. The Merton and KMV models fall into this category. Finally, there are reduced-form models. These use the credit rating derived using some other quantitative or qualitative approach to derive a probability of default.

Probit and logit models

As described above, probit and logit models are types of general linear models that are used when the dependent variable can take a value only between zero and one. This makes them particularly suitable for modelling the probability of default for a firm.

Both types of model start by considering the numerical characteristics of firms that have defaulted or remained solvent over a particular period, together with coefficients for these characteristics. For example, the independent variables might include accounting ratios such as financial leverage and income cover. The dependent variable will be one for a firm that has defaulted and zero for a firm that has remained solvent. The coefficients from the regression can then be applied to a set of accounting ratios for a new firm to give a figure representing the probability of default.

Probit and logit models both offer highly effective approaches to determining credit risk. Whilst they do not allow for the inclusion of qualitative factors – a factor common to all quantitative models – they are among the most commonly used type of model around today.

Discriminant analysis

Whilst probit and logit models use accounting ratios to arrive at probabilities of default, linear discriminant analysis has been more widely used in practice. The most familiar credit modelling approach using this technique is Altman's Z-score (Altman, 1968). As discussed earlier, this uses linear discriminant analysis to give each firm a score that indicates whether it is likely to become insolvent or not. There are two reasons that the Z score uses financial ratios. The first is that ratios allow firms of different sizes to be compared on a consistent basis – a firm's level of earnings is less important than, say, the earnings as a proportion of the firm's assets. However, as well as giving consistency across firms, ratios also allow sensible comparisons to be made over time. For example, a firm's earnings would be expected to drift upwards over time in line with price inflation, rendering any analysis based on this measure redundant. However, since asset prices are similarly affected by inflation, a measure such as earnings over assets should be more stable over time.

The original Z-score was calibrated using publicly quoted manufacturing firms. If there are N firms, $1, 2, \ldots, n, \ldots, N$, then the score Z_n for a particular firm n is calculated as:

$$Z_n = 0.012 X_{1,X} + 0.014 X_{2,n} + 0.033 X_{3,n} + 0.006 X_{4,n} + 0.999 X_{5,n}, \quad (14.30)$$

where:

- $X_{1,n}$ is the ratio of working capital to total assets;
- $X_{2,n}$ is the ratio of retained earnings to total assets;
- $X_{3,n}$ is the ratio of earnings before interest and taxes to total assets;
- $X_{4,n}$ is the ratio of the market value of equity to the book value of total liabilities; and
- $X_{5,n}$ is the ratio of sales to total assets,

each for firm n. Each of the ratios is entered into this model in the form of a percentage. For example, if the ratio of retained earnings to total assets is 5.5%, then the a value of 5.5 is used, not 0.055.

A value of Z_n above 2.99 indicates that a firm is 'safe', whilst a score below 1.80 indicates that the firm is at risk of distress. If Z_n is between 1.80 and 2.99, then the firm falls in the zone of uncertainty.

A subsequent version of the model replaced $X_{4,n}$ with the ratio of the book value of equity to the book value of total liabilities so that the approach could be applied to private companies that did not have equity market values. Then, a later model was parameterised without $X_{1,n}$, since this formulation is more appropriate for non-manufacturing firms where the necessary working capital can differ greatly from industry to industry.

A key assumption underlying discriminant analysis is that the independent variables used are normally distributed. However, this is often not the case for financial ratios. Even within industry groups, they can have complex distributions, sometimes even being u-shaped with observations taking either very low or very high values. Also, if the items used to calculate ratios can take only positive values, then the ratios themselves will also be positive. This means that it is more likely that the distribution of ratios will be skewed, due to the lower bound of zero. Despite this, Altman's Z-score and its descendants have been widely used in credit scoring for a number of years.

The k-nearest neighbour approach

The probit and logit approaches can both be described as parametric, as can Altman's Z-score. However, non-parametric approaches can also be applied to credit modelling. One such approach is the k-nearest neighbour (kNN) approach. This approach has already been discussed in general, and in relation to credit modelling the characteristics could easily be the same accounting ratios that were used in probit and logit models, or in Altman's Z-score.

As discussed earlier, the choice of the number of neighbours is not straightforward, but techniques do exist. More difficult is the choice of dimensions, which requires judgement. There are also issues around the volume of calculations required. In particular, if the number of credits is large – as it would be for a bank considering its loan portfolio – then it might be impossible to run the model in a reasonable time frame.

Support vector machines

Another non-parametric approach described earlier is the support vector machine (SVM). The technical aspects of SVMs have been described, and each dimension in such a model could represent a different accounting ratio in the context of credit modelling.

This is a flexible approach, but care should be taken not to over-fit such models if non-linear versions are used.

The Merton model

A different approach to modelling credit risk is to use an equity-based approach, such as the contingent claims model of Merton (1974). This is more

Table 14.3. *Option pay-offs*

Price (X_t)	Call option (buyer)	Call option (writer)	Put option (buyer)	Put option (writer)
4	0	0	6	−6
8	0	0	2	−2
12	2	−2	0	0
16	6	−6	0	0

Table 14.4. *Pay-offs to investors*

Price (X_t)	Shareholders	Bondholders
4	0	4
8	0	8
12	2	10
16	6	10

appropriate for larger borrowers with liquid, frequently traded equity stock, since an accurate number for the volatility of the corporate equity is needed. The core assumption with this method is that the value of the firm as a whole follows a lognormal random walk and that insolvency occurs when the value of the firm falls below the level of debt outstanding. This means that the debt is being treated as a call option on the firm, and in this way the Merton model is closely related to the Black–Scholes model for option pricing.

To appreciate this link further, it is worth considering the pay-offs to various parties under financial options and in relation to a firm and its bond- and shareholders. First, recall the pay-offs excluding the option prices for put and call options exercised at time t with an exercise price, E, of ten units, where the price of the underlying asset is X_t, as shown in Table 14.3

Now consider instead a firm with a total value at time t of X_t and total debt, B, of ten units. The value of the firm to the bondholders, who own the debt, and to the shareholder, who own whatever is left, is given in Table 14.4.

From this, it should be clear that shareholders effectively have a call option on the underlying value of the firm, with an exercise price of the level of the firm's debt. The bondholders, on the other hand, are entitled to a fixed amount – the level of debt – but they have also written a put option on that debt, and the option is held by the firm. If the firm cannot repay the debt in full, it is essentially exercising the put option that it holds.

The values of these options can be calculated using the Black–Scholes model. However, in relation to credit risk it is sometimes helpful simply to know the probability that a variable will cross a particular level, without needing to know the degree to which that level is exceeded. The Merton model considers the probability that the value of an asset X at a fixed time T in the future, X_T, will be below some fixed level, B at the same time T, where the current value of the asset is X_0. The probability that X_T is lower than B is given by:

$$\Pr(X_T \leq B) = \Phi\left(\frac{\ln(B/X_0) - (r_X - \sigma_X^2/2)T}{\sigma_X\sqrt{T}}\right), \qquad (14.31)$$

where r_X is the expected increase in X_0. It should be clear that if r^* is substituted for r_X and E is substituted for B, then this expression can be written in terms from the Black–Scholes formula as $\Phi(-d2)$.

Looking at what happens as the parameters change is helpful to validate the model intuitively. As X_0 increases relative to B, $\Pr(X_T \leq B)$ falls. This is as would be expected. Similarly, a higher rate of growth in X_0 reduces $\Pr(X_T \leq B)$. However, increasing either σ_X or T results in greater uncertainty and, therefore, a higher probability.

As for the Black–Scholes model, an absence of transaction costs is assumed. There is also an assumption that X_t, where $0 \leq t \leq T$, increases in line with a lognormal random walk with a fixed rate of growth and volatility. However, as with the Black–Scholes model, these assumptions are not necessarily valid.

Another issue is that both the rate of growth of X_t and the volatility of this growth could be linked to the degree of leverage that exists within a firm – in other words, how much debt and how much equity a firm has. If the level of debt, B, is fixed, then the firm's leverage will change, increasing as X_t falls and falling as X_t rises. Such changes could have an impact on the profitability of a firm, and thus the pattern of its future growth. The Merton model does not allow for this.

Example 14.4 A firm has a total asset value of 500. The expected rate of growth of this asset value is 10% per annum, whilst its volatility is 30% per annum. If the firm's total borrowing consists of a fixed repayment of 300 that must be made in exactly one year's time, what is the probability that the firm will be insolvent at this point?

Merton's model gives the probability of default at time T as:

$$\Pr(X_T \leq B) = \Phi\left(\frac{\ln(B/X_0) - (r_X - \sigma_X^2/2)T}{\sigma_X\sqrt{T}}\right).$$

For this firm, $B = 300$, $X_0 = 500$, $r_X = 0.10$, $\sigma_X = 0.30$ and $T = 1$
Substituting these values into the above equation gives:

$$\Pr(X_1 \leq 300) = \Phi\left(\frac{\ln(300/500) - [0.10 - (0.30^2/2)]}{0.30}\right).$$

$$= 0.0296.$$

The probability of insolvency is therefore 2.96%.

The KMV model

Many subsequent authors have expanded on Merton's initial insight, including additions such as an allowance for coupons, more elaborate capital structures and negotiation between equity- and bondholders. However, the most commercially successful change came in the form of the KMV model (Kealhofer, 2003a, 2003b).

The KMV model was developed by the company of the same name founded by Stephen Kealhofer, John 'Mac' McQuown and Olrich Vasicek. KMV is now owned by Moody's, who are therefore able to apply both qualitative and quantitative ratings to firms.

Whilst the Merton model goes straight from the data to a probability of default, the KMV model uses an indirect route. The first stage is to replace B, the level of a firm's debt, with a variable $\tilde{B}$, which better represents the structure of these debts. In particular, it considers the term structure of these liabilities, allowing for the fact that insolvency over the next year will occur if payment dues to bondholders cannot be made over that period.

The KMV model also derives values for X_0 and σ_X from the quoted value of a firm's equity rather than assuming that they are directly observable. It does this by defining two equations that take advantage of the fact that the price of an equity can be regarded as a call option on the underlying assets of the firm. The first of the equations gives an expression for the value of the firm's equity, whilst the second gives an expression for volatility of that equity. Both items are observable. Each equation gives the dependent variable as a function of:

- the asset value of the firm, X_0;
- the asset volatility, σ_X;

- the capital structure, a function of X_0 and $\tilde{B}$; and
- the interest rate, r_X.

Since there are two equations and only X_0 and σ_X are unknown, it is possible to solve the two equations to find these terms.

Having found these two variables, the KMV model does not go straight to a probability of default. Instead, it calculates an interim measure, the distance to default, DD:

$$DD = \frac{X_0 - \tilde{B}}{X_0 \sigma_X}. \qquad (14.32)$$

This represents the number of standard deviations the firm value is from default. Distances to default are calculated for thousands of companies, solvent and insolvent, and calibrated with this data to give a default probability.

Credit migration models

Structural models are attractive because they give the probability that a firm will default. However, changes in market value can reflect changes in market sentiment as much as a view on the level or certainty of a firm's cash flows. This means that the results from the Merton and KMV models can change significantly, despite there being no real change in a firm's prospects. An alternative approach to arriving at default probabilities is to use a credit migration model, such as CreditMetrics.

Credit migration models use transition matrices to infer default probabilities. Most credit rating agencies produce transition matrices, which give the proportion of entities with a particular credit rating at the start of each year having various credit ratings at the end of that year. For example, Table 14.5 gives a transition matrix produced by Moody's Investor Services for default changes that took place in 2009. Tables 14.6 and 14.7 give similar information calculated over longer time periods. Moody's places each entity that it rates into one of nine categories, ranging from Aaa down to C, with those entities rated Aaa being the most secure, whilst those rated C have typically already defaulted on payments. Many entities rated Ca are also either in or near default, so in these tables the ratings Ca and C are combined. Each of the ratings from Aaa down to Caa is also subdivided into categories 1, 2 and 3, with the modifier 1 indicating that an entity is at the higher end of its category and the modifier 3 indicating that it is at the lower end.

Standard and Poor's has a similar rating structure, with ten ratings from AAA down to D. Here, those entities with a rating of AAA are considered the most secure with D denoting an entity in default. The Standard and Poor's system also has within-rating modifiers denoting more and less secure entities, in

Table 14.5. *Moody's 2009 one-year global migration rates (%)*

		\multicolumn{10}{c}{Year-end rating}
		Aaa	Aa	A	Baa	Ba	B	Caa	Ca-C	Default	Unrated
	Aaa	62.42	33.76	0.00	0.00	0.00	0.00	0.00	0.00	0.00	3.82
	Aa	0.00	70.98	22.62	1.04	0.15	0.00	0.00	0.00	0.00	5.21
	A	0.00	0.18	80.20	12.61	0.44	0.53	0.00	0.00	0.18	5.86
Initial	Baa	0.00	0.09	0.93	85.38	5.12	0.84	0.09	0.00	0.74	6.80
rating	Ba	0.00	0.00	0.00	3.85	71.54	13.27	0.77	0.58	2.31	7.69
	B	0.00	0.00	0.00	0.00	2.88	68.35	13.46	0.41	6.99	7.91
	Caa	0.00	0.00	0.00	0.00	0.00	7.59	48.81	6.51	28.20	8.89
	Ca-C	0.00	0.00	0.00	0.00	0.00	0.00	4.76	20.63	65.08	9.52

Source: Moody's Investor Services: 'Corporate Default and Recovery Rates, 1920–2009,' *Moody's Special Comment* (2010).

Table 14.6. *Moody's average one-year global migration rates, 1970–2009 (%)*

		\multicolumn{10}{c}{Year-end rating}
		Aaa	Aa	A	Baa	Ba	B	Caa	Ca-C	Default	Unrated
	Aaa	87.65	8.48	0.61	0.01	0.03	0.00	0.00	0.00	0.00	3.22
	Aa	1.01	86.26	7.82	0.34	0.05	0.02	0.01	0.00	0.02	4.47
	A	0.06	2.78	87.05	5.21	0.48	0.09	0.03	0.00	0.05	4.24
Initial	Baa	0.04	0.19	4.65	84.40	4.20	0.79	0.18	0.02	0.17	5.35
rating	Ba	0.01	0.06	0.38	5.66	75.74	7.25	0.53	0.08	1.13	9.16
	B	0.01	0.04	0.13	0.35	4.81	73.50	5.66	0.70	4.37	10.43
	Caa	0.00	0.02	0.02	0.16	0.44	8.17	59.90	4.25	14.72	12.32
	Ca-C	0.00	0.00	0.00	0.00	0.32	2.24	8.65	38.48	33.28	17.03

Source: Moody's Investor Services: 'Corporate Default and Recovery Rates, 1920–2009', *Moody's Special Comment* (2010).

this case through the addition of a '+' or a '−'. Transition matrices for Standard and Poor's are given as Tables 14.8 and 14.9.

The one-year default probability for a firm with a particular credit rating is therefore simply given by the number in the final column for the rating shown at the start of the row. For example, in 2009 the probability of default according to Moody's for an A-rated bond was 0.18%; for an B-rated bond, it was slightly higher at 6.99%; however, for a bond with either a Ca or a C rating, it was 65.08%. This highlights two features: that firms with higher credit ratings have lower default probabilities, but also that the relationship between

Table 14.7. *Moody's average one-year global migration rates,*
1920–2009 (%)

		Year-end rating									
		Aaa	Aa	A	Baa	Ba	B	Caa	Ca-C	Default	Unrated
	Aaa	86.82	8.06	0.81	0.16	0.03	0.00	0.00	0.00	0.00	4.11
	Aa	1.22	84.63	7.09	0.73	0.17	0.04	0.01	0.00	0.07	6.06
	A	0.08	2.96	84.84	5.47	0.67	0.11	0.03	0.01	0.09	5.74
Initial	Baa	0.04	0.29	4.50	81.30	5.01	0.79	0.13	0.02	0.29	7.63
rating	Ba	0.01	0.08	0.48	5.89	73.65	6.77	0.56	0.07	1.34	11.16
	B	0.01	0.05	0.16	0.60	5.79	71.60	5.45	0.55	3.91	11.90
	Caa	0.00	0.02	0.03	0.19	0.74	7.73	63.37	3.94	12.48	11.49
	Ca-C	0.00	0.00	0.11	0.00	0.44	2.97	7.48	54.35	22.15	12.51

Source: Moody's Investor Services: 'Corporate Default and Recovery Rates,
1920–2009', *Moody's Special Comment* (2010).

Table 14.8. *Standard and Poor's 2009 one-year global migration rates (%)*

		Year-end rating								
		AAA	AA	A	BBB	BB	B	CCC-C	Default	Unrated
	AAA	87.65	8.64	0.00	0.00	0.00	0.00	0.00	0.00	3.70
	AA	0.00	76.17	15.96	0.64	0.21	0.00	0.00	0.00	7.02
Initial	A	0.00	0.36	84.67	7.74	0.43	0.29	0.00	0.21	6.30
rating	BBB	0.00	0.00	2.00	83.71	5.94	0.80	0.20	0.53	6.81
	BB	0.00	0.00	0.00	3.09	72.95	11.48	0.60	0.70	11.18
	B	0.00	0.00	0.16	0.00	2.29	69.34	8.42	10.14	9.65
	CCC-C	0.00	0.00	0.00	0.00	0.00	6.32	27.37	48.42	17.89

Source: Standard and Poor's: '2009 Annual Global Corporate Default Study
and Rating Transitions', *Standard and Poor's Global Fixed Income Research*
(2010).

credit ratings and default probabilities is non-linear, increasing rapidly as credit
quality declines.

These figures also highlight a third feature – that 2009 was a particularly
bad year for defaults. This means that 2009 default rates might not necessarily
be a good indicator of the rates expected in 2010, 2011 or 2012, since default
rates vary over the economic cycle. One approach is to use the average default
rates over a longer period. Standard and Poor's calculate averages from 1981
to the current day as shown in Table 14.9, whilst Moody's give numbers from
1970 or even 1920 as shown in Tables 14.6 and 14.7. However, this replaces

Table 14.9. *Standard and Poor's average one-year global migration rates, 1981–2009 (%)*

		\multicolumn{9}{c}{Year-end rating}
		AAA	AA	A	BBB	BB	B	CCC-C	Default	Unrated
	AAA	88.21	7.73	0.52	0.06	0.08	0.03	0.06	0.00	3.31
	AA	0.56	86.60	8.10	0.55	0.06	0.09	0.02	0.02	4.00
Initial	A	0.04	1.95	87.05	5.47	0.40	0.16	0.02	0.08	4.83
rating	BBB	0.01	0.14	3.76	84.16	4.13	0.70	0.16	0.26	6.68
	BB	0.02	0.05	0.18	5.17	75.52	7.48	0.79	0.97	9.82
	B	0.00	0.04	0.15	0.24	5.43	72.73	4.65	4.93	11.83
	CCC-C	0.00	0.00	0.21	0.31	0.88	11.28	44.98	27.98	14.37

Source: Standard and Poor's: '2009 Annual Global Corporate Default Study and Rating Transitions', *Standard and Poor's Global Fixed Income Research* (2010).

the problem of excessive volatility, seen in structural models, with one of a complete lack of response to current economic conditions. It is always possible to look at default rates from a similar economic climate to that expected over the coming year, but such an approach is subjective and relies on firms being affected by similar conditions the same way at different points in time.

Perhaps a more sensible approach is to use credit migration models to calculate default probabilities over longer periods, preferably covering an economic cycle. This can be done if it is assumed that credit migrations follow a Markov chain process. In other words, this approach requires the assumption that the probability of a firm having a particular credit rating or indeed defaulting at time $t + 1$ depends only on what its credit rating is at time t, and is completely independent of its credit rating at time $t - 1$ or any prior time. One complication with this approach is that a number of issuers have their ratings withdrawn each year. This can happen for a number of reasons. Some are benign, such as the maturity of all rated bonds, or because of a merger or acquisition. However, ratings are also withdrawn if an issuer fails to provide information requested by the rating agency, or if the issuer decides it no longer wishes to be rated. There are a number of ways in which rating withdrawals can be dealt with, but the simplest is to assume that issuers whose ratings are withdrawn would have the same future patterns of changes to creditworthiness as those who retained their ratings. This means that each migration probability should be scaled up such that the total migrations excluding rating withdrawals sum to 100%.

Both Moody's and Standard and Poor's provide default probabilities calculated over a number of years as shown in Tables 14.10 to 14.12, so it is

Table 14.10. *Moody's average cumulative issuer-weighted global default rates, 1970–2009 (%)*

		Time horizon						
		1	2	3	4	5	10	15
	Aaa	0.00	0.01	0.01	0.04	0.11	0.50	0.93
	Aa	0.02	0.06	0.09	0.16	0.23	0.54	1.15
	A	0.05	0.17	0.34	0.52	0.72	2.05	3.57
Rating	Baa	0.18	0.49	0.91	1.40	1.93	4.85	8.75
	Ba	1.17	3.19	5.58	8.12	10.40	19.96	29.70
	B	4.55	10.43	16.19	21.26	25.90	44.38	56.10
	Caa-C	17.72	29.38	38.68	46.09	52.29	71.38	77.55

Source: Moody's Investor Services: 'Corporate Default and Recovery Rates, 1920–2009', *Moody's Special Comment* (2010).

Table 14.11. *Moody's average cumulative issuer-weighted global default rates, 1920–2009 (%)*

		Time horizon						
		1	2	3	4	5	10	15
	Aaa	0.00	0.01	0.03	0.08	0.16	0.85	1.36
	Aa	0.07	0.20	0.31	0.47	0.72	2.22	4.13
	A	0.09	0.28	0.57	0.91	1.26	3.30	5.51
Rating	Baa	0.29	0.84	1.55	2.32	3.14	7.21	10.93
	Ba	1.36	3.29	5.47	7.74	9.90	19.22	26.65
	B	4.03	9.05	14.05	18.50	22.42	36.37	44.75
	Caa-C	14.28	24.03	31.37	36.89	41.18	52.80	62.36

Source: Moody's Investor Services: 'Corporate Default and Recovery Rates, 1920–2009', *Moody's Special Comment* (2010).

possible to test whether the Markov chain assumption holds in practice – and it does not appear to. This is not surprising, since a firm being down-graded to a lower credit rating is unlikely to have the same characteristics as a long-term holder of that rating. In particular, it is likely either to be experiencing some temporary difficulties, or to be at the start of a continuing downward trend. However, the approximation is not bad.

An even simpler approach is to assume that credit migration follows a martingale process. This means that the expected credit rating at time $t + 1$ is the same as the credit rating at time t. This can clearly not be the case for the highest credit rating, since if the only way is down, the expected rating at time $t + 1$

Table 14.12. *Standard and poor's average cumulative issuer-weighted global default rates, 1981–2009 (%)*

		Time horizon						
		1	2	3	4	5	10	15
	AAA	0.00	0.03	0.14	0.26	0.39	0.82	1.14
	AA	0.02	0.07	0.14	0.24	0.33	0.74	1.02
	A	0.08	0.21	0.35	0.53	0.72	1.97	2.99
Rating	BBB	0.26	0.72	1.23	1.86	2.53	5.60	8.36
	BB	0.97	2.94	5.27	7.49	9.51	17.45	21.57
	B	4.93	10.76	15.65	19.46	22.30	30.82	35.74
	CCC-C	27.98	36.95	42.40	45.57	48.05	53.41	57.28

Source: Standard and Poor's: '2009 Annual Global Corporate Default Study and Rating Transitions', *Standard and Poor's Global Fixed Income Research* (2010).

must be lower than the credit rating at time t. Furthermore, since default probabilities can be no higher than one, an approach that simply scales a one-year probability will inevitably produce impossible answers given a long enough time horizon; however, it can be used to give a very rough estimate of default over a multi-year period.

> **Example 14.5** A firm has Standard and Poor's credit rating of A. Using the credit migration rates averaged over 1981 to 2009, what is the probability that the firm will have defaulted in two years time? What is the answer using an N times one-year approximation? How do these results compare to the actual two-year default rates for 1981 to 2009?
>
> The firm has a 0.08% chance of defaulting before the end of the first year. Scaling this up to allow for the 4.83% of issuers losing their rating leaves the result at 0.08%. If the firm survives the first year, then the following probabilities can be calculated, again allowing for the issuers losing their ratings:
>
> - the probability that the firm will be promoted to AAA and then default is $(0.04\%/(1-4.83\%)) \times (0.00\%/(1-3.31\%)) = 0.00\%$;
> - the probability that the firm will be promoted to AA and then default is $(1.95\%/(1-4.83\%)) \times (0.00\%/(1-4.00\%)) = 0.00\%$;
> - the probability that the firm will remain at A and then default is $(87.05\%/(1-4.83\%)) \times (0.08\%/(1-4.83\%)) = 0.08\%$;
> - the probability that the firm will be demoted to BBB and then default is $(5.47\%/(1-4.83\%)) \times (0.26\%/(1-6.68\%)) = 0.02\%$;

- the probability that the firm will be demoted to BB and then default is $(0.40\%/(1-4.83\%) \times (0.97\%/(1-9.82\%) = 0.00\%$;
- the probability that the firm will be demoted to B and then default is $(0.16\%/(1-4.83\%) \times (4.93\%/(1-11.83\%) = 0.01\%$; and
- the probability that the firm will be demoted to CCC-C and then default is $(0.02\%/(1-4.83\%) \times (27.98\%/(1-14.37\%) = 0.01\%$.

Summing these probabilities gives a total of 0.20% over two years.

If the two-year default probability is simply taken to be $2 \times 0.08\%$, then the result is instead 0.16%, which is close to the value obtained using the migration approach.

However, both of these results are lower than the two-year default rate calculated directly from the data, which is 0.21%.

There are also a number of practical issues with credit migration models. Credit ratings do not give a high level of granularity – the number of available ratings is small compared with the number of rated firms. Having said this, credit ratings are unavailable for the vast majority of firms. Obtaining a rating is not free, and given that the main purpose of being rated is to reduce the cost of borrowing, the level of borrowing needs to be sufficient to justify the expense of obtaining a rating. A further issue is that different agencies can also produce different ratings for the same firm, particularly financials.

CreditMetrics uses the probability of a change in rating together with estimated recovery rates and volatility in credit spreads to estimate the standard deviation of the value of a corporate bond due to credit quality changes. This is done using the following approach:

- calculate the value of a bond in one year's time for each potential credit rating;
- multiply each bond value by the probability of having that credit rating;
- sum these items to get the expected bond value;
- deduct this from the bond value at each potential credit rating;
- square each result and multiply it by the probability of having that credit rating; and
- sum these items to get the variance of the bond value.

Example 14.6 A firm has a bond in issue that has a Standard and Poor's credit rating of BBB. The projected values of the bond, allowing for changes in gross redemption yield and, when relevant, default and recovery, are given below for each credit rating. Using a credit migration approach based on Standard and Poor's migration rates from 1981 to 2009, what are the expected value and variance of the value of the bond?

Year-end rating	Value given rating
AAA	104.27
AA	103.18
A	102.10
BBB	100.00
BB	94.98
B	90.29
CCC-C	81.78
Default	61.97
Unrated	–

Using the process outline above, an additional column is needed giving the probability of migration to the various credit ratings. A second column is then added giving the probabilities adjusted for rating withdrawals. This column is then multiplied by the value given the credit rating to arrive at a probability weighted value. The sum of these values gives the mean, which can then be deducted from the values given the credit ratings to arrive at the difference from the mean. Each of these values is then squared and multiplied by the probability of occurrence. The sum of these results gives the variance of returns.

Year-end rating	Value given rating	Pr of rating (%)	Adjusted Pr of rating (%)	Pr-weighted value	Value less mean	Pr-weighted squared value
AAA	104.27	0.01	0.01	0.01	4.61	0.0021
AA	103.18	0.14	0.15	0.15	3.52	0.0173
A	102.10	3.76	4.03	4.11	2.45	0.2250
BBB	100.00	84.16	90.18	90.18	0.34	0.0985
BB	94.98	4.13	4.43	4.20	−4.68	0.9033
B	90.29	0.70	0.75	0.68	−9.37	0.6146
CCC-C	81.78	0.16	0.17	0.14	−17.87	0.5111
Default	61.97	0.26	0.28	0.17	−37.69	3.6931
Unrated	–	6.68	–	–	–	–
Total				99.66		6.0651

Therefore the expected value of the bond is 99.66 with a variance of 6.0651.

14.5.4 Credit portfolio models

It is important to be able to quantify groups of credit risks together. This might be for portfolios of loans or mortgages that a bank is keeping on its books, but it might also be to price credit derivatives. These derivatives are discussed in more detail in Chapter 16 as ways in which credit risk can be managed, but a key feature of these derivatives is that the impact of defaults is magnified for investors in particular classes of investor. Since the overall pattern of defaults is heavily dependent on the relationships between the underlying securities, it is important that these relationships are modelled accurately.

A key issue with portfolios of credits is that the relationship between the underlying securities or loans will change with the economic climate. In particular, the distributions have *jointly fat tails*. For example, whilst the price movements of corporate bonds might appear to be relatively independent when those movements are small, the correlations will often increase substantially when price movements are large and negative.

Multivariate structural models

A simple way to allow for the portfolio aspect of credit modelling is to construct a multivariate version of the Merton model. This involves modelling the values of the firms under consideration using some sort of multivariate model. The most obvious would be a multivariate lognormal model, linked by a matrix of correlations between the firm values. However, the logarithm of firm values could be modelled using a multivariate t-distribution, or an explicit copula could be used to model the relationship between the asset values.

Example 14.7 The firm X has a total asset value of £500m, whilst the firm Y has a total asset value of £800m. The expected rate of growth of X's asset value is 10% per annum, whilst its volatility is 30% per annum. For Y, the expected rate of growth is 5% per annum with a volatility of 10% per annum. The returns of the two firms are linked by a Frank copula with a parameter, α, of 2.5. If the total borrowing for firm X consists of a fixed repayment of £300m, whilst for Y it is £750m, in each case repayable in exactly one year's time, what is the probability that the both firms will be insolvent at this point?

Merton's model gives the probability of default at time T as:

$$\Pr(X_T \leq B) = \Phi\left(\frac{\ln(B/X_0) - (r_X - \sigma_X^2/2)T}{\sigma_X\sqrt{T}}\right).$$

The default probability for firm X was established in Example 14.5.3 as 2.96%. For firm Y, $B = 750$, $Y_0 = 800$, $r_Y = 0.05$, $\sigma_Y = 0.10$ and $T = 1$ Substituting these values into the above equation gives:

$$\Pr(Y_1 \leq 800) = \Phi\left(\frac{\ln(750/800) - [0.05 - (0.10^2/2)]}{0.10}\right)$$

$$= 0.1367.$$

The joint probability under a Frank copula in terms of firm values at time T is given by:

$$\Pr(X_T \leq x \text{ and } Y_T \leq y) = -\frac{1}{\alpha}\ln\left[1 + \frac{(e^{-\alpha F(x)} - 1)(e^{-\alpha F(y)} - 1)}{e^{-\alpha} - 1}\right].$$

Here, for $T = 1$, the values needed are $F(x) = 0.0296$, $F(y) = 0.1367$ and $\alpha = 2.5$. This gives:

$$\Pr(X_1 \leq 300 \text{ and } Y_1 \leq 750) = -\frac{1}{2.5}\ln\left[1 + \frac{-0.0713 \times -0.2895}{-0.9179}\right]$$

$$= 0.0091.$$

The probability that both firms will be insolvent is, therefore, 0.91%.

Similarly, the derived asset values and volatilities in the KMV model can be parameterised and simulated in a multivariate context, either through a multivariate distribution or with an explicit copula.

Multivariate credit migration models

CreditMetrics, mentioned above, is actually a multivariate credit migration model, used to determine various risk measures for portfolios of corporate bonds. However, to move from the single-bond approach described above, a number of additional steps are needed.

The first stage taken by CreditMetrics is to use the Merton model to link the migration probabilities to changes in the underlying asset value of a firm, meaning that the change in the value of a firm's assets is assumed to have a normal distribution. Each credit rating has its own probability of default, with that probability being higher for lower credit ratings. This means that a change in rating can be regarded as a change in the underlying value of the firm. In particular, the change in rating is a function of the change in the value of a firm's assets and the volatility of those assets.

The next stage is to consider the correlation between these asset values for different firms. Since correlations are dimensionless, the exact value of a firm's volatility does not matter so long as it is fixed, meaning one fewer parameter that is needed for each firm. To calculate the correlations, the firm's equity value is used as a proxy for its asset value, the rationale being that most of the volatility in a firm's value will be reflected in the equity price rather than the bond price.

Correlations between the equity values are not calculated directly; instead, equity returns are modelled by a range of country-specific industry indices, with the unexplained variation defined as independent firm-specific volatility.

Simulations of the indices and the independent firm-specific factors are then produced, giving consistent simulations of the values of the firms. These simulations are in terms of the number of standard deviations moved by each firm in each simulation, which means that they can be mapped back to a change in rating for each firm. This can itself be converted to a change in bond value, meaning that when the results are aggregated over all firms for each simulation, the change in portfolio value is given. These changes in portfolio value can then be converted to the desired measure of risk.

Common shock models

A simple way to model default is to assume that bond defaults are linked by Poisson processes, and are subject to shocks affecting one, several or all of the bonds. This means that the probability of receiving all of the payments due can be modelled using a multivariate Marshall–Olkin copula.

However, if each bond itself defaults according to a Poisson process and a common time horizon for all bonds is considered, then the probability the N firms subject to M shocks survive to time T can be simplified to:

$$\text{Pr(no defaults)} = e^{-\sum_{m=1}^{M} \lambda_m T}. \tag{14.33}$$

The probability of exactly one default can be obtained by looking at the ways in which such a default could occur. In particular, there will be only one default when a shock occurs that affects only one bond, and no other shocks occur. For N firms there will be a maximum of N of the $M = 2^N - 1$ shocks that can produce this outcome. Consider, for example, the situation where $N = 3$, so $M = 7$. This means that:

- three Poisson shocks, λ_1, λ_2 and λ_3 affect only one firm each;

- three Poisson shocks, λ_{12}, λ_{13} and λ_{23} affect two firms each; and
- a single Poisson shock, λ_{123}, affects all three firms.

When the Poisson probability of default is λ, the probability of a bond staying out of default over a time period T is $e^{-\lambda T}$. This means that the probability of the bond defaulting is $1 - e^{-\lambda T}$. Therefore the total probability of a single default is:

$$\text{Pr(exactly one default)} = (1 - e^{-\lambda_1 T})e^{-(\lambda_2 + \lambda_3 + \lambda_1 2 + \lambda_1 3 + \lambda_2 3 + \lambda_1 23)T} +$$
$$(1 - e^{-\lambda_2 T})e^{-(\lambda_1 + \lambda_3 + \lambda_1 2 + \lambda_1 3 + \lambda_2 3 + \lambda_1 23)T} +$$
$$(1 - e^{-\lambda_3 T})e^{-(\lambda_1 + \lambda_2 + \lambda_1 2 + \lambda_1 3 + \lambda_2 3 + \lambda_1 23)T}.$$

$$(14.34)$$

Generalising this for N bonds with M shocks, where $\lambda_1, \lambda_2, \ldots, \lambda_N$ are the Poisson means for the shocks affecting only single bonds, whilst the parameters $\lambda_{N+1}, \lambda_{N+2}, \ldots, \lambda_M$ are the Poisson means for the shocks affecting more than one bond, gives the following expression for the probability of a single default:

$$\text{Pr(exactly one default)} = \sum_{n=1}^{N}(1 - e^{-\lambda_n T})e^{-\left(\sum_{m=1}^{M} \lambda_m - \lambda_n\right)T}. \qquad (14.35)$$

This approach can be extended to calculate the probabilities of more than one default occurring, but the number of combinations of defaulting and non-defaulting bonds can soon become very large.

Time-until-default models

Time-until-default or survival models describe the defaults in a portfolio of bonds in terms of copulas linking the time at which a bond defaults.

A survival function $\bar{F}(t)$ is defined for each bond. This gives the probability that a bond will not have defaulted by time t, and it can be expressed in terms of a hazard rate function, $h(t)$, as follows:

$$\bar{F}(t) = e^{-\int_0^t h(s)\,ds}. \qquad (14.36)$$

If $h(t)$ is taken to be a constant, h, then this expression becomes:

$$\bar{F}(t) = e^{-ht}. \qquad (14.37)$$

The probability that a bond will have defaulted by time t is given by the distribution function, $F(t) = 1 - \bar{F}(t)$. Since the density function $f(t)$ is given by the first differential of $F(t)$ with respect to t, this means that

$$f(t) = he^{-ht}. \tag{14.38}$$

In other words, if the hazard rate is constant, then the survival time has an exponential distribution with parameter h.

The hazard rate can be estimated from a number of sources – the Merton model, published credit ratings and historical default information can all be used – but the way in which these sources are employed is the same. This involves looking at the implied default probability, α, over a defined time horizon, setting the distribution function equal to this probability and solving for h:

$$F(t) = 1 - e^{-ht} = \alpha, \text{ so } h = -\frac{\ln(1-\alpha)}{t}. \tag{14.39}$$

The next stage is to link these default times. This can be done using copulas, parameterised by some measure of correlation between the default times. The normal copula has been widely used, but there is no reason why another copula could not be used instead. Indeed, given the fact that defaults are likely to occur more widely in poor credit environments, perhaps a copula with higher tail dependence is more appropriate. Such a model can then be used to calculate the likelihood of a particular aggregate default rate for a portfolio of bonds.

14.5.5 The extent of loss

Most of the analysis of credit risk concentrates on the probability of loss. However, the extent of loss must also be assessed. In practice, the recovery rate rather than the proportion of lost is modelled. Two distinct measures of recovery are the price after default and the ultimate recovery. The former is a short-term measure and the latter has a longer time horizon. The ultimate recovery is often significantly larger than the price after default.

A number of factors can affect the expected recovery, including (de Servigny and Renault, 2004):

- the seniority of the obligation;
- the industry;
- the point in the economic cycle;
- the degree and type of collateralisation;

- the jurisdiction; and
- the composition of the creditors.

The impact of these factors is often modelled using historical data. The results can be translated into a deterministic expectation of recovery rate applied to all debt, or stochastic recovery rates can be modelled, allowing for the volatility of recovery rates calculated from historical data as well as their expected values.

If the recovery rate is to be parameterised, a distribution bounded by zero and one, such as the beta distribution, is most appropriate. However, non-parametric approaches using kernel estimation are useful for more complex distributions, including bi- or polymodal distributions.

14.5.6 Credit risk and market risk

One issue with some credit risk portfolio models is that, whilst they model the credit risks in a portfolio sense, they sometimes ignore other risks which will be closely linked to credit risk. Most institutions are exposed to a range of risks, of which credit is only one. In particular, the market risk of any asset portfolio may well be linked to the credit risks. Pension schemes provide a prime example. They generally have a disproportionally large exposure to the credit risk of the sponsoring employer, but are often subject to significant market risks which are not independent of the credit risk borne. The relationship between various credit risks, and between credit and other financial risks, therefore needs to be considered. If credit and market risks have each been measured independently using sophisticated methods, then this effort will have been wasted if the risks are assumed to be independent or linked using a crude measure such as correlation.

Another fundamental way in which credit risk and market risk are linked is when the credit risks have duration, such as with long-term fixed-rate loans, or corporate bonds. In this case, valuing the credit risk involves linking the risks to a yield curve and modelling the yield curve risk as well, as discussed in the interest rate risk section above.

14.6 Liquidity risk

Both funding and market liquidity risks need to be assessed by financial institutions. This involves analysing the potential outflows and ensuring that the assets held are sufficiently liquid or provide sufficient cash flows to provide the required liquidity with an acceptable degree of confidence. Whilst it is tempting to apply the quantitative techniques discussed above to liquidity risk, this is rarely possible. The data on liquidity crises are limited, and it is crises that are the focus of liquidity risk modelling. Furthermore, liquidity risk will occur in every organisation in a different way, so industry information on liquidity problems is of little used from a modelling point of view. Instead, stress testing is more commonly used. This involve projecting cash inflows and outflows under a range of scenarios. However, before considering the range of scenarios it is important to understand the nature of the cash inflows and outflows faced by an organisation.

The starting point is to gain an understanding of liabilities, in particular their term and potential variation in this term. Pension schemes offer the least variation, since there are few options to accelerate or postpone payments, except before retirement for the former and at retirement for the latter. For insurance companies, there are more options, with some products offering early withdrawal – often with a penalty – or the option to lapse. Banks, however, face the biggest issues, with even the longer-term products often having early withdrawal options and many accounts being instant access.

Having understood the liabilities, it is of course important to understand the assets, including the timing and certainty of payment streams, the potential ability to sell assets within particular time frames and the price at which such assets might be realised. This final point is an important issue addressed again later.

Each scenario involves the projection of both asset and liability cash flows. Ideally, there should be no scenarios for which cash cannot be found to meet outgoings. A variety of short- and long-term scenarios should be considered, covering periods as short as a few days and as long as a week. They should also take into account both institution-specific and market-wide stresses. A range of scenarios might comprise the following:

- rising interest rates;
- ratings downgrade;
- large operational loss;
- loss of control over a key distribution channel;
- impaired capital markets;
- large insurance claim from a single or related events; and
- sudden termination of a large reinsurance contract.

The final two items relate exclusively to insurance companies, but the others affect banks as well. A rise in interest rates could see holders of bank accounts or insurance contracts withdrawing their assets in search of higher returns elsewhere. Conversely, money could be taken away following a ratings downgrade if the credit worthiness led these people to seek a more secure home for their assets. Rather than a loss of customer assets, an operational loss could cause a drain on funds, or for an insurance company a single large loss or a series of losses with a common cause could require larger than expected payments. Rather than losing monies already held, future cash flows could be disrupted if a distribution channel closed, reducing the amount of new income. Similarly, if capital market liquidity fell, then raising capital from that source could also be difficult. Finally, for insurance companies, the termination of a large reinsurance contract could leave an institution exposed to large cash outflows in the case of a large claim.

When modelling sources of liquidity, it is important to allow for potential limits on transfers of assets between legal entities. In particular there may be legal, regulatory and operational issues that limit the extent to which liquidity in one part of an organisation can be used to provide liquidity elsewhere in the group.

As mentioned earlier, it is important to allow for interactions between liquidity and other risks, in particular market and interest rate risks in the scenario specifications. When liquidity is low, many asset values may well be depressed, limiting the amount their sale would raise.

14.7 Systemic risks

Systemic risks are usually – but not always – extensions of market risk. This means that they require the model used to contain particular features.

Feedback risk implies that returns exhibit some degree of serial correlation. This implies that the model used to project a series of returns should include this feature if feedback risk is thought to be relevant. One potential issue is that serial correlation implies that returns in a period can in part be derived from returns in past periods. This in turn suggests the possibility of arbitrage. However, if arbitrage opportunities exist, then there is also the possibility that arbitrageurs will try to exploit the opportunities, moving prices to the extent that the opportunity vanishes. For this reason, the possibility of arbitrage is often excluded from financial models.

Contagion risks relate to the interaction between different financial series. In particular, they suggest that certain series might be more closely linked for extreme negative values. This means that the linkages between series are perhaps better modelled using copulas. This assumes that sufficient information is available to parameterise such copulas.

14.8 Demographic risk

14.8.1 Types of demographic risk

As mentioned earlier, there are many types of demographic risk. However, mortality and longevity risk usually receive more attention. There are four types of mortality risk:

- level;
- volatility;
- catastrophe; and
- trend.

The same risks also exist for longevity, except for catastrophe risk – there is no possibility of a one-year-only fall in underlying mortality rates, whilst one-off spikes can occur as a result of wars and pandemics. I discuss the way in which each of these risks is modelled below.

14.8.2 Level risk

There are two main ways in which the current underlying level of mortality can be determined:

- from past mortality rates for a group of lives; and
- from the underlying characteristics of those lives.

The first of these approaches is known as experience rating, whilst the second is known as risk rating. Both approaches are often used together, the relative contribution of each measure being determined by a measure of credibility attached to the experience.

Experience rating

Experience rating involves looking at the number of deaths that have occurred in a portfolio of lives to determine the mortality rate at each age. Data can be used to calculate two rates of mortality that might be of interest: the central rate of mortality and the initial rate of mortality.

The central rate of mortality gives the number of deaths as a proportion of the average number of lives over a particular period. The result can be used as an approximation for the force of mortality, which is the instantaneous rate of mortality applying at any point in time, analogous to the force of interest.

The initial rate of mortality gives the number of deaths as a proportion of the number of lives present at the start of a particular period. This is usually a more practical measure since it is generally the number of lives at the start of a period that is known rather than the average number of lives over a period.

If the number of deaths in a particular period – usually a year – for a group of lives aged x is d_x, and the number of lives aged x at the start of the period is l_x, then the central mortality rate, m_x, is:

$$m_x = \frac{d_x}{(l_x + (l_x - d_x))/2} = \frac{d_x}{l_x - (d_x/2)}. \tag{14.40}$$

In other words, the denominator is calculated assuming that the deaths occur uniformly over the period. The calculation of the initial mortality rate, q_x, is more straightforward:

$$q_x = \frac{d_x}{l_x}. \tag{14.41}$$

When calculating the mortality rates for a group of lives, it is important to divide the data into homogeneous groups, where possible. At the most basic level, this means means calculating different variables for males and females. However, for a life insurance company, writing different classes of business, it will usually be desirable to calculate separate mortality rates for each class. For pension schemes, it might be possible to calculate different rates for different types of employee, such as managers and staff. However, there is a trade-off between ensuring that groups are as homogeneous as possible whilst making sure that no group is so small that the differences in mortality are hidden by random variation.

A similar compromise is needed when deciding the period of time from which data should be taken. Using raw data covering a long period of time gives more deaths and a larger effective population from which rates can be calculated. However, the earlier data are less relevant to current mortality rates, and the final rate calculated may hide an underlying trend in the rates.

Risk rating

The alternative approach to estimating the current underlying mortality profile of a group of lives is to use risk rating. The broad process behind such an approach is as follows:

- divide the population as a whole into a number of homogeneous groups;
- derive expressions for the mortality of each of the groups in terms of a range of risk factors;

- analyse the structure of the group of lives of interest – for example, a portfolio of annuitants – in terms of these risk factors;
- use these risk factor exposures to infer the underlying mortality of the group of interest.

The risk factor analysis can be carried out using generalised linear models (GLMs), in particular logit or probit models. This involves using the mortality rate as the dependent variable, and items such as socio-economic group as the independent variables. The result is a formula which means that, if the independent variables are known for a particular portfolio of lives, then the underlying mortality rate can be calculated. Survivor models can also be used to reflect the impact of these factors on a broader function of mortality.

A recent innovation that aggregates the effect of a number of different underlying factors is postcode rating. This involves grouping postcodes by the type of population that lives there, using marketing classifications, and calculating the mortality rates for those classifications. This means that, if the underlying mortality to which an individual is exposed is needed, the individual's postcode can provide that information.

However, as attractive as risk rating is, it cannot allow for the fact that individuals will not necessarily conform to their risk factor stereotype. This is particularly important if a group of lives has a particular characteristic that is not picked up by the risk factors, meaning that experience rating is always helpful. What is needed, then, is a way of linking the results of the experience and risk rating approaches.

Credibility
Credibility, described earlier, can be used to combine experience and risk rating information. This involves choosing what credibility weighting, Z, is applied to the mortality rate calculated using experience rating. The balance of the estimate, coming from risk rating, is weighted by $(1 - Z)$.

14.8.3 Volatility risk (*Process risk*)

Volatility risk occurs because the number of individuals in a pension scheme or insurance portfolio is finite. This means that even if the nature of the underlying population is correctly identified, the number of deaths occurring could easily differ from that predicted.

Volatility risk can be modelled stochastically by assuming that deaths occur according to some statistical process. The most obvious is a binomial process, but assuming a Poisson distribution can give a good approximation when mortality rates are low. In either case, simulated future populations can be obtained

Table 14.13. *Selected probabilities for the binomial distribution* $(n = 100, p = 0.5)$

x	$\Pr(X = x)$	$\Pr(x \leq X)$
1	0.0059	0.0059
2	0.0312	0.0371
3	0.0812	0.1183
4	0.1396	0.2578
5	0.1781	0.4360
⋮	⋮	⋮
10	0.0167	0.9885
⋮	⋮	⋮
100	0.0000	1.0000

by projecting the underlying mortality rates forward and using these rates as the input for a binomial or Poisson process. This involves deriving the cumulative probability distribution, generating a series of random numbers between zero and one, and reading off the number of deaths that each random number infers. For example, if there is a population of one hundred at a particular age and the underlying probability of death for each individual is 5% per annum, then the expected number of deaths over the next year would be five. However, the distribution of deaths is quite broad, as shown in Table 14.13.

A random number of deaths can therefore be generated by simulating a uniform random number between zero and one, U, and determining the greatest number deaths for which the cumulative probability is less than or equal to U. A similar approach can be used if deaths are assumed to follow a Poisson distribution.

Volatility risk is also important when fitting mortality models, since the level of volatility risk differs at different ages. For this reason many mortality models are fitted not through least squares optimisation, but through Poisson maximum likelihood estimation. The first stage in this process is to define the expected number of deaths as a function of age, time or some other variable. This becomes the Poisson mean. The probability of the observed number of deaths at each age and in each period is then calculated in terms of this function. These probabilities are then multiplied together to give a likelihood function, and the parameters in the function giving the expected number of deaths are calculated such that the likelihood function is maximised.

Example 14.8 The numbers of deaths in a given year and the initial population sizes for that year at ages 80, 90 and 100 are shown below:

Age (x)	Initial population (l_x)	Deaths (d_x)
80	250	20
90	80	14
100	7	3

It has been suggested that the initial mortality rate for age x, q_x, could be modelled as a log-linear function of age, with the estimated initial mortality rate, $\hat{q}_x$, being equal to $a + bx$. Show that if such a model is fitted using a Poisson maximum likelihood approach, $a = -9.05$ and $b = 0.08$.

The Poisson probability of there being d_x deaths at age x is given by $f(d_x) = e^{-\lambda_x} \lambda_x^{d_x} / d_x!$, where $\lambda_x = \hat{q}_x l_x$ and l_x is the initial population at age x. Inputting the values of $a = -9.05$ and $b = 0.0815$, and values either side, into this formula gives the results shown in Table 14.8.

Age (x)	Initial population (l_x)	Deaths (d_x)	$f(d_x)$ a b	$f(d_x)$ a $b_{-1\%}$	$f(d_x)$ a $b_{+1\%}$	$f(d_x)$ $a_{-1\%}$ b	$f(d_x)$ $a_{+1\%}$ b
80	250	20	0.0888	0.0847	0.0855	0.0823	0.0814
90	80	14	0.1054	0.1045	0.0984	0.0957	0.1032
100	7	3	0.2231	0.2183	0.2238	0.2236	0.2176
Likelihood			0.0021	0.0019	0.0019	0.0018	0.0018

where the subscripts -1% and $+1\%$ represent deviations of 1% either side of the two parameters. The values given in the final row show that the values of $a = -9.05$ and $b = 0.0815$ maximise the likelihood of the observed deaths occurring.

14.8.4 Catastrophe risk

Catastrophe risk occurs when there is a large, temporary increase in mortality rates. This can be due to wars, pandemics or some other common risk factor.

There are a number of ways that catastrophe risk can be modelled. Scenario analysis can be used to determine the effect of particular changes to mortality rates, for example a 20% increase in mortality across all age groups. However, it is also possible to model more complex dependencies between individual lives by linking them with copulas.

14.8.5 Trend risk

Trend risk is the risk that mortality rates will change in such a way that causes financial loss. For pension schemes, this means that mortality will improve more quickly than expected; for term assurance portfolios, it means that improvements will not be as fast as in the past.

There are two aspects to trend risk that are important. The first is determining the expected levels of mortality rates in the future, whilst the second is assessing the uncertainty in these predictions. There are also two broad types of approach that can be used: parametric and non-parametric.

The most common non-parametric approach used to project mortality rates is the P-spline approach. This uses penalised splines to smooth historical mortality rates, and then to project rates into the future. However, as with all non-parametric methods, this approach cannot be used to simulate large numbers of potential outcomes, so gives no indication of the uncertainty of mortality projections. For this, parametric mortality models are needed.

Parametric models describe mortality rates as a function of a range of factors. These factors can be projected stochastically, and therefore used to generate simulated future mortality rates.

Most parametric mortality models are aggregate or all-cause models, which consider the mortality rates from all causes of death in a single rate. However, cause-of-death models are being used increasingly to project mortality rates. This can be important if falls in aggregate rates of mortality are due to large reductions in mortality from a particular cause of death.

Parametric mortality models typically use two or more of the following factors to describe and project mortality rates:

- the age to which the rate applies, x;
- the calendar year or period in which the rate applies, t; and
- the year of birth – or cohort – to which the rate applies, $c = t - x$.

The importance of age is clear. Mortality rates tend to be different at different ages. In particular, they tend to rise at an increasing rate with age, apart from an initial fall in the months after birth, and the presence of a 'mortality hump' particularly for males – around the late teens and early twenties. The first of these effects reflects the fact that surviving the first few months after birth is more difficult than surviving the subsequent few years, whilst the second reflects the higher propensity of young men to take risks.

The importance of time – the period effect – is also usually clear. Mortality rates for a particular age tend to fall with time. However, this is not always the case, as has been seen in Russia following the collapse of the Soviet Union.

Finally, there is the cohort effect. This allows for the fact that people born in particular years can experience heavier or lighter mortality than those born before or later, the effect being additional to the age and period effects.

The 'raw' mortality function modelled will generally be either an initial mortality rate, $q_{x,t}$, or a central mortality rate, $m_{x,t}$. Whilst the latter has theoretical attractions, being more closely linked to the force of mortality, the former is of more practical use. However, both of these rates of mortality tend to have non-linear relationships with age, increasing rapidly as the population ages. For this reason, either the natural logarithm or the logit of the mortality rate is typically used as the dependent variable.

The Lee–Carter model

A simple and popular approach to describing and projecting mortality is found in the Lee–Carter model (Lee and Carter, 1992). This models mortality rates using two age-related parameters, one of which is constant and the other of which varies by age. The dependent variable used in the original model is the natural logarithm of the central mortality rate. This means that the model can be written as:

$$\ln m_{x,t} - \alpha_{0,x} + \alpha_{1,x}\beta_{1,t} + \epsilon_{x,t}. \tag{14.42}$$

There is no unique solution to this model, so some restrictions are needed. These are that:

- the sum over x of $\alpha_{1,x}$ is equal to one; and
- the sum over T of $\beta_{1,t}$ is zero.

This means that $\alpha_{0,x}$ is the average value of $\ln m_{x,t}$ over all t for each x.

The Lee–Carter model was originally fitted by applying singular value decomposition (SVD) to a table of the natural logarithms of central mortality rates. However, Poisson maximum likelihood approaches have also been used. These combine the rates implied by the model with the population sizes at each age in each period to give a series of Poisson means. These are then used to generate the probabilities of deaths actually observed, the probabilities are combined into a likelihood function, and the parameters are chosen to maximise this likelihood function.

This model can be used to project mortality rates by taking the age-specific variables, $\alpha_{0,x}$ and $\alpha_{1,x}$, and applying them to projected values of the time-specific variable, $\beta_{1,t}$. A typical approach for producing simulated values of $\beta_{1,t}$ is as follows:

- calculate the series $\Delta\beta_{1,t} = \beta_{1,t} - \beta_{1,t-1}$;
- calculate $\mu_{\Delta\beta}$ and $\sigma_{\Delta\beta}^2$ as the mean and variance of $\Delta\beta_{1,t}$;
- generate a series of normally random variables with mean $\mu_{\Delta\beta}$ and variance $\sigma_{\Delta\beta}^2$; and
- add these variables to the most recent fitted estimate of $\beta_{1,t}$ to give projected values of $\beta_{1,t}$.

The Renshaw–Haberman model

The Renshaw–Haberman model (Renshaw and Haberman, 2006) uses the same broad approach as the Lee–Carter model, but with the addition of an age-related cohort parameter:

$$\ln m_{x,t} = \alpha_{0,x} + \alpha_{1,x}\beta_{1,t} + \alpha_{2,x}\gamma_{2,t-x} + \epsilon_{x,t}. \tag{14.43}$$

As with the Lee–Carter model, there is no unique solution here, so some restrictions are again needed. These are that:

- the sum over x of $\alpha_{1,x}$ is equal to one;
- the sum over x of $\beta_{1,t}$ is equal to zero;
- the sum over x of $\alpha_{2,x}$ is equal to one; and
- that the sum over t of $\gamma_{2,t-x}$ is zero.

This means again that $\alpha_{0,x}$ is the average value of $\ln m_{x,t}$ over all t for each x.

This model is more difficult to parameterise than the Lee–Carter model. One approach is to use an iterative process, alternately holding values of $\beta_{1,t}$ and $\gamma_{2,t-x}$ constant, whilst the model is fitted by adjusting the variable that is left free to change.

Simulated mortality rates can again be derived by producing projected values for $\beta_{1,t}$. However, whilst this can be used to give values for all the ages included in the dataset for the Lee–Carter model, the minimum age for which projections can be produced under the Renshaw–Haberman model increases by one each year into the future if only $\beta_{1,t}$ is projected. To be precise, projections can be produced only for age and time combinations where $t - x$ is no greater than the largest value of $c = t - x$ from the data. To counter this, projected values for $\gamma_{2,t-x}$ must also be derived – in other words, the nature of future cohorts must be predicted. This can be achieved using the same projection approach as is applied to $\beta_{1,t}$ if there is thought to be a pattern to the development of future cohorts.

The Cairns–Blake–Dowd models

The original Cairns–Blake–Dowd model (Cairns *et al.*, 2006) uses a different approach to those described above, in that it assumes a linear relationship between the logit of the mortality rate and age in each calendar year. Projections can then be derived by modelling the parameters of this fit and projecting these parameters into the future. The assumption of a linear relationship means that unlike earlier models, the Cairns–Blake–Dowd model can be used only for older ages – certainly no lower than age fifty – and cannot be used to model the full mortality curve.

The model is described as follows:

$$\ln\left(\frac{q_{x,t}}{1-q_{x,t}}\right) = \beta_{0,t} + \beta_{1,t}(x-x^*) + \epsilon_{x,t}. \tag{14.44}$$

In the original model, x^* was taken to be equal to zero, but later formulations set x^* equal to $\bar{x}$, the average of the ages used. The model is fitted using Poisson maximum likelihood estimation, as described above.

Simulated values of $\beta_{0,t}$ and $\beta_{1,t}$ can be calculated in the same way as described for the Lee–Carter model, but the correlation between the changes in $\beta_{0,t}$ and $\beta_{1,t}$ is also calculated. This is so that when $\beta_{0,t}$ and $\beta_{1,t}$ are projected, the links between them can be taken into account through the use of correlated normal distributions.

Several changes to the original model have been published. These involve:

- a flat cohort effect;
- an age-related cohort effect; and
- a component of the age squared.

In Equations (14.45), (14.46) and (14.47), x^* is again equal to $\bar{x}$, the average age used in the analysis. In Equation (14.46), x^{**} is a constant that is estimated as part of the fitting process, and in Equation (14.47), x^{***} is the average value of $(x-\bar{x})^2$. Combinations of the effects noted above have also been considered.

$$\ln\left(\frac{q_{x,t}}{1-q_{x,t}}\right) = \beta_{0,t} + \beta_{1,t}(x-x^*) + \gamma_{2,t-x} + c_{x,t} \tag{14.45}$$

$$\ln\left(\frac{q_{x,t}}{1-q_{x,t}}\right) = \beta_{0,t} + \beta_{1,t}(x-x^*) + \gamma_{2,t-x}(x-x^{**}) + \epsilon_{x,t} \tag{14.46}$$

$$\ln\left(\frac{q_{x,t}}{1-q_{x,t}}\right) = \beta_{0,t} + \beta_{1,t}(x-x^*) + \beta_{2,t}((x-x^*)^2 - x^{***}) + \epsilon_{x,t} \tag{14.47}$$

14.8.6 Other demographic risks

Most other demographic risks are of less importance than mortality and longevity risk. In particular, the proportion of pension scheme members who are married, the age differences of spouses and the number and ages of children will often be known, so pose minimal risks. Even if these details are unknown, it is straightforward to recalculate the value of liabilities making conservative assumptions for unknown variables which will generally be uncorrelated with other risks.

Some other demographic risks, however, require more thought. The number of lapses in relation to insurance policies and the number of pension scheme members either retiring early or leaving a firm before their retirement date can have a meaningful impact on the profitability of an insurance policy or the size of pension scheme liabilities. These items are also related to other factors affecting financial institutions. In an economic downturn, policy lapses are likely to be higher. Redundancies are also likely to rise, meaning that pension scheme withdrawals and early retirements might be more common. Given that the state of the economy is also linked to the performance of investments and the rate of interest used to discount long-term liabilities, the importance of allowing for these interactions should be clear.

However, there is often insufficient information to derive a useful statistical distribution for any of these additional demographic risks, not least because each pension scheme and insurance portfolio will be slightly different. This means that scenario analysis can be particularly helpful in assessing the impact of particular demographic outcomes.

14.9 Non-life insurance risk

As discussed above, non-life insurance claims contain two aspects: incidence and intensity. This means that there is an additional complication compared with mortality risk since for most defined benefit pension schemes and insurance policies the intensity – more commonly known as the benefit – is known, either in absolute terms or is exactly defined by some other variables. The way in which the intensity of insurance claims is estimated is discussed below.

The incidence of non-life insurance claims is similar in nature to mortality and longevity risk. Volatility risk exists, because portfolios are of finite size. There is also level risk, which is dealt with by a combination of experience and risk rating. Catastrophe risk is also present, arising in terms of incidence from concentrations of risk. However, trend risk is more difficult to define and usually less important. The change in the incidence of claims over time is more likely to follow the economic cycle than to trend in a particular direction. This means that changes should be modelled – probably through scenario analysis – consistently with other economically sensitive risks. However, the short period of exposure for many insurance policies means that changes in the underlying risk are much less important than identifying the risk correctly in the first place.

There are two broad groups into which insurance classes can be placed. The first group includes those classes where there is a relatively high frequency of claims, such as motor or household contents insurance. Conversely, the second group includes classes where the frequency of claims is very low and the size of claims is greatly variable. Excess-of-loss reinsurance – where payments are made only if aggregate claims exceed a particular level – is a good example of this type of insurance.

14.9.1 Pricing high claim frequency classes

Classes of insurance where the claim frequency is high tend to produce a significant volume of data, which is relatively straightforward to analyse and model.

The most common way to model the incidence of claims in these classes of insurance for rating purposes is to construct multi-way tables covering all of the risk factors, so that the proportion of claims for any given combination of risk factors can be calculated. For example, the risk factors for motor insurance might include items such as the engine size and the use to which the vehicle would be put, as shown in Table 14.14

A major drawback with this approach is as the number of rating factors increases, the number of observations in each 'cell' falls to levels that make statistical judgements difficult. For these reasons, generalised linear models (GLMs) are now often used to model the impact and interaction of the various risk factors.

When analysing the claim frequencies for a range of policies, there are a number of statistical issues that need to be addressed. The first relates to the fact that the data set will often span a number of years. Given that different external factors such as the weather and the state of the economy will influence claims to different degrees in different years, the use of dummy variables will often be appropriate.

A more subtle statistical issue arises from the fact that some policyholders will be included in the dataset for each year of the data set, whilst others will not. In particular, policyholders who move to a competitor in later years will be present only at the start of the period, whilst policyholders joining from a competitor will be present only at the end. This means that the data comprise what is known as an 'unbalanced panel'. This is important, because if this factor is ignored, then the policyholder-specific nature of claim frequencies will also be ignored, meaning that the standard errors for some explanatory variables might

Table 14.14. *Hypothetical car insurance claim probability for two risk factors*

		Social, domestic and pleasure (SDP)	SDP and commuting	SDP and work
	< 1,000cc	0.10	0.15	0.22
Engine	1,000cc – 1,499cc	0.15	0.16	0.24
size	1,500cc – 1,999cc	0.17	0.19	0.28
	2,000cc – 2,999cc	0.23	0.24	0.33
	> 2,999cc	0.15	0.17	0.25

Table 14.15. *Claims per year*

		Year				
		2005	2006	2007	2008	2009
	A	1	0	0	1	0
	B	n/a	n/a	2	3	2
Policyholder	C	1	0	3	n/a	n/a
	D	n/a	0	0	0	n/a
	E	3	0	1	0	n/a

be too low. Consider, for example, a portfolio of five policyholders, with claim frequencies tracked over five years.

Whilst there are a total of eighteen observations in Table 14.15, these observations are clearly not independent, and ignoring this fact in a regression will mean that policyholder-specific factors will artificially lower the variability in claim frequencies.

The way to counter this is to include an additional item of data in any regression, an indicator of the policyholder to whom the observations apply, which is used in the calculation of robust standard errors. Whilst the statistical methods behind this calculation are complex, the option to calculate robust standard errors exists in most statistical packages.

Intensity for premium rating in these classes of insurance is generally modelled through multiple regression approaches. Clearly, if the distribution of claim amounts is not normal (and for many classes it will not be), ordinary least squares regression is inappropriate and an alternative approach must be used.

As with claim incidence, there are statistical issues that should be faced when modelling claim intensity. As before, there are issues of seasonality and the fact that claim amounts will differ due to economic and environmental factors. Clustering is also an issue. However, there is another statistical issue faced here – the issue of censoring. It is tempting, and it seems logical, to calculate the influence of independent variables on claim amounts using the information only from policies where claims have occurred. However, there is also useful information in relation to policies where there have been no claims. This information can be used by carrying out what is known as a censored regression. As above, the statistics underlying this approach are complex, but the option to carry out censored regressions exists in most statistical packages.

Experience rating is also carried out in non-life insurance. This can be for an individual client (such as a car insurance policyholder), a corporate client (such as a firm with employer liability insurance) or for the entire portfolio of an insurance company. Particularly in this final case, it is important to consider the level of risk within a number of homogeneous subgroups, so that any changes in the mix of risk types over time is properly reflected. As with life insurance, the results from experience rating and risk rating can be combined through the use of credibility.

14.9.2 Reserving for high claim frequency classes

An important aspect of all insurance is calculating the amount of money needed to cover future outgoings. However, a particular issue for non-life insurance, especially in relation to high claim frequency classes, is that there is a delay between claims being incurred and these claims being reported. Ignoring incurred-but-not-reported (IBNR) claims can significantly understate the reserves that must be held to cover future outgoings. There are a number of approaches that can be used to determine the level of outstanding claims. Three of the most common are:

- the total loss ratio method;
- the chain ladder method; and
- the Bornhuetter–Ferguson method.

All of these methods can be applied to aggregate claims or to loss ratios, and can be adjusted to allow for claim inflation.

The total loss ratio method
The total loss ratio method simply looks at the total premium that has been earned for a particular year and assumes that a particular proportion of those

premiums will result in claims. The premium earned is the part of any premium covering risk in a given. So, for example, if a premium of £300 was received in respect of cover from 1 September 2008 to 31 August 2009, the premium earned in 2008 would be £100 and the premium earned in 2009 would be £200.

The loss ratio can be determined from historical data and adjusted as appropriate. It can also be adjusted for changes in the underlying mix of business to give more accurate results.

Example 14.9 The table below gives the history of claims occurring in the last three years. All claims are notified no more than three years after happening:

		Earned premium	Development year (d)		
			1	2	3
Year	2007	250	150	160	200
of	2008	300	150	200	–
claim (c)	2009	350	200	–	–

What are the total estimated claims for 2007, 2008 and 2009 under the total loss ratio approach, assuming a total loss ratio of 80%?

The loss ratio of 80% is consistent with that observed for claims occurring in 2007, being equal to 200/250. The total projected claims for 2008 and 2009 are therefore $80\% \times 300 = 240$ and $80\% \times 350 = 280$ respectively. The claim development can therefore be completed as follows:

		Earned premium	Development year (d)		
			1	2	3
Year	2007	250	150	160	200
of	2008	300	150	200	**240**
claim (c)	2009	350	200	–	**280**

The attraction of this approach is that it is simple, and can be applied with limited data. This also means that it can be applied to new classes of business. However, if historical loss ratios are used it is not clear how the method should be adjusted, if it becomes clear that the rate of losses emerging is higher than has been experienced in the past.

The Chain ladder method

The chain ladder method is still the dominant approach for calculating the total projected number of claims. This considers the claims that have already been reported and uses the historical pattern of claim development to project reported claims forward. This is done by calculating link ratios, the change in the total proportion of claims notified over subsequent periods historically and applying these to years where the development of claims is incomplete. The approach can be applied to the cumulative value of claims or to the incremental value of claims in each year. The former approach places much greater weight on earlier claims, whilst the latter can become volatile when the period since claiming is great. The best way to explain the chain ladder method is by example.

Example 14.10 Considering the case in Example 14.9.2, what are the total estimated claims for 2008 and 2009 under the chain ladder approach?

		Earned premium	Development year (d)		
			1	2	3
Year	2007	250	150	160	200
of	2008	300	150	200	–
Claim (c)	2009	350	200	–	–

Each cell, $X_{c,d}$, contains the number of claims that occurred in year c and were reported by the end of the dth year. To calculate an estimate for $X_{2008,3}$, the link ratio must be calculated using data from the claims that occurred in 2007. In particular, this link ratio, $l_{2,3}$ is calculated as $200/160 = 1.25$. The estimated value of $X_{2008,3}$ is therefore $1.25 \times 200 = 250$.

To calculate $X_{2008,3}$, two years' worth of data are available, so the link ratio, $l_{1,2}$ is calculated as $(160+200)/(150+150) = 1.20$. This means that the estimated value of $X_{2009,2}$ is $1.20 \times 200 = 240$. The link ratio $l_{2,3}$ can then be applied to this value to estimate $X_{2009,3}$ is $1.25 \times 240 = 300$. The claim development can therefore be completed as follows:

		Earned premium	Development year (d)		
			1	2	3
Year	2007	250	150	160	200
of	2008	300	150	200	**250**
Claim (c)	2009	350	200	240	**300**

One recent augmentation to the chain ladder approach is the treatment of the development as stochastic. Such an approach could be carried out by considering the statistical distribution of the link ratios in each year, obtaining not just an average ratio but also the extent to which they vary from year to year. In order to do this, a number of years of claim development history would be needed, and as the length of the history rises its relevance falls. However, a potential advantage of this approach is that it allows the linkage of claim levels to economic, market and other variables.

The Bornhuetter–Ferguson method

Another way of assessing the ultimate claim level is to use the Bornhuetter–Ferguson method. This is essentially the total loss ratio approach adjusted for claims reported to date. The chain ladder approach is used to derive link ratios, and these are used to calculate the proportion of the expected loss ratio that should develop in each period. Then, for any period that the loss is actually known, the prediction from the combined loss ratio and chain ladder approach is replaced with the known figure. Again, this is best demonstrated by example.

Example 14.11 Considering the case in Example 14.9.2 and assuming a loss ratio of 80%, what are the total estimated claims for 2008 and 2009 under the Bornhuetter–Ferguson approach?

		Earned premium	Development year (d)		
			1	2	3
Year	2007	250	150	160	200
of	2008	300	150	200	–
claim (c)	2009	350	200	–	–

The link ratios $l_{1,2}$ and $l_{2,3}$ have already been calculated as 1.20 and 1.25 respectively. This means that the claims reported by the end of the third year – which are also the total claims – are 1.25 times the claims reported by the second year, and they are $1.25 \times 1.20 = 1.50$ the claims reported by the end of the first year. This means that $1/1.50 = 0.67$ claims are paid by the end of the first year, whilst $1/1.25 = 0.80$ are paid by the end of the second year.

The first part of the Bornhuetter–Ferguson estimate for the claims arising in 2008 is therefore the value of claims already reported, 200. The second part is the product of the premiums earned, the loss ratio and the chain

---

ladder proportion of claims outstanding, $300 \times 0.80 \times (1 - 0.80) = 48$. This means that the 2008 claims estimate is $200 + 48 = 248$.

Using the same approach, the first part of the Bornhuetter–Ferguson estimate for the claims arising in 2009 is the value of claims already reported, again 200. The second part is the product of the premiums earned, the loss ratio and the chain ladder proportion of claims outstanding, $350 \times 0.80 \times (1 - 0.67) = 93$. This means that the 2009 claims estimate is $200 + 93 = 293$. The table can therefore be completed as follows:

		Earned premium	Development year (d)		
			1	2	3
Year	2007	250	150	160	200
of	2008	300	150	200	**248**
claim (c)	2009	350	200	**240**	**293**

14.9.3 High claim frequency classes

The second group of classes is that where the frequency of claims is very low and the size greatly variable. This means that these classes of insurance are particularly amenable to modelling using extreme value theory.

Copulas can also play an important part in the modelling of these asset classes, since many large aggregate claims will arise as a result of some sort of concentration of risk. For example, hurricane damage will affect a large number of properties in a particular area. This means that if an insurer receives a large claim in respect of a particular property and large claims are more likely to arise from hurricane damage, then it is also likely that more large claims will be received.

One aspect of catastrophe-type modelling that is less of an issue for low-frequency high-value claims is the reserve for IBNR claims. Since catastrophes are generally covered in the press, most insurers have an idea of the amount of claims they face even before the claims are made.

14.10 Operational risks

Operational risks can seem daunting to quantify, but it should be borne in mind that not all operational failures are large, infrequent, enterprise-threatening events. For example, a bank carrying out millions of transactions each day will inevitably make small mistakes on a regular basis. This means that the frequency and size of these events can be modelled. For these types of claims, non-life reserving techniques for high frequency classes may well be appropriate.

However, financial institutions may also have rare but very costly operational losses, either from one-off causes such as fraud or the cumulative effects of poor project management. Here, extreme value theory can be used.

The nature of operational losses means that their distribution is skewed to the right and fat-tailed in terms of amounts lost, which can influence the distribution used. It is also important to consider potential links between operational and other risks. For example, fraud is more likely to occur in an economic downturn. As a result, scenario analysis can be useful, although if the purpose of modelling is to arrive at an amount of capital required, then stochastic techniques might be more appropriate.

Such methods would be consistent with Basel II's advanced measurement approach, discussed later. However, a simpler approach would be to model operation risk by multiplying the income received by a fixed percentage, either in aggregate or by business line. These are the basic indicator and standardised approaches under Basel II.

These approaches could all be classified as bottom–up methods. However, it is also possible to use top–down methods to assess the exposure to operational risks. For example, the total income volatility could be measured, as could the income volatility arising from credit risk, market risk, mortality risk and any other material sources. The excess of the first item over the sum of the other two could be regarded as income volatility arising from operational risk. However, as an historical measure it does not necessarily capture the forward-looking nature of operational risk. It also looks only at the impact on income rather than value, which could well differ due to issues such as reputational damage or, conversely, an increase in the value of a brand.

The value issue can be addressed by looking instead at changes in the market price of a firm, if it is listed. Using a model such as the capital asset pricing model (CAPM), it is possible to strip out the changes in the value of a share due to overall market movements, and to concentrate on the firm-specific changes in value. From this, it should be possible to see the impact of past operational losses on the value of the firm. This means that issues such as reputation are directly included in the assessment. However, it is difficult to disaggregate the effects of various factors from the change in firm values. This is important from a risk prevention standpoint, as it means that it is impossible to focus on the impact of individual events.

Another approach is to consider the risk capital of an organisation. The total risk capital can be estimated and whatever number is left, after the risk capital for credit, market and other risks has been deducted, is the operational risk capital. This is a forward-looking approach so it is more relevant as an indication of risk. However, assessing the total risk capital is not straightforward. Furthermore, the interactions between the risks are ignored with this approach.

14.11 Further reading

There are a number of texts that described many of the risks here in greater detail. Market risk is covered by McNeil *et al.* (2005), and derivatives are described more fully by Hull (2009). Another popular derivatives-based text is Wilmott (2000), which discusses many practical issues related to the use of the various mathematical techniques. Interest rate risk is covered by both of these books, although greater coverage is given by Rebonato (1998), who covers this topic alone. Cairns (2004) provides an accessible introduction to the topic of interest rate risk, whilst a comprehensive and up-to-date analysis is given in the three volumes by Andersen and Piterbarg (2010a,b,c).

As mentioned earlier, de Servigny and Renault (2004) is a good resource for the analysis of credit risk, but the rating agencies also provide a great deal of information, much of it at no cost.

Chief Risk Officer Forum [2008] and Basel Committee on Banking Supervision [2008] give good, detailed advice on liquidity risk, whilst Bas (2003) describes operational risk in detail. Operational risk is also covered in detail by Lam (2003).

Demographic and non-life insurance risks are still better dealt with in journal articles than by books. Cairns *et al.* (2009) give a comparison of a number of different mortality models, but the original model by Lee and Carter (1992) is still worth looking at. Int (2008) is also helpful, describing as it does the various mortality risks in more detail. An overview of current issues in longevity risk is given in McWilliam (2011). No comparable book exists for non-life risk, but Würthrich and Merz (2008) discuss a number of more advanced approaches to dealing with issues in this area.

15

Risk assessment

15.1 Introduction

Once risks have been analysed, the results must be assessed. This is true whether considering a project to be initiated, a product to be launched or an asset allocation to be adopted. Such analysis will generally involve trying to maximise (or minimise) one variable subject to a maximum (or minimum) permissible level of another variable.

Creating these variables will often involve applying particular risk and return measures to particular items. The different types of measures are described below, and choosing the appropriate one involves careful consideration.

The item to which risk and return measures are applied also requires some thought. These might be income or capital measures, and they might be prospective or retrospective. Income measures might be profit or earnings related, but cash flow might also be important, as liquidity problems can result in the closure of otherwise-profitable firms. Capital measures might relate to the share price of a firm, or the relationship between some other measure of assets and liabilities.

As well as determining the measures of risk and return to be used, and the items to which they should be applied, the level of risk that can be tolerated must be determined. This means visiting the concept of risk appetite. However, it is also important that risk appetite is placed in the context of other risk-related terminology.

There are many different classifications in this regard, and the terminology here is not intended to be definitive. However, it is intended to be unambiguous and to give an idea of the range of considerations an organisation will have in respect of risk.

15.2 Risk appetite

Once an organisation has identified, described and, where appropriate, quantified all of the risks to which it is exposed, the resulting summary is known as its *risk profile*. This is an assessment of the risk that an organisation is currently taking.

However, for this to be useful, it needs to be compared with the organisation's *risk appetite*. This is itself a combination of two things: *risk tolerance* and *risk capacity*.

15.2.1 Risk tolerance

Risk tolerance is a cultural issue, part of the organisation's internal risk management context, and is about the subjective decision a firm has taken on where it would like to be in the risk spectrum. Different stakeholders may well have different risk tolerances. For example, bondholders and other creditors will have lower risk tolerances than equity investors and with-profits policyholders, due to the share that each has in the potential benefits arising from higher risk activities. An individual's risk tolerance can be assessed in conversations with an investment advisor, through questionnaires or through evidence of past decisions; however, determining the risk tolerance for an organisation is more difficult, since it develops over time. As a philosophy, it is best identified by the strategic decisions made by the board, but in this respect it will be as much based on the perception of investors as the behaviour of the directors. However, the directors of a firm can define their risk tolerance in terms of some measure of solvency, a target credit rating or volatility of earnings and the ability to pay dividends.

Risk tolerance can be expressed mathematically in terms of a utility or preference function. A utility function, $u(W)$, combines measures of risk and return based on a given level of wealth, W, into a single measure of utility, or 'happiness'. There are particular features that utility functions should have if they are to be realistic. First, $u(W)$ should be monotonically increasing in W. This means that having more of something is always better. Secondly, utility functions should be concave. Mathematically speaking, this means that for all $\delta_2 > \delta_1$, $[u(W + \delta_1) - u(W)]/\delta_1 > [u(W + \delta_2) - u(W)]/\delta_2$).

What this means is that, whilst having more of something increases utility, the proportional increase in utility falls as the amount possessed rises. Concavity also implies risk aversion, which is also important for a utility function to be sensible. The level of risk aversion can be quantified in terms of the first and second differentials of $u(W)$ with respect to W, denoted $u'(W)$

and $u'(W)$, as:

$$a(W) = -\frac{u'(W)}{u'(W)}, \qquad (15.1)$$

this expression being positive if a function implies risk aversion.

There are several commonly used utility functions. One which has been much-used in financial economics is the quadratic utility function which has the following form:

$$u(W) = \alpha E(W) - \frac{1}{2} E(W^2), \qquad (15.2)$$

where $W \leq \alpha$. This is essentially the utility function behind mean–variance optimisation, which says that investors should try to maximise the expected value of their investments subject to the volatility of those investments being constrained to a particular level. One drawback of this function is that if $W > \alpha$, $u(W)$ starts to decrease. In most financial scenarios, this is not realistic. Furthermore, it implies increasing absolute risk aversion, since $a(W)$ can be shown to be increasing in W. Risk aversion is more likely to be decreasing in W, since the more an individual or institution has, the smaller the impact of a particular fixed monetary loss is.

The exponential utility function, shown below, has also been used:

$$u(W) = -\frac{e^{-\alpha W}}{\alpha}, \qquad (15.3)$$

where $\alpha > 0$. This exhibits constant absolute risk aversion, with $a(W) = \alpha$, which is more likely than increasing absolute risk aversion, but still not ideal.

Finally, there is the power utility function which has the following form:

$$u(W) = \begin{cases} \dfrac{W^{1-\alpha}}{1-\alpha} & \text{if } \alpha > 0 \text{ and } \alpha \neq 1; \\[2ex] \ln W & \text{if } \alpha = 1. \end{cases} \qquad (15.4)$$

This has decreasing absolute risk aversion, and constant relative risk aversion, with $a(W) = \alpha/W$. These features mean that it is an intuitively attractive utility function. All three utility functions are shown together in Figure 15.1, scaled and shifted so they follow similar paths and pass through the origin.

However, a more recent innovation is the distinction between the utility function and the prospect function. The key difference between the two is that, whilst the utility function simply maps the combinations of risk and reward that are acceptable, the prospect function considers the combinations that an

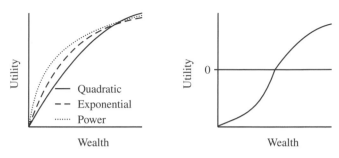

Figure 15.1 Utility (left) and prospect (right) functions

investor would choose *given a particular starting point*. The prospect function is so named because it assumes that participants consider their prospective wealth.

The shape of the prospect function for levels of W greater than W_0, the starting level of wealth, is similar to a utility function, in that it is concave in W. This means that for this part of the function investors are assumed to be risk averse, preferring guaranteed to possible gains.

However, there is a discontinuity at W_0, where the prospect function kinks. Below W_0, the prospect function is convex, suggesting that if a loss is to be made, a possible loss is preferred to a guaranteed one. Gains are still preferred to losses, though, by a ratio of two to one. This can be seen by the relative gradients of the prospect function above and below W_0.

Finally, the prospect function seems to tend to zero risk aversion – indifference to whether additional risks are taken or not – for very large gains or losses. This can be seen by the fact that at the extremes the prospect function tends to straight lines.

The current level of wealth, W_0, serves as an anchor relative to which decisions are made. However, the anchoring power of W_0 is not constant, and the way in which this anchor changes has an impact on the way in which risks are viewed.

In a sense, prospect theory can be regarded as a positive version of the normative theory of utility in that it reflect how people actually behave rather than how they ought to behave. This means that there is merit in developing strategies to 'dislodge' any mental anchors when developing strategies so that views of risk are as rational as possible.

As well as considering risk tolerance in aggregate, it can also be expressed in relation to individual risks such as investment or liquidity restrictions. Each individual restriction constitutes a well-defined *risk limit*. These risk limits are important, as they give the implications of the risk tolerance for each individual department. As such, they must be clear and unambiguous.

15.2.2 Risk capacity

The risk tolerance of an organisation (or an individual) is tempered by the capacity an organisation has to take on risk. Risk capacity is a function of the resources that are available. For financial institutions, it is even more a function of regulatory and legislative limits, and as such part of the organisation's external risk management context. Expressions of risk capacity can be made in the same way as for risk tolerance, and can again be expressed in aggregate or as an individual risk.

This means that risk appetite can be expressed as the more restrictive of the risk tolerance and the risk capacity. However, it is important to recognise that organisations should consider the risks to which they believe they are exposed as well as just considering the risks that they are obliged to manage – just because there are no regulatory limits in a particular area, it does not mean that risks should develop unchecked.

4 15.3 Upside and downside risk

When considering risk, it is important to recognise that both positive and negative outcomes are of interest. Unexpected positive outcomes, or upside risks, are the reasons for accepting exposure to unexpected negative outcomes, or downside risks, in the first place. For example, when investing in equities, there is exposure to downside risk from the potential fall in share prices, but upside risk from the potential rise.

It is important that the downside risks to which an organisation is exposed are consistent with the potential upside risks available, as well as with the risk appetite. In particular, downside risks that do not present any potential upside risk are not desirable. In some cases, this is unambiguous – returning to the example of equity investments, the level of exposure to downside and upside risks is a function of the risk appetite. Conversely, there is no potential upside from having inadequate systems and processes for the settlement of derivatives, whilst there is significant potential downside. If it is virtually cost-free to improve these systems and processes, then they should be improved.

However, it is not always so unambiguous. What if the cost of improvement is significant? Failing to make changes effectively results in a guaranteed upside of not spending money. In the case of systems and processes, the cost of improvements will almost always be justified by the reduction in risk, although a very good system will often suffice rather than a more expensive excellent system. A more difficult situation is faced when insurance is considered. Taking out insurance will remove a downside risk – whilst often leaving residual risks – but this is at the cost of the insurance premium.

This is similar to an issue faced by many financial institutions: the extent to which underwriting should take place. For example, when an insurance company offers a life insurance policy or a bank offers a loan, how much money should it spend to avoid potential adverse selection by the customer? A detailed medical would given an insurance company a lot of information on the appropriate price for a life insurance policy, but would the price differential be larger than the cost of the medical? The broad principle here – which applies equally to any other risk reduction measure – should be that any expenditure to reduce risk should be consistent with the expected saving arising from the measure being put in place. *cost - benefit*

The concept of downside risk is also particularly important when risk quantification is considered, as it brings the shape of the statistical distribution into focus, as discussed below.

15.4 Risk measures

There are a number of ways in which risks can be measured, from the very simple to the very complex. The more simplistic approaches tend to use broad-brush approaches, meaning that at best key risks can be overlooked, and at worst that there is active regulatory arbitrage in order to maximise the genuine level of risk for a stated value of a risk metric. However, broad-brush approaches can at least be recognised as flawed, only giving a broad indication of the level of risk taken.

The more complex approaches can cover a wider range of risks and can allow more accurately for the different levels of risk between firms. However, this complexity can lead to a false sense of security in models. This is particularly true if the risk being considered is in relation to extreme events, which most models are very poor at assessing.

One issue with all of these measures is the time horizon or holding period used in the calculation. For a liquid security, in an environment where the measure is being used to assess the risk of holding particular positions, a shorter holding period can be used since positions can be closed out quickly; however, for analysis including less liquid assets, such as loans to small businesses for a retail bank, or holdings in illiquid assets such as property or private equity, a longer time horizon is more appropriate.

It is also worth noting that scaling of risk measures from one holding period to another (such as monthly to annual) is not always possible, particularly if the underlying statistical distribution is non-normal. Also, if there is non-linearity in any of the investments being analysed – options being the prime example – then separate analysis is needed for different holding periods.

15.4.1 Deterministic approaches

The broad-brush risk measures are essentially the deterministic approaches. These all involve taking the item to be measured and performing a simple transformation of it in order to get to the item to be assessed. Such approaches are popular with regulators as a basic test of solvency. Some examples are given below.

Notional amount

This approach is best described by example. Consider an institution with a fixed value of liabilities backed by a portfolio of assets. Whilst the market value of assets might exceed the value of the liabilities, such a comparison ignores the risk inherent in those assets. A way of dealing with this would be to apply a multiple of between 0% and 100% to each of the assets depending on its properties to arrive at a notional amount of the assets. For example, the notional value of government bonds could be taken as 100% of their market value, the notional value of domestic equities could be taken as 80% of their market value and the notional value of overseas equities at 60% of their market value.

This has the advantage of being very easy to implement and interpret across a wide range of organisations. However, it has a number of shortcomings. First, it can be used only if the asset class is defined. A 'catch all' multiple can be defined to apply to asset classes not otherwise covered, but such an approach is not ideal. In particular, if a low notional value is applied to the catch-all asset class, then this might distort the market, leading to an increase in prices for assets regarded as high quality by the regulator.

This approach also fails to distinguish between long and short positions. For example, if the investments included a portfolio of UK equities and a short position on the FTSE 100 Future (a UK index future), then both positions would be risk weighted even though the effect of holding the future is to reduce the risk of the equity investments. Similarly, there is no allowance for diversification, since the multiples used take no account of what other investments are held. Finally, there is no allowance for concentration. If the multiple for equities is 80%, then this could apply if the only investment was a holding in a single firm's securities. This risk could be limited by having admissibility rules as well. For example, the notional holding would be 80% of any equity holdings, with no equity holding in any one firm making up more than 5% of the notional assets of the firm. However, this is another blunt tool.

Factor sensitivity

Continuing with the same example, a factor sensitivity approach produces a revised value of assets and, possibly, liabilities based on a change in a single underlying risk factor. For example, with an insurance company the effect on bond investments and long-term liabilities of a 1% fall in interest rates might be considered, with a firm being considered solvent only if the stressed value of assets exceeded the stressed value of liabilities.

As this approach considers the change in a single underlying risk factor, it is not very good at assessing a broader risk profile. In particular, it is difficult to aggregate over different risk factors.

Scenario sensitivity

One way of solving the problem of combining factor sensitivities is to combine various stresses into scenarios, so for example combine a 1% fall in interest rates with a 20% fall in equity markets. This is more robust than considering individual factors, but all the earlier points about scenario analysis should still be considered.

15.4.2 Probabilistic approaches

The more complex approaches are generally probabilistic. These involve measuring risk by applying some sort of statistical distribution and measuring a feature of that distribution.

Standard deviation

The standard deviation of returns is often used as a broad indication of the level of risk being taken, and is used in a number of guises. The most obvious is portfolio volatility, which is simply the standard deviation of returns. This can be calculated one of three ways:

- retrospectively, from the past volatility of the portfolio;
- semi-prospectively, from the past covariances of the individual asset classes but the current asset allocation; or
- fully prospectively, from estimated future covariances of the individual asset classes and the current asset allocation.

With the fully prospective approach, future covariance may be adapted from historical data, derived from market information such as option prices, or arrived at in some other way.

Volatility also arises in the calculation of tracking error. This is a measure of the difference between the actual returns and the performance benchmark of an investment manager. It is calculated as:

$$TE = \sqrt{\frac{1}{T} \sum_{t=1}^{T} (r_{X,t} - r_{B,t})^2}, \tag{15.5}$$

where $r_{X,t}$ is the manager's return in period t where $t = 1 \ldots T$ and $r_{B,t}$ is the benchmark return in that period. This is often approximated as the standard deviation of $r_{X,t} - r_{B,t}$; however, if the average excess return is significantly different from zero, the standard deviation approach can seriously understate the true tracking error.

The variable TE is more properly known as the *ex post* tracking error, as it records the level of deviation that occurred. It is also possible to estimate a level of *ex ante* tracking error by considering the difference between the holdings in the portfolio and in the benchmark. However, whilst the *ex post* tracking error is unambiguous, the *ex ante* tracking error requires a number of assumptions regarding the behaviour of the components of the portfolio. It is usually calculated by simulating the performance of a portfolio relative to a benchmark using a factor-based stochastic model.

The *ex post* tracking error is used as part of the information ratio. This is calculated as:

$$IR = \frac{ER}{TE}, \tag{15.6}$$

"Sharp" ratio" like

where:

$$ER = \frac{1}{T} \sum_{t=1}^{T} (r_{X,t} - r_{B,t}), \tag{15.7}$$

the average excess return.

The standard deviation is also commonly used in pension scheme analysis when comparing the efficiency of different asset allocations. Here it may be used both to derive the set of efficient portfolios (through mean–variance optimisation) and to highlight the risk of the actual and proposed asset allocations.

Using the standard deviation in a dimensionless measure, such as the information ratio, is potentially useful as a ranking tool. However, there are those who question the usefulness of the information ratio because it can lead to 'closet tracking' – claiming to be an active manager but making few if any active decisions. The standard deviation also has value as a broad measure of risk, since it is relatively straightforward to calculate for a wide number of financial risks; indeed, and if the correlations are known, it is straightforward to calculate an aggregate standard deviation without having to resort to stochastic simulations. However, unless the underlying distributions are normally distributed, this information cannot be used to derive accurate percentile statistics. It is, though, arguable that the standard deviation is less than clear as a measure of risk in its own right. If a particular asset allocation gives an expected funding level of 100% with a standard deviation of 10%, how clear is it to clients (or consultants) what the 10% means? Clearly, it is better than 11% and worse than 9%, but beyond this, it is less useful.

The standard deviation is similarly opaque if extreme events are the concern. It requires additional calculations to be carried out to show the risk of extreme events (these calculations are described below), but it also gives misleading results if the underlying distributions are skewed. Another way of thinking of this is that a symmetrical risk measure is only useful if the underlying distributions are symmetrical. Similarly, the standard deviation underestimates risk if the underlying distribution is leptokurtic. For extreme event analysis it is necessary to move away from measures of dispersion to measures of tail risk.

Value at Risk

A commonly used measure in the world of finance, to the extent that it is the measure of choice in most banking organisations, is the Value at Risk (VaR). It can be defined as the maximum amount that will be lost over a particular holding period with a particular degree of confidence. For example, a 95% one-month VaR of 250 tells us that the maximum loss for a one-month is 250 with a 95% level of confidence. VaR can also be expressed in terms of standard deviations, so a two sigma daily VaR of 100 tells us that the maximum daily loss is 100 with level of confidence of around 96%, if returns are assumed to be normally distributed. The VaR can also be given as a percentage of capital, so a 95% one-month VaR of 3.2% tells us that 3.2% is the maximum loss over a one-month period with a probability of 95%.

Confusingly, the terminology around VaR is inconsistent, so whilst a 95% VaR gives the maximum loss expected with a 95% level of confidence, the same figure is expressed by others as a 5% VaR, being the point below which the worst 5% of losses are expected to occur. This confusion is not helped by the fact that losses are sometimes referred to as positive ('a loss of 250') and sometimes as negative ('a −3.2% return'). In this analysis, the convention will be to refer to losses as positive, defining the loss in period t for portfolio X as:

$$L_{X,t} = -(X_t - X_{t-1}), \qquad (15.8)$$

where X_t is the portfolio value at time t.

The time horizon over which VaR is calculated is an important part of the calculation. In particular, it should reflect the time for which an institution is committed to hold a portfolio. It should be recognised that this can change over time as levels of liquidity rise and fall.

VaR also feature in many pension scheme asset allocation presentations, being calculated over increasing holding periods and presented as the percentiles in a 'funnel of doubt'.

There are three broad approaches to calculating VaR:

- empirical;
- parametric; and
- stochastic.

The empirical approach is the most straightforward and intuitive. It involves recording daily (or weekly, or monthly) profits and losses within a portfolio. The worst 1% of results (so the 1st centile) represents the 99% VaR; the worst 5% or results (so the 5th centile) represents the 95% VaR; and so on.

The derivation of the VaR from empirical data is shown in Figure 15.2, but the results can also be expressed as a formula. If VaR_α is the VaR at a level of confidence of α and the losses $L_{X,t}$, where $t = 1, 2, \ldots, T$, are ranked such that $L_{x,1}$ is the smallest loss and $L_{X,T}$ is the greatest, then:

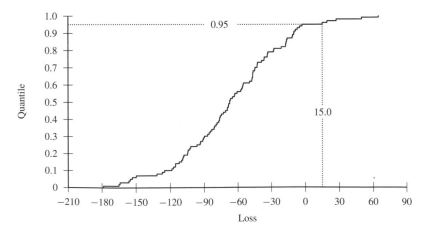

Figure 15.2 Calculating 95% VaR from historical loss data

$$VaR_\alpha = L_{X,T\alpha}.^1 \qquad (15.9)$$

If T_α is not an integer, then $L_{X,T\alpha}$ can be calculated by linearly interpolating using the values of $L_{X,t}$ for the values of t immediately above and below $T\alpha$, t_+ and t_- respectively, as:

$$L_{X,T\alpha} = (T\alpha - t_-)L_{X,t+} + (t_+ - T\alpha)L_{X,t-}. \qquad (15.10)$$

This approach has a number of advantages. It is simple and it is also realistic, as it allows for major market movements (provided these occur during period analysed). It also avoids the need for assumptions of the distribution of returns.

However, it has potentially more disadvantages, although many of these can be overcome. First, it is unsuitable if the composition of the portfolio changes over time. This problem is easily overcome by modifying using the returns on individual asset classes or business lines and combining them in the proportions of the current business mix to give a simulated historical return series.

A more serious problem is that it relies on the suitability of past data in describing future volatility, so it is unsuitable if economic circumstances change significantly. Furthermore, even if past data are reliable, the results will not reflect the full range of possible future scenarios. It is also difficult to use scenario testing to analyse robustness to changes in assumptions.

The parametric approach assumes that price changes in the underlying assets follow a simple statistical distribution. The VaR is then simply the quantile of the distribution that corresponds to the level of confidence to which the VaR is being calculated. A common assumption is that the losses follow a multivariate normal distribution. Under this assumption, it is possible to calculate the portfolio standard deviation from:

- the variance of the losses for each asset class;
- the proportion invested in each asset class; and
- the correlations between losses for each asset class.

The proportions and expected returns for each asset can also be used to calculate the portfolio's expected return. In particular, if:

- the expected loss for asset class n is μ_n;
- the standard deviation of losses for asset class n is σ_n;

---

[1] Dowd (2005) notes that the level of loss for VaR_α could be taken to be $L_{X,T\alpha}$, $L_{X,T\alpha+1}$ or some where in between, but as this is an approximation the exact choice should not be of too great a concern. Similar considerations apply to related measures such as tail value at risk.

- the correlation between the losses for asset classes m and n is $\rho_{m,n}$; and
- the proportion of assets in asset class n is w_n where $\sum_{n=1}^{N} w_n = 1$;

then the expected loss for the portfolio is:

$$\mu = \sum_{n=1}^{N} w_n \mu_n, \qquad (15.11)$$

and the variance of losses for the portfolio is:

$$\sigma^2 = \sum_{m=1}^{N} \sum_{n=1}^{N} w_m w_n \sigma_m \sigma_n \rho_{m,n}. \qquad (15.12)$$

Once the expected loss of the portfolio and its volatility have been determined, it is possible to use the standard normal distribution to calculate the loss at the desired probability level as follows:

$$VaR_\alpha = \mu + \sigma \Phi^{-1}(\alpha), \qquad (15.13)$$

where $\Phi^{-1}(\alpha)$ is the inverse standard normal distribution evaluated for the probability, α. This calculation is shown graphically in relation to the density function in Figure 15.3 and the distribution function in Figure 15.4.

The parameter estimates are often obtained from historical data or using implied volatilities calculated from option prices. They are usually calculated with daily or weekly time horizons. In this case, where the time frame is short, it is often common to assume that the expected return on each asset is zero

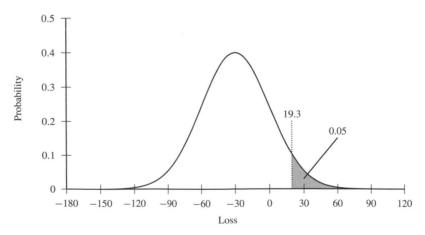

Figure 15.3 95% VaR from a normal density functions

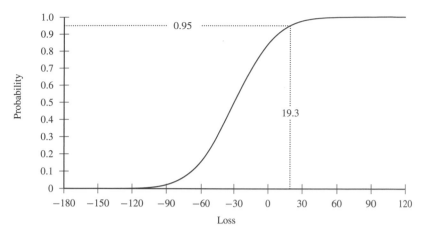

Figure 15.4 95% VaR from a cumulative normal distribution

over the period. This also reduces the number of parameters that need to be estimated.

If VaR is being calculated over longer time horizons, such as monthly or annual, then the assumption that the loss distribution is normally distributed becomes less appropriate. A better approximation is that returns, as defined by the difference between the natural logarithms of successive asset values, are normally distributed. This means that the loss distribution can be redefined as follows:

$$L_{X,t} = -(\ln X_t - \ln X_{t-1}) = -r_{X,t}, \tag{15.14}$$

where $r_{X,t}$ is the return on asset X for time t. The result is that the VaR under this approach is quoted as a return rather than a loss.

The main advantage of the parametric approach to VaR calculation is the ease of computation. It also reduces the dependence on actual historical profits and losses (although the choice of parameter values may depend on historic data). Furthermore, even if historical data is used, the variance and covariance parameters can readily be adjusted if past data are felt to be unreliable.

However, this approach is still far from perfect. First, a consistent set of parameters must be chosen. The approach is also more difficult to explain than the historical method, and it can be unwieldy if there are many assets involved. The relationships between the returns on the different assets may not be stable, making correlation estimates unreliable. In response to this, some practitioners use undiversified VaR (which assumes that all correlation coefficients are equal to one, to simulate crash conditions). A final – and important – criticism is that the normal distribution is often inappropriate for modelling investment returns. In particular, equity returns are leptokurtic over short time horizons.

The next level of complexity is stochastic VaR. This is similar to the historical method, except that the profits and losses generated are simulated. A common approach is to use a multivariate probability distribution to simulate future investment returns and interactions. This means that using the multivariate normal distribution would give the same results as the variance–covariance approach – but the key feature of this approach is that the multivariate normal distribution need not be used. Another approach is to draw randomly from historical returns (bootstrapping) to avoid having to come up with a returns distribution, although any inter-temporal links between returns in one period and the next will be lost with this approach. Typically, results are based on thousands of simulations. The VaR is calculated from the simulated data using the same method as for historical VaR, by sorting the results by size.

The key advantage of this approach is that more complex underlying statistical distributions can be used if appropriate, in particular if skew or leptokurtosis is present. This approach is potentially more realistic than the historical method, as the full range of possible future outcomes can be considered. Results can also be analysed for sensitivity to chosen probability distributions and parameter values. However this approach can be difficult to explain to lay investors. Furthermore, the choice of probability distribution function and parameter values is subjective and can be very difficult. This means that the results may be unreliable. Finally, it can be very time-consuming, particularly for large portfolios, since the calculations required are significant.

VaR itself has a number of advantages. First, it provides a measure of risk that can be applied across any asset class, allowing the comparison of risks across different portfolios (such as equity and fixed income). Other measures are more closely tied to particular asset classes, duration and fixed-income being a prime example. Indeed, VaR can be used to aggregate all types of risk, not just market risk.

VaR also enables the aggregation of risks taking account of the ways in which risk factors are associated with each other. Furthermore, it gives a result that is easily translated into a risk benchmark, so judging 'pass' and 'fail' are straightforward. Finally, VaR can be expressed in the most transparent of terms, 'money lost'.

However, VaR is not always appropriate. If it is being used to determine the amount of capital that must be held (thus limiting the probability of insolvency to that used in the VaR calculation), or to determine some other trigger point at which action must be taken, then no assessment of the events in the tail are needed; however, in many instances, it is useful to know something about the distribution of extreme events. VaR gives only the point at which loss is expected to occur with a predetermined probability, and gives no indication of how much is likely to be lost if a loss is incurred.

Parametric VaR is also potentially misleading if the assumed distribution does not reflect the risks being borne. A prime example is if normality is assumed for risks with leptokurtic or skewed outcomes. Furthermore, if there is significant tail dependence between risks, and correlations are used to describe the dependence structure rather than copulas, then there is a risk that a VaR calculation will underestimate the risk, since it involves an assessment of extreme scenarios. These risks are particularly great where all models are at their least reliable – when extreme observations are being considered. Furthermore, when tail events occur, they often come with other risks such as decreased liquidity. This can invalidate the VaR calculation, since the time horizon for which a portfolio must be held has by definition increased as liquidity has fallen. There is also the risk that the results of any calculation can be very sensitive to changes in the underlying parameters. If this is the case, it is necessary to ask whether a VaR result that changes significantly over time really reflects changes in the underlying risks.

There is also a risk that if VaR is used in regulation, then it might encourage similar hedging behaviour for similar firms, leading to systemic risk.

A final theoretical problem with VaR is that it does not constitute a coherent risk measure as it is not sub-additive. This means that the combined VaR for a number of portfolios is not necessarily less than or equal to the sum of the VaRs of the individual portfolios. As a result, it is not appropriate to determine the VaR for an organisation by aggregating the VaRs for the organisation's constituent departments.

Probability of ruin

The reciprocal of VaR is the probability of ruin. Whereas VaR sets the level of confidence (usually 95%) and then considers the maximum loss, the probability of ruin looks at the loss that would bring insolvency and looks at how likely this is. Ruin probabilities suffer from many of the limitations of VaR. However, provided they are used to assess the probability of insolvency (rather than the capital needed to meet a particular probability of insolvency), the assessment of loss if it occurs is not such a high priority – if ruin occurs, the extent of ruin is at most a second-order consideration.

Tail VaR

Another important measure of risk is the tail Value at Risk, or tail VaR. This measure has a wide number of other names, including expected tail loss, tail conditional expectation and expected shortfall, although an alternative definition of expected shortfall is given below. There are also a number of expressions used for evaluating tail VaR, although they generally reduce to the same formula. Tail VaR can be defined as the expected loss given that a loss beyond some critical value has occurred.

As with VaR, tail VaR can be calculated using empirical, parametric or stochastic approaches. If $TVaR_\alpha$ is the tail VaR at a critical level of α and the losses $L_{X,t}$ are again ranked such that $L_{X,1}$ is the smallest and $L_{X,T}$ the greatest, then the tail VaR can be calculated as:

$$TVaR_\alpha = \frac{\sum_{t=T\alpha}^{T} L_{X,t}}{\sum_{t=T\alpha}^{T} I(t \geq T\alpha)}, \quad (15.15)$$

where $I(t \geq T\alpha)$ is an indicator function that is equal to one if $t \geq T\alpha$ and zero otherwise. If T_α is not an integer, then the contribution of L_{X,t_-}, the loss for the value of t immediately below $T\alpha$, is calculated as $L_{X,t_-}(t_+ - T\alpha)$, so Equation (15.15) can be rewritten as:

$$TVaR_\alpha = \frac{\sum_{t=\lceil T\alpha \rceil}^{T} L_{X,t} + L_{X,t_-}(t_+ - T\alpha)}{\sum_{t=T\alpha}^{T} I(t \geq T\alpha) + (\lceil T\alpha \rceil - T\alpha)}, \quad (15.16)$$

where $\lceil T\alpha \rceil$ represents $T\alpha$ rounded up to the next integer.

The parametric calculation of the tail VaR involves choosing an appropriate statistical distribution to reflect the nature of the loss distribution, and integrating to find the area under the upper tail. This can be expressed in terms of the VaR as:

$$TVaR_\alpha = \frac{1}{1-\alpha} \int_\alpha^1 VaR_a \mathrm{d}a. \quad (15.17)$$

If, as before, the loss distribution is assumed to have a normal distribution, then using the parameters derived earlier, the tail VaR can be calculated as:

$$TVaR_\alpha = \mu + \sigma \frac{\phi(\Phi^{-1}(\alpha))}{1-\alpha}. \quad (15.18)$$

The results of such a calculation are shown in Figure 15.5

As with VaR, a stochastic approach can also be used. To calculate the tail VaR, the empirical approach is simply applied to the output from the stochastic model.

The tail VaR has a number of advantages over the VaR. First, it considers not only whether a particular likelihood of loss would result in insolvency, but also the distribution of losses beyond this point. This is not necessarily that important if the only issue is whether or not insolvency occurs, but if the

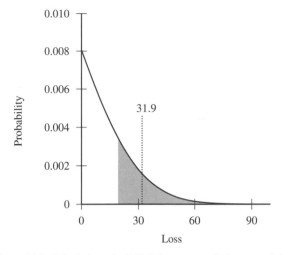

Figure 15.5 Calculation of tail VaR from a cumulative normal distribution

question instead relates to how bad things are beyond a particular point, then this is a vital aspect of the loss distribution.

Unlike VaR, the tail VaR is also coherent, as described below. This means that it has a number of mathematical properties that are intuitively attractive. In particular, if the tail VaR is calculated for a number of lines of business within an organisation, the results can be aggregated to give an overall tail VaR for the organisation. If needed, this can then be converted back to the VaR if this is the measure by which an organisation is judged.

Expected shortfall

The expected shortfall is closely related to the tail VaR. However, rather than being just the average value in the tail, it is defined as the probability of loss multiplied by the expected loss given that a loss has occurred.

As with VaR and tail VaR, expected shortfall can be calculated using empirical, parametric or stochastic approaches. If ES_α is the expected shortfall at a level of confidence of α and the losses $L_{X,t}$ are again ranked such that $L_{X,1}$ is the smallest and $L_{X,T}$ the greatest, then the expected shortfall can be calculated as:

$$ES_\alpha = \frac{\sum_{t=T\alpha}^{T} L_{X,t}}{\sum_{t=T\alpha}^{T} I(t \geq T\alpha)} \frac{\sum_{t=T\alpha}^{T} I(t \geq T\alpha)}{T}$$

$$= \frac{\sum_{t=T\alpha}^{T} L_{X,t}}{T}. \qquad (15.19)$$

In other words, it can be calculated as the sum of the losses in the tail divided by the total number of observations. If T_α is not an integer, then Equation (15.19) can be rewritten as:

$$ES_\alpha = \frac{\sum_{t=\lceil T\alpha \rceil}^{T} L_{X,t} + L_{X,t_-}(t_+ - T\alpha)}{T}. \qquad (15.20)$$

The parametric calculation of the expected shortfall can be expressed in terms of the VaR and tail VaR as:

$$ES_\alpha = \int_\alpha^1 VaR_a da$$
$$= (1-\alpha)TVaR_\alpha. \qquad (15.21)$$

If the loss distribution is assumed to have a normal distribution, then the expected shortfall can be calculated as:

$$ES_\alpha = (1-\alpha)\mu + \sigma\phi(\Phi^{-1}(\alpha)). \qquad (15.22)$$

A stochastic approach can also be used, applying the empirical approach to the output from the stochastic model as before.

Like the tail VaR, expected shortfall considers not only whether a particular likelihood of loss would result in insolvency, but also the distribution of losses beyond this point. However, unlike VaR and tail VaR the expected shortfall has little intuitive meaning. Whilst the VaR and tail VaR give results that are easy to relate to the current value of a portfolio, this is not the case for the expected shortfall.

Coherent risk measures

As mentioned earlier, VaR is not a coherent risk measure, whilst tail VaR is. It is therefore worth explaining exactly what makes a risk measure coherent. In simple terms, coherence implies that when loss distributions are altered or combined, the risk measure used behaves sensibly. Consider two loss distributions, L_X and L_Y, where $L = L_X + L_Y$. Consider also a risk measure F, which is calculated for L_X, L_Y and L as $F(L_X)$, $F(L_Y)$ and $F(L)$. For F to be a coherent risk measure, it must have the following properties:

- monotonicity – if $L_X \le L_Y$, then $F(L_X) \le F(L_Y)$;
- sub-additivity – $F(L_X + L_Y) \le F(L_X) + F(L_Y)$;

- positive homogeneity – $F(kL) = kF(L)$, where k is a constant amount; and
- translational invariance – $F(L+k) = F(L)+k$, where k is, again, a constant amount.

It is worth considering what these properties actually mean.

For a risk measure to be monotonic, it should increase if the potential losses increase. Note that monotonicity does not specify by how much a risk measure should grow, only that it should not fall.

Sub-additivity – the feature that VaR cannot guarantee – implies that combining two risks cannot create any additional risk; on the contrary, the total amount of risk according to a coherent measure may fall due to the effects of diversification.

Positive homogeneity implies that if a risk is scaled by some factor n, then the risk measure increases by the same factor. It also implies that if you aggregate a number of identical risks, then a coherent risk measure does not give any credit for diversification which does not exist.

Finally, translational invariance implies that if you reduce your risk of loss by fixed amount, then your measure of risk falls by the same amount.

Convex risk measures

Another feature that risk measures ought to have is convexity. Mathematically, this means that for $0 \le \lambda \le 1$:

$$F[\lambda L_X + (1-\lambda)L_Y] \le \lambda F(L_X) + (1-\lambda)F(L_Y). \qquad (15.23)$$

In other words, the a convex risk measure should give credit for diversification between risks where such a benefit exists. This can be thought of as a generalisation of the sub-additivity criterion discussed above.

15.5 Unquantifiable risks

It is important to recognise that, whilst quantification is an important tool, not all risks can be quantified. This might be because the potential losses are difficult to assess with any degree of certainty, as with reputational risk and the negative impact of poor publicity on future sales. Many types of operational risk relating to issues such as fraud and business continuity also fall into this category

The issue here is not necessarily with the potential size of the loss. Whilst this can be difficult to assess, it is possible to make educated guesses or to consider worst-case scenarios. It is also possible to put in place a framework to ensure that the cost is measured in direct terms, but also that the indirect costs are allowed for in terms of lost time, damaged reputation and the impact on sales and so on.

However, it is often very difficult to assess the likelihood of many events. Whilst this does make modelling the risk difficult, it still allows for the inclusion of these risks in scenario analyses. It is also possible to classify such risks in broad terms. These terms can be qualitative or quantitative. An example of a qualitative assessment would be to classify a risk as being very likely, moderately likely or very unlikely. On the other hand, broad percentage ranges could be used, for example with risks being given a probability of less than 25%, between 25% and 75% or over 75%.

It is also true to say that if the potential size of a loss is great enough, then no matter how unlikely the loss is – providing it is feasible – action should be taken to mitigate the risk.

One way of assessing these unquantifiable risks is to use a risk map, as shown in Figure 15.6. This is a diagram which maps the likelihood and impact of various risks onto a two-dimensional chart, so that their relative importance can be assessed. On this chart, both likelihood and impact are scored from one (unlikely/low impact) to five (very likely/high impact). The same chart can be shown after risks have been treated – the result would be a residual risk map, although in this context 'residual' refers to the new levels of exposure to the original risk rather than different risk arising following the risk treatment.

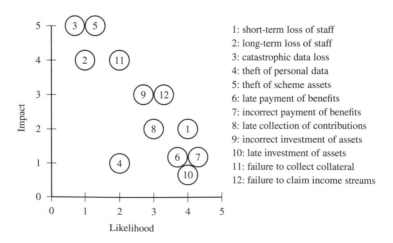

Figure 15.6 A sample risk map for pension scheme operational risk

Risks may also be hard to quantify even if they are fundamentally quantifiable. This may be because the risk relates to a new or heterogeneous asset class, or one where the amount of publicly available data is limited. In these cases, using another source of data as a proxy can help. Quantification may also be difficult if past losses have occurred infrequently, but extreme value theory is designed to deal with these situations.

/○ 15.6 Return measures

Once the measure of risk has been determined, the measure of return must be agreed. In this way, strategies can be compared and the results narrowed down to a set of efficient opportunities. With these two measures, two questions can be answered: what is expected to happen; and what are the risks of this not happening. This suggests that the return measure is a measure of central tendency, such as the mean, median or, less commonly, the mode.

Return measures, though they differ across types of institution, are generally more straightforward than risk measures because they are often linear, additive measures. The expected return on a portfolio invested in equities and bonds is simply a linear combination of the return on an equity portfolio and the return on a bond portfolio. Expected values have a key role in this kind of two-dimensional analysis in that they are fundamental to the concept of mean-variance optimisation, discussed below.

A type of return measure that relates to the previous section is the generic risk-adjusted performance measure. There are a large number of these, including:

- the return on risk-adjusted assets ('RORAA');
- the risk-adjusted return on assets ('RAROA');
- the return on risk-adjusted capital ('RORAC');
- the risk-adjusted return on assets ('RAROC'); and
- the risk-adjusted return on risk adjusted capital ('RARORAC').

These seek to embody the risk being taken in the return measure itself. The Sharpe ratio could be regarded as a simplistic version of risk-adjusted return, being calculated as:

$$SR = \frac{r_X - r^*}{\sigma_X}, \tag{15.24}$$

where r_X is the return on investment X, σ_X is its volatility and r^* is the return on a risk-free investment. This can be evaluated in terms of historical averages or prospective estimates. However, prospective assumptions can be difficult to estimate. This means that the usefulness of the above statistics, which are sensitive to the expected return parameter, is limited.

15.7 Optimisation

Having decided on measures of risk and return, the next stage is to use these measures to choose an investment or business strategy that provides an optimal combination of these measures.

15.7.1 Mean–variance optimisation

The classic approach to finding an optimal asset allocation is mean–variance optimisation. This involves finding a set of portfolios for which no higher expected return, measured by the mean, is possible given a particular level of risk, as measured by the variance – or, more commonly, the standard deviation – of returns. Such portfolios are described as mean–variance efficient, and together they form the efficient frontier, as shown in Figure 15.7 together with a range of possible portfolios. The asset allocations that are implied by the points on the frontier are shown in Figure 15.8.

The basic form of this model involves a group of assets each of which has returns that are normally distributed and considered over a single period. In this case, the mean and variance of all possible asset allocations can be calculated analytically from the means, standard deviations, correlations and weightings of the underlying asset classes as follows:

$$\mu = \sum_{n=1}^{N} w_n \mu_n, \tag{15.25}$$

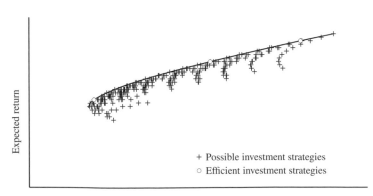

Figure 15.7 Efficient frontier

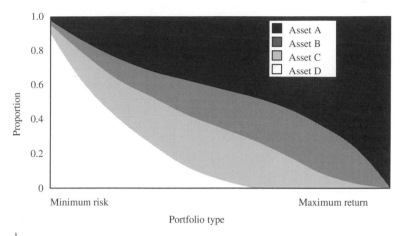

Figure 15.8 Composition of portfolios on the efficient frontier

and:

$$\sigma^2 = \sum_{m=1}^{N} \sum_{n=1}^{N} w_m w_n \sigma_m \sigma_n \rho_{m,n}. \qquad (15.26)$$

Here, μ and σ^2 are the mean and variance of the returns rather than the losses, as was the case in the calculation of VaR. The other variables are:

- the expected return for asset class n, μ_n;
- the standard deviation of returns for asset class n, σ_n;
- the correlation between the returns for asset classes m and n, $\rho_{m,n}$; and
- the proportion of assets in asset class n, w_n where $\sum_{n=1}^{N} w_n = 1$.

The portfolios that form the efficient frontier can then be found by varying the values of w_n. If there are only two asset classes, then every combination giving a return greater than that available from the minimum risk portfolio is efficient. However, for more than two asset classes optimisation algorithms such as those built into statistical or spreadsheet packages are needed.

Example 15.1 You have two asset classes, 1 and 2. Asset class 1 has an expected return of 8% per annum with a standard deviation of 15%, whilst asset class 2 has an expected return of 5% with a standard devation of 6.5%. The correlation between the asset classes is –20%. Find the expected risk and return for a portfolio consisting of 20% of asset 1 and 80% of asset 2, and show that this is the minimum risk portfolio.

The expected return for this portfolio is:

$$\mu = \sum_{n=1}^{2} w_n \mu_n$$

$$= (0.2 \times 0.08) + (0.8 \times 0.05)$$

$$= 0.05600,$$

or 5.600%.

The variance of the portfolio is given by:

$$\sigma^2 = \sum_{m=1}^{2} \sum_{n=1}^{2} w_m w_n \sigma_m \sigma_n \rho_{m,n}$$

$$= (0.2 \times 0.2 \times 0.15 \times 0.15 \times 1)$$

$$+ (0.2 \times 0.8 \times 0.15 \times 0.065 \times -0.2)$$

$$+ (0.8 \times 0.2 \times 0.065 \times 0.15 \times -0.2)$$

$$+ (0.8 \times 0.8 \times 0.065 \times 0.065 \times 1)$$

$$= (0.2 \times 0.15)^2 + (0.8 \times 0.065)^2 + (2 \times 0.2 \times 0.8 \times 0.15 \times 0.065 \times -0.2)$$

$$= 0.00298,$$

or 0.298%. The standard deviation is the square root of this amount, 5.459%.

Reworking this calculation with asset allocations of 19% for asset class 1 and 81% for asset class 2 gives a standard deviation of 5.463%; using asset allocations of 21% and 79% gives a standard deviation of 5.461%. Since both of these are higher than 5.459%, the allocation of 20% to asset class 1 and 80% to asset class 2 is the minimum risk asset allocation.

15.7.2 Separation theorem

One particular portfolio arising from mean–variance optimisation is of particular interest since it is particularly efficient. To see why, one additional asset is needed, one that has a fixed return over the period under consideration and no risk. Since it is risk free, the standard deviation of a portfolio consisting of a proportion α of a risky asset and $(1 - \alpha)$ of the risk-free asset is simply α times the standard deviation of the risky asset. This means that the most efficient

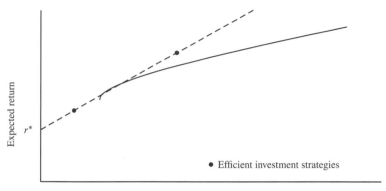

Figure 15.9 The separation theorem – single rate of interest

portfolios are those consisting of combinations of the efficient portfolio and the risk-free asset. This can include combinations where $\alpha > 1$, implying that additional money has been borrowed at the risk-free rate of interest to invest in this portfolio. The efficient portfolio is defined as the one where a line drawn from a point of complete investment in the risk-free asset is at a tangent to the efficient frontier, as shown in Figure 15.9.

If all investors have the same view of market risk and return, then everyone should want to hold only combinations of this portfolio and a risk-free asset or liability. This means that this efficient portfolio is actually the market portfolio, consisting of all assets in proportion to their market capitalisation. The tangent to the efficient frontier shown in Figure 15.9 is then known as the capital market line.

If the return on the risk-free asset is r^*, the return on the market portfolio is r_U and the volatility of this return is σ_U, then the return on a portfolio consisting of a proportion α of a risky asset and $(1 - \alpha)$ of the risk-free asset is $\alpha r_U + (1 - \alpha)r^*$, and the volatility of this portfolio is $\alpha\sigma_U$

However, it is often the case that a higher rate of interest is paid on money borrowed than is received on money invested. This means that there is a discontinuity, in that it is possible to mix investment in a risk-free asset and one efficient portfolio, to mix borrowing at the risk-free rate and another efficient portfolio, or to invest in the range of efficient portfolios in between.

If the rate at which money can be borrowed is $r_{F'}$ and the rate of return on the higher risk of the two portfolios is r_V, then the return on a portfolio consisting of a proportion α of a risky asset and $(1 - \alpha)$ of the risk-free borrowing is $\alpha r_V + (1 - \alpha)r_{F'}$, and the volatility of this portfolio is $\alpha\sigma_V$. For levels of risk

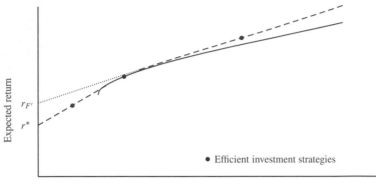

Figure 15.10 The separation theorem – differential lending and borrowing rates of interest

between those offered by portfolios with expected returns of r_U and r_V, the optimal strategy is simply to hold a portfolio on the original efficient frontier, as shown in Figure 15.10.

15.7.3 Issues with mean–variance optimisation

The reasons why the normal distribution might not be suitable for modelling investment returns have been discussed in detail. However, it is not usually appropriate to simply substitute another distribution into the approach described above. Using a joint distribution that is not elliptical means that the portfolio standard deviation will not necessarily reflect the full nature of the risked faced. This is partly because the standard deviation will not capture the impact of skew and kurtosis, and since the correlations used to combine the standard deviation do not give a full picture of the extent to which the various asset classes are linked. In this case, a different measure of risk can be used instead, but this may well need to be evaluated using stochastic simulation.

Stochastic simulation is also required if the efficiency being considered is over several periods with decisions being made at the end of each period. For example, if the contributions paid into a pension scheme depend on the solvency at the end of each period, then the outcome after a number of periods cannot be calculated analytically.

Even if all of the criteria for mean–variance (or similar) analysis are met, this approach to optimisation has limitations. In relation to the separation theorem, which assumes a market portfolio, it is not clear exactly what should count as 'the market'. Listed equities and bonds will be included, but for all markets or just domestic ones? This issue has already been discussed in relation to using the CAPM to choose an equity risk premium, but it also exists here. If global assets are chosen, should difficult-to-access classes like hedge funds and private equity be included? What 'free float' adjustments should be made for large investments that are held on a long-term basis by investors? There are no easy answers to these questions.

Even when a market portfolio is agreed, issues still remain. One of the foremost is that mean–variance optimisation can lead to efficient portfolios that appear unrealistic or impractical. An important example is when the two asset classes have similar expected volatilities, have similar correlations with other asset classes and are highly correlated with each other, but one has a slightly higher expected return than the other. In this case, the asset class with the higher return will tend to feature in the efficient frontier, whereas the asset class with the lower return will not.

One solution to this issue is to manually choose more 'acceptable' alternatives that lie close to but not on the efficient frontier; another is to place upper (and perhaps lower) limits on the allocations to 'difficult' asset classes. Both of these approaches seem too subjective. A third approach is to consider asset classes in broad groups, so optimising using global equities rather than regional equity weights. Whilst this results in subjectivity in arriving at the allocation within such a group, a bigger issue is that it provides no solution for standalone asset classes such as commodities.

15.7.4 The Black–Litterman approach

A more analytical approach to deal with this problem is the Black–Litterman approach (Black and Litterman, 1992). This is essentially a Bayesian approach where an investor's assumed asset returns are combined with the asset returns implied by the market. The market-implied returns for each asset class are those that would result in the market portfolio being the efficient portfolio, as described earlier, given the volatilities and correlations of the individual asset classes. The more confidence an investor has in his or her own assumptions, the greater the weight these are given; the less confidence there is, the more the assumptions will tend towards those implied by the market.

15.7.5 Resampling

This solves the issue of assets being excluded from the efficient frontier, but one issue remains. This is that the portfolio giving the maximum expected return is always an investment in a single asset class. This implies that high-risk investors should put all their eggs in one basket, which is not generally how even these investors behave. A solution which does address this issue is re-sampling.

The first stage in the re-sampling approach involves calculating the asset allocations for a single efficient frontier based on a relatively small number of simulations, representing the projection period. For example, if monthly data were used with a projection period of ten years, this part of the process would involve producing 120 simulated returns from each asset class. From these simulations, an efficient frontier could be created with, say, the asset allocation for ten portfolios being highlighted. These portfolios would be the minimum risk, the maximum return (a single asset class) and eight in between, equally spaced by the level of risk in the portfolio.

This process is then repeated many times to give a large number of candidate efficient frontiers and sets of ten asset allocations. A re-sampled efficient frontier is then calculated by averaging the asset allocation for each risk point. This means that the asset allocation for the minimum risk portfolio in the re-sampled frontier is the average of the asset allocations over all of the minimum risk portfolios, the asset allocation for the second portfolio is the average over that for all second portfolios and so on, up to the maximum return portfolio.

Michaud (1998) describes a patented bootstrapping version of this approach using historical data, but the approach can also be implemented using forward-looking simulated data.

This approach does address all of the issues discussed above. However, there are a number of issues with re-sampling. On a practical level, it involves significantly more work than more 'traditional' approaches and can only be implemented using simulations, either historical or forward-looking. On a theoretical level, the statistical properties of the points on the re-sampled efficient frontier are not clear. In particular, it is not obvious that, say, the asset allocations on the ninth point of a series of ten-point efficient frontiers should be considered to be sufficiently related to be combined into a single re-sampled point.

One aspect of re-sampling which is more robust is the maximum return point. It is interesting, for example, to consider the asset allocation that would give the maximum expected return allowing for uncertainty in those expectations over various periods. As the time horizon gets smaller, the allocation tends towards an equal weight in each asset class, whilst as it gets longer the allocation tends towards an total investment in the asset class with the highest expected return.

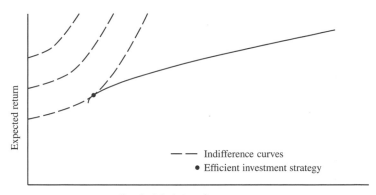

Figure 15.11 Portfolio selection

15.7.6 Choosing an efficient portfolio

Most of the above analysis is concerned with describing the range of efficient portfolios. However, in practice, a particular portfolio or strategy must
be chosen. There are a number of ways in which this can be done.

If the risk appetite is known in absolute terms – for example, a VaR beyond a
certain limit is unacceptable – then the strategy of choice is the one that simply
maximises the expected return for a given VaR. However, in many cases, the
constraints are not so obvious.

One approach in this situation is to turn the risk preference of investors into
quantitative limits. In this case, the process above can be used; however, they
will often be expressed as trade-offs, and so be more appropriate for conversion
to preference or utility functions.

In this case, a series of lines can be drawn, each representing the combination of risk and return that gives a particular level of utility. Since each point
on this line represents combinations of risk and return to which an investor
is equally attracted, these lines are known as indifference curves. If these are
plotted on the same chart as an efficient frontier, then the point at which an
indifference curve is tangential to the efficient frontier defines the optimal portfolio, as shown in Figure 15.11. This approach can also be extended to allow
for the separation theorem, as shown in Figures 15.12 and 15.13.

15.8 Further reading

Many of the issues in this chapter are described in a range of finance textbooks
such as Copeland *et al.* (2004). Elton *et al.* (2003) also includes a discussion

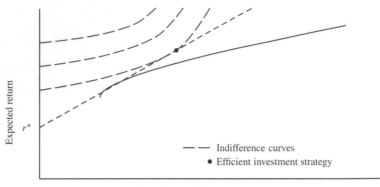

Figure 15.12 Portfolio selection using the separation theorem – single rate of interest

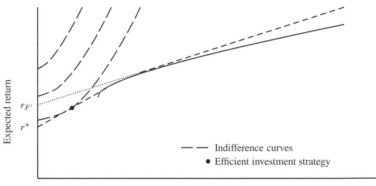

Figure 15.13 Portfolio selection using the separation theorem – differential lending and borrowing rates of interest

of utility theory, with even more information being given in Eeckhoudt *et al.* (2005). Market risk assessment is dealt with by Dowd (2005). Meucci (2009) covers much of the same ground more formally, and also includes discussion of areas such as the Black–Litterman model. Whilst both of these books discuss coherent risk measures, McNeil *et al.* (2005) explores them in more detail. The definitive reference for this topic is Artzner *et al.* (1999). Michaud (1998), on the other hand, considers exclusively the subject of re-sampling.

16

Responses to risk

16.1 Introduction

Having not only identified and analysed risks but also compared the risks faced with the stated risk appetite, the next stage is to respond to those risks. The responses to risk are generally placed into one of four categories:

- reduce;
- remove;
- transfer; or
- accept.

There is little point in trying to fit every potential risk response into one of these categories, since there is often ambiguity over which category a particular treatment should be put. The main purpose of detailing these four groups is to ensure that all potential responses are considered in relation to a risk as it arises.

16.1.1 Risk reduction

Risk reduction involves taking active steps to limit the impact of a risk occurring. This group includes approaches such as diversification. This involves combining a risk with other uncorrelated risks, or at least with risk where the correlation is less than one. At the extreme, it can involve taking on risks which have a high negative correlation with the risk faced, in which case it becomes hedging rather than just risk reduction. Whilst this approach is most obviously connected to investments, it can also relate to the choice of projects on which a firm embarks. Risk reduction can also involve the creation of more robust systems and processes, in order to reduce the chance of a risk emerging, or to limit the impact of a risk.

413

16.1.2 Risk removal

Removing a risk means ensuring that an institution is no longer exposed to that risk at all. To achieve this, a firm can choose to avoid a project or an investment altogether, or can decide to achieve its aims differently. For example, a firm concerned about counter-party risk from OTC swaps could instead use exchange traded derivatives.

16.1.3 Risk transfer

Risk transfer is a key response to risk. This involves changing the exposure to a risk by transferring the consequences of a risk event to another party. Two important categories are non-capital market and capital market risk transfer.

Non-capital market risk transfer

The most common form of non-capital market risk transfer is insurance – the payment of a premium to buy protection from a risk. This itself can take several forms. The traditional route is for a firm wishing to transfer a risk to pay a premium to another firm – the insurer – in exchange for protection. However, some firms choose to self-insure, either through setting aside assets or through setting up a wholly owned captive insurance company. Captives tend to be set up in tax-beneficial offshore locations so that they can be used as tax-efficient ways of setting aside reserves as a cushion against adverse events. Formal and informal captives can also be set up by groups of firms in order to achieve an element of diversification between them.

The types of policies can also vary hugely. Proportional or quota share insurance (or reinsurance, if bought by an insurance company) transfers a proportion of each policy sold to a third party, allowing the firm to take on more business and therefore to build a more diversified portfolio; excess-of-loss (re)insurance, on the other hand, pays out only if losses exceed a certain level. If the level is very high, then this becomes catastrophe insurance. Insurance policies can also average the loss events over a number of years, to smooth profits and lower premiums, or can require a range of events to occur before payout is made. This can be helpful if the desire is to protect against concentrations of risk.

Capital market risk transfer

Capital market risk transfer – also known as securitisation – is a way of turning risk exposure into an investment that can be bought and sold, investors taking exposure to the risk but earning a risk premium for doing so.

One of the most common formats is to package risks in a bond where the payments to investors are reduced if losses rise above a certain level. However, a broader approach is to issue a put option that allows a firm to raise capital at a predetermined price in the event of a pre-specified catastrophe.

One attractive feature of capital market risk transfer is that if the security bearing the risk is traded, then its price can be used to provide a market-based price for the risk. This means a market price can be determined for any risks which are of a similar nature to those transferred but are retained by the firm. Such marking-to-market is an important part of risk frameworks such as Basel II and Solvency II.

Capital market risk transfer can also provide a quicker way of raising capital to cover risks than the more indirect route of issuing equity before taking on the risk, either through a rights issue or through the creation of a new firm.

16.1.4 Risk acceptance

Accepting, retaining or taking a risk, rather than reducing, removing or transferring it, implies that no action is taken to respond to the risk. This can be done because the risk is of trivial – either because the potential severity of the risk is small or the probability of occurrence is vanishingly unlikely – but large risks can also be retained. This might be done if the cost of removal is greater than the exposure to that risk. If a risk is retained that would often be transferred, then this is sometimes known as self-insurance. This can happen if a risk is very large, so insurers would require an additional margin to cover this risk, or if claims would be so frequent that the amount claimed would often be similar in magnitude to the size of premiums. However, risks are also retained when the taking of a particular risk is part of the business plan. An example might be mortality risk taken by a life insurer. It is important to note that just because a risk is retained, it does not mean that it is not analysed. Indeed, the analysis of a risk is often an important part of the decision on how to deal with that risk.

16.1.5 Good risk responses

There are a number of features that a good risk response should have. First, it should be economical. This not only means that the solutions chosen should be the least costly way of achieving the results, but it should also cost less than the amount saved in the reduction of risk. In some instances, this is easy to quantify. For example, if a new expense monitoring system is introduced to reduce the number of fraudulent expense claims, then the cost of the system can easily be compared with the reduction in the total volume of expenses. However, if a strategy is put in place to reduce the chances of reputational damage, then it is much more difficult to assess whether that strategy has been cost effective.

It is also important to ensure that risk responses match as closely as possible the risks that they are intended to control. However, this can involve a compromise with the principal of economy. For example, if trying to limit the downside risk of investments in a portfolio of mid-cap shares, options on that portfolio of shares might be thinly traded and therefore expensive. Even though an options on the corresponding large-cap index might not match the liability as well, the lower cost might compensate for the higher basis risk.

Linked to this point, responses should also be as simple as possible, since the more complex a solution is, the greater the chance that a mistake will be made. This does not mean that no complex solutions should be used – sometimes the only ways of dealing with complex risks are themselves complex; however, it is important to consider the full range of possible solutions.

Risk responses should also be active, not just informative. Whilst it is important that key personnel are notified when a risk limit is close to being breached, it is more important that action is taken to avoid the breach. For example, if equity markets fall to a level where solvency is threatened, a good risk treatment would ensure that management were aware of this fact; however, a better system would also implement a change in investment strategy, either through the prior purchase of options, programmed trades or some other approach. However, this is not to say that solutions should be rigid, and it is important that the flexibility to change risk responses remains.

Risks should often be retained unless they are significant. This does not necessarily mean that the expected value of the risk should be large. In particular, low frequency/high severity risks should almost always be mitigated if the potential damage from such a risk is large enough.

16.2 Market and economic risk

Market risk is an important risk for all financial institutions, and is often the most important. All firms should have clear strategies and policies on market risk. It is also important to recognise the way in which market risk is linked to other risks. For example, operational failures can often be highlighted in extreme market conditions, so it is important to consider the extent of market risk exposure when designing systems to limit operational risk. Market risk is also closely linked to credit risk. Not only does credit risk tend to be higher when markets are subdued, but many derivative-based responses to market risk can expose a firm to counter-party risk. In particular, OTC derivatives expose each counter-party to the risk that the other will fail before the end of the contract, whilst owing money on it on the date of failure. One way to deal with this is collateralisation, which is discussed later.

16.2.1 Policies, procedures and limits

The most fundamental aspect of managing market risk is to have clear policies. At a high level, this can include policies on the overall level of market risk that is acceptable by some measure such as VaR. However, it should also include details of what constitutes an acceptable investment, and what limits there are to investments in particular asset classes, individual securities or with individual counter-parties. In this way, policies, procedures and limits are closely linked to diversification – discussed below – and counter-party risk.

A firm's policies should also include a statement of who can make various investment decisions, and the financial limits on such decisions. This provides the link between market risk and operational risk.

16.2.2 Diversification

A key way to manage market risk is through diversification. By holding a range of investments, exposure to the poor performance of one is limited. Diversification can be measured by the extent to which a portfolio holds assets in different asset classes, geographic regions and economic sectors, either in absolute terms or relative to benchmarks. Factor analysis can also be used to to determine the extent to which particular economic and financial variables influence a portfolio of stocks. If the exposure to one or more factors is thought to be too great, then this implies that the portfolio should be diversified further.

16.2.3 Investment strategy

This is arguably the easiest way to manage market risk, although the scope for change and the effect of that change will vary across the different types of firm. For banks, the effect is reasonably important, but market risk is not generally the greatest risk faced. For insurers, the scope for change is controlled by the degree to which the assets held are admissible from a regulatory point of view. This can mean that assets that are relatively similar from a risk point of view are treated in very different ways from a regulatory point of view. The market risk aspect of the investments is secondary to the admissibility aspect for insurance companies. Market risk is often the key risk for pension schemes, so the investment strategy is a key way of controlling the risk taken, although it is only one aspect and should be considered in the light of the various other 'levers'.

Investment strategy is often determined using stochastic asset-liability modelling. This helps determine the appropriate investment strategy by maximising the return by some measure such as shareholder earnings subject to some maximum level of risk such as a VaR target. In reality, there may well be a number of risk limits that are applied.

16.2.4 Hedging against uncertainty

Rather than changing an investment strategy directly, derivatives may instead be used. One approach is to use derivatives to hedge against uncertainty. This means that both losses and gains are reduced. The easiest way to do this is using a future or a forward. Each of these is an agreement to buy or sell a fixed amount of some asset for a fixed price at some fixed date in the future, the delivery date.

Futures or forwards can be used as an alternative to buying and selling securities if the investment strategy is being changed. They might be used if there is a desire to leave a particular stock selection strategy in place, in terms of the actual investments held, whilst changing the underlying asset allocation. Futures and forwards can also be used to change the asset allocation more quickly and cheaply than can sometimes be achieved by trading the underlying securities.

An important point to note about this type of hedging is that it means that profits as well as losses are neutralised. This should not be a problem if this issue is understood by all parties, but even if offset by a large profit in an underlying asset, a large loss on a derivative contract can be unsettling. This is particularly true if the department carrying out the hedging constitutes a separate cost centre to the department holding the underlying asset. Communication is therefore key in these circumstances.

It is also important to recognise that no matter how good a hedge might be in theory, uncertainty over the amount of hedging required can reduce the effectiveness of a hedge. For example, a pension scheme might want to hedge a future sale of assets, but may still be in receipt of contributions that are based on the total payroll and so are uncertain.

Differences between futures and forwards

Whilst futures and forwards have similar underlying properties, they differ in some important ways. The most fundamental is that futures are traded on exchanges, whilst forwards are OTC contracts. Anyone wishing to trade a future must be a member of an exchange or must trade through a broker who is a member. Each futures trade involves matching a party who wishes to take a long position in a future with one who wishes to take a short position, since each future is a contract. However, even though each trade will match these two parties, the parties do not contract with each other. Instead, all parties contract directly with the exchange.

Forwards, on the other hand, are simply OTC agreements directly between the two parties wishing to trade. The details of each contract are set out in an ISDA (International Swaps and Derivatives Association) agreement. This is a very detailed document outlining all aspects of how the contract works.

As OTC contracts, forwards are very flexible and can be provided on virtually any underlying asset with any delivery date. However, the bank providing the forward will itself want to mitigate this risk, either through other positions held or with other banks, and the more unusual a forward is, the more difficult this will be. More importantly, the more difficult it is to pass on the risk in the forward, the more risk capital a bank will need to write the forward. Since this cost is passed on to the investor, there is a real cost to pay for demanding an unusual forward contract.

Exchange-traded contracts such as futures have virtually no flexibility – they are highly standardised in terms of the nature of the underlying asset and the delivery date. However, this level of standardisation means that exchange-traded contracts tend to be very liquid, meaning that large transactions can be effected very quickly with a minimal impact on the price of the contract.

Counter-party risk

The nature of exchange-traded and OTC contracts has an impact on the credit risk faced by the various counter-parties. Looking first at exchange-traded contracts, counter-party risk is reduced by the pooling of contracts – since each party has a contract directly with the exchange, the failure of a single counter-party does not directly affect the payment of any single futures contract. However, since this means that the exchange is underwriting all contracts, the exchange needs to protect itself from the failure of any of its counter-parties – in other words, those holding futures contracts.

Exchanges protect themselves through the use of margins. These are deposits that members of an exchange post with the exchange to ensure that if a member becomes insolvent, there are assets available to cover any losses they have made on their contracts.

There are several types of margin that might be required, the most common being:

- initial margin;
- maintenance margin; and
- variation margin.

The initial margin is the value of assets transferred to a margin account once a contract is opened. This will be some proportion of the contract size, with the proportion depending on the volatility of the contract. At the end of each day – and sometimes during the day – the cost to the member of closing out a position at the current price of the future is calculated by the exchange. A futures contract is closed by taking an opposite position in the same contract. If the cost of closing the position is greater than the initial cost of the future, then the difference is deducted from the margin account; if it is lower, then the difference is added to the margin account. This process is known as marking to market.

If the margin account drops below a specified level – the maintenance margin – then the member is required to transfer assets to the margin account to top it back up to the level of the initial margin. This amount is known as the variation margin.

Margins can be reduced if members hold diversifying positions in similar futures – that is, if they hold spread rather than naked positions. At the extreme, this can involve margins being calculated taking into account the individual construction of each member's portfolio with the exchange. This is the case with the Standard Portfolio Analysis of Risk (SPAN) developed by the Chicago Mercantile Exchange.

Each exchange will specify what assets can be counted as collateral. They will also specify the extent to which each asset counts. For example, high-quality government bonds might be counted at 90% of their face value, whilst shares might only be counted at 50% of their face value.

With the absence of pooling, counter-parties to OTC derivatives such as forwards face a higher degree of counter-party risk. This is often dealt with using collateralisation. Collateralisation involves the transfer in response to the marking to market of a contract in a similar way to margin requirements. The obligations of both counter-parties in relation to collateral are outlined in the credit support annex (CSA) of the ISDA agreement. In particular, whilst the ISDA agreement covers all aspects of the structure of the derivative and the calculation of its price, the CSA covers issues such as the types of security that can be used as collateral and when the required amount of collateral is calculated. It will also specify the minimum transfer amount – the level below which no transfer of collateral will be needed. This is to avoid counter-parties making very small transfers of assets when marking to market reveals that only a small change in the collateral required is needed.

Not all OTC contracts will involve collateralisation; however, collateralisation can reduce the cost of a transaction for a counter-party whose risk of default is regarded as significant.

Even if adequate collateral has been posted, the failure of a counter-party can leave a firm exposed to a risk that it had hoped to deal with. If the failure has occurred at a time of more general difficulties in the market, then putting a replacement derivative contract in place might take some time, leaving the firm exposed to risk for longer than it would prefer. This must be borne in mind when considering how to deal with market risk.

Pricing futures and forwards

If costs are ignored and the asset on which the future is based pays no income, then the price of a future or forward has a simple relationship to the spot price of the underlying asset. In particular, the price at time zero of a future or forward with a delivery time T, F_0, is related to the spot price at time zero, X_0, and the continuously compounded risk-free rate of interest, r^*, as follows:

$$F_0 = X_0 e^{r^* T}. \tag{16.1}$$

In other words, the price of the future – which represents the price at which an investor is agreeing at time zero to buy or sell an asset at time T – is simply equal to the current spot price rolled up at the risk-free rate of interest. The rationale for this formula can best be seen by considering two equivalent ways of owning an asset at time T. The first is simply to pay the spot price for the asset, X_0, at time zero; the second is to enter into a futures contract at time zero to pay F_0 for the asset at time T, and to invest sufficient assets in an account paying a risk-free rate of interest to accumulate to F_0 at time T. This would require an investment of $F_0 e^{-r^* T}$. Since the two transactions must have the same price – otherwise an arbitrage opportunity would exist – this means that $F_0 e^{-r^* T} = X_0$, which, after rearrangement, is equivalent to Equation (16.1).

In practice, there are often complications. In particular, there may be:

- a fixed amount of income available from the underlying asset;
- a fixed rate of income available from the underlying asset;
- a fixed amount of benefit associated with the underlying asset;
- a fixed rate of benefit associated with the underlying asset;
- a fixed amount of cost associated with the underlying asset;
- a fixed rate of cost associated with the underlying asset; or
- a differential rates of interest for inter-currency contracts.

If a fixed amount of income is payable on the underlying asset, then this is normally foregone if the investment position is replicated by a future or forward – such a derivative commits the holder to trade in the asset, but does not result in a transfer of the asset's income in the period before transfer. To allow for the lack of an amount of income with a present value at time zero of D, this amount must be deducted from the spot price of the asset. This means that the price of the forward becomes:

$$F_0 = (X_0 - D)e^{r^*T}. \qquad (16.2)$$

If income is instead received at some fixed rate, r_D, then this rate is instead deducted from the rate at which assets would need to accumulate:

$$F_0 = e^{(r^* - r_D)T}. \qquad (16.3)$$

As well as income, holding an asset can provide other benefits. For example, holding a physical asset rather than obtaining the position synthetically might result in a reduction in capital requirements. If such an effect is market-wide, then it could have an effect on the price of a future and can be reflected in Equations (16.2) and (16.3), with the benefit being converted to a fixed amount, D, or a rate, r_D. The rate of benefit is known as the convenience yield.

Assets can also generate explicit costs, which are not borne if exposure is gained through a futures contract. For commodities, this can be storage costs, but it might also be a cost of financing. Such costs can be regarded as negative income in Equations (16.2) and (16.3).

As a result, the price of a future can be above or below the current spot price. It can also be above or below the *expected future spot price*. If the price of a future is lower than the expected future spot price, then the situation is described as *normal backwardation*. This occurs when any income produced by an underlying asset together with its convenience yield exceed any storage or financing costs. However, it also occurs if the main reason for the existence of the market is for producers of a commodity to hedge against future falls in the commodity's price – the volume of demand for short positions in futures drives the price of the future down.

The opposite situation occurs if the market is driven by a desire to gain exposure to an underlying asset synthetically, so there is a high volume of demand for long positions in futures. This can happen if the reason for the creation of a market was to allow users of particular commodities to hedge their input costs, or if prices are driven by investors trying to gain exposure to particular commodities. This also means that the effect can be exacerbated

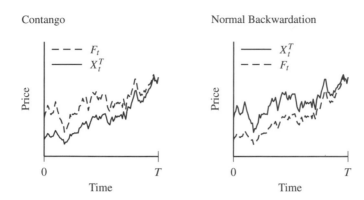

Figure 16.1 Markets in contango and normal backwardation

if storage costs are particularly high. In this case, the market is said to be in *contango*. Both markets are shown in Figure 16.1, with X_t^T denoting the expectation at time t of the spot price at time T.

These terms should not be confused with normal and inverted market. A normal market is one where, on a particular day, the futures prices increase with the expiry date of the future. This is what might be expected when there is no predictable seasonality in the availability of the underlying asset. However, for some commodities there might be an expectation of increased availability of the underlying asset at some future date. This could be expected to lead to a fall in the spot price *at that time*, which would be reflected in a lower price for futures of that term.

Basis risk in futures

The point of a forward is that it can be used to hedge exactly the risk faced, with the size of the contract being equal to the size of the risk. However, the fact that futures contracts are standardised means that they might not provide an exact hedge. In particular:

- the futures position might need to be closed before the expiry date of the future;
- the future may expire before the planned date of the asset's sale or purchase, requiring that the future be rolled over into another position;
- the date of sale or purchase for the asset might be uncertain;
- the asset on which the future is based might not be the same as the asset being hedged; or

- items excluded from the future such as dividend income from the underlying
  asset or costs associated with investment in this asset might not be known
  accurately in advance.

All of these issues can give rise to basis risk, which is defined as uncertainty
in the basis at the point at which the futures position is closed. The basis at time
t, B_t, is the difference between the spot price of the asset, X_t, and the futures
price, F_t:

$$B_t = X_t - F_t. \tag{16.4}$$

As can be deduced from the earlier comments on discounting, storage costs,
convenience yield and so on, the basis can be positive or negative. It can also
be defined as $F_t - X_t$, particularly in the context of financial futures.

The basis at the time a futures contract is effected is known, since both the
price of the future and the spot price of the asset are known. Furthermore, if
a hedge is required until the exact date of expiry, T, the asset being hedged is
exactly the same as that underlying the future and there are no uncertain cash
flows in the period between which the hedge is effected and the expiry date,
then there is no basis risk. This is because in this case the basis at the time of
expiry is zero – at expiry, the spot price is equal to the futures price. However,
if any of the conditions above hold, then the basis at the time of sale, expiry or
roll into a new position will be unknown. It is the uncertainty around the basis
at this time that gives rise to basis risk.

Consider a situation where a portfolio of equities held at time $t = 1$
must be sold at time $t = 2$ in the future, as shown in Figure 16.2. A

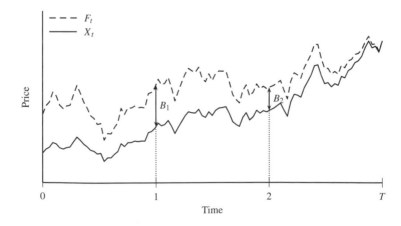

Figure 16.2 Basis risk – early sale

way to hedge this sale would be to take a short position in a futures contract at time $t = 1$ based on the same underlying equity portfolio. However, if the future expires at some future time $t = T$ where $T >= 2$, then the contract must be closed out early. In particular, an an offsetting, long position in the contract would be taken at time $t = 2$ when the portfolio was sold.

Taking the short position in the futures contract at time $t = 1$ means that a price at time $t = T$ of F_1 is being guaranteed on the sale of the portfolio at time $t = T$. The offsetting contract taken at time $t = 2$ means that a price of F_2 for the purchase of the portfolio at time $t = T$ is also guaranteed. This means that the profit (or loss) on the two contracts at time $t = T$ is $F_1 - F_2$ – essentially the guaranteed sale price less the guaranteed purchase price. At time $t = 2$, the equity portfolio will also be sold, realising an income of X_2. This means that the total income is $X_2 + F_1 - F_2$. However, since $B_t = X_t - F_t$, this can be rewritten $F_1 + B_2$. In other words, the total return is a function of just the futures price at time $t = 1$ and the basis at time $t = 2$.

As can be seen in Figure 16.2, the basis reduces to zero as t approaches T, verifying that if the hedge is held until the expiry of the futures contract, then there is no basis risk.

However, if the expiry date of the future occurs before a hedged asset must be sold, then the hedge must be rolled into a new future. This means that, at expiry of the first future, a new short position in another futures contract is entered into. This means that, whilst there has been no basis risk in the first future, there is basis risk at the time the new futures contract is taken out and, if it is to be closed out before expiry, at this point as well. In fact, if a contract is rolled N times, then there are N opportunities for basis risk if the final contract expires when the underlying asset is sold, and $N + 1$ if the final contract is closed out before expiry.

Uncertainty over the time at which an underlying asset must be sold can lead to any of the situations above, primarily because it would be difficult to choose a future with the correct expiry date.

A different type of basis risk occurs when the issue is that the underlying asset on which the future is based differs from the asset being hedged. In this case, the hedge is actually a cross-hedge. Here, even if the hedge is held until the expiry of the future, basis risk arises, as shown in Figure 16.3. If X_t is the spot price of the asset underlying the future and Y_t is the spot price of the asset being hedged, then the basis at time t can be split into two parts:

- the difference between the spot price on the asset being hedged and the spot price of the asset underlying the future $(Y_t - X_t)$; and

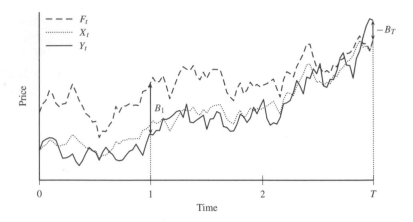

Figure 16.3 Basis risk – cross-hedging

- the difference between the spot price of the asset underlying the future and the price of the future $(X_t - F_t)$.

  Putting these together gives:

$$B_t = (Y_t - X_t) + (X_t - F_t) = Y_t - F_t. \qquad (16.5)$$

Consider the situation where a portfolio of equities must be hedged until time $t = T$, the expiry date of the future, but where the portfolio underlying the future differs from the portfolio being hedged as shown in Figure 16.3. If this hedge is transacted by holding a short position in the future until expiry, then the total return is $F_1 + Y_T - X_T$ – in other words, the price of the future at time $t = 1$ plus the difference between the two spot prices at time $t = T$. Since $X_T = F_T$, Equation (16.5) reduces to $B_T = Y_T - X_T$ at time $t = T$. This means that the return can also be written $F_1 + B_T$.

This shows that if there is a hedge where the futures position must be closed before expiry *and* the asset underlying the future is not the same as the asset being hedged, then the return will remain $F_1 + B_t$, but with B_t being defined by Equation (16.5) rather than Equation (16.4).

Hedging with futures
Because futures are standardised, it is necessary to determine the number of contracts needed to hedge a particular position. There are two approaches to this calculation, depending on whether the hedge is described in terms of an amount of exposure – for example, barrels of oil – or whether it is described in terms of financial exposure.

In the first situation, there are two parts to the calculation. First, an optimal hedge ratio, h, must be calculated. This gives the units of the futures required to hedge each unit of exposure. Three items are needed to calculate the optimal hedge ratio:

- the volatility of the per-unit price of the asset to be hedged, σ_Y;
- the volatility of the per-unit price of the future over the term of the hedge, σ_F; and
- the correlation between these two amounts, $\rho Y, F$.

It can be shown that the optimal hedge ratio, which minimises the volatility of the hedged position, is given by:

$$h = \rho_{Y,F} \frac{\sigma_Y}{\sigma_F}. \tag{16.6}$$

The next stage involves using this figure to calculate the total number of contracts required for the hedge, N_h. This involves two further items:

- the number of units being hedged, N_Y; and
- the number of units each futures contract represents, N_F.

These are combined as follows:

$$N_h = h \frac{N_Y}{N_F} = \rho_{Y,F} \frac{\sigma_Y N_Y}{\sigma_F N_F}. \tag{16.7}$$

If the price of the future and the asset being hedged are perfectly correlated, then this reduces to the volatility-adjusted ratio of the size of the position to be hedged and the contract size; if the volatilities are also equal – as would be the case if the asset being hedged were the same as the asset underlying the contract – then the sizes of the position to be hedged and the contract would be the only items required.

For financial assets, a similar approach is used. However, if the asset underlying the future is regarded as the market portfolio, then the optimal hedge ratio can be regarded as the CAPM beta of the portfolio being hedged, β_Y. Then all that is needed is the value of the portfolio, Y, and the notional value of the futures contract, X. If the future is on an index, then X is defined as the current index in points multiplied by the change in the value of a contract for a one-point move; if the future is on a single share, then X is the current share price multiplied by the number of shares per contract. The number of contracts needed to hedge the portfolio, N_h, is then given by:

$$N_h = \beta_Y \frac{Y}{X}. \tag{16.8}$$

Example 16.1 You are managing a portfolio of equities for a pension scheme. The portfolio is actively managed with a benchmark of the FTSE All-Share Index, and its current value is £120 million. The scheme has decided it wishes to disinvest from this portfolio as quickly as possible, but selling all of the equities could cause a fall in the price of some of the assets. You therefore decide to sell futures on the FTSE 100 Index to hedge price movements in your portfolio. The size of each FTSE 100 futures contract is £10 per point. The current FTSE 100 index value is 6,000. The volatility of the FTSE 100 Index, on which the future is based, is 15% per annum; the beta of the portfolio relative to the FTSE 100 index is 1.2. Calculate the number of contacts required to hedge this position.

Using Equation (16.8), the portfolio size, Y, is £120,000,000. The notional value of a futures contract, X, is the index value, 6,000, multiplied by the change in value for a one-point move, £10. This gives a current notional contract value of £60,000. Combining these with a value of 1.2 for β_Y gives:

$$N_h = \beta_Y \frac{Y}{X} = 1.2 \frac{120,000,000}{60,000} = 2,400. \tag{16.9}$$

16.2.5 Hedging against loss

Whilst these derivatives offer an alternative to trading in the underlying securities, options offer a way of changing the return profile of a portfolio in a more fundamental way. In particular, if a put option on an investment is bought, then this can be used to protect against falls in that investment below the strike price. This is because, below the strike price, the fall in the value of the underlying investment will be offset by the increase in the pay-off from the option.

Whilst futures and forwards are free apart from the dealing costs, options require a premium to be paid – they are essentially a form of insurance. This means they can be less attractive than futures and forwards for many scenarios. However, if downside risk is the main concern, options can offer a good way of limiting this risk.

One other limitation of options should also be noted. An option can be used to limits the loss faced in absolute terms. Therefore, if an option is used in a portfolio of assets that are held to meet liabilities whose values are changing – as would be the case for a pension scheme, where the liability value is sensitive to changes in interest rates – only the asset risk will be addressed. Whilst more complex out-performance options can be bought to deal with these risks together, these are often more costly.

Another type of derivative that can be used to provide protection against loss is a credit default swap (CDS). This provides a payment on the default of a named bond or index, and thus can be used to hedge against falls in prices. CDSs are described in more detail in the section on credit risk.

Many options are traded on exchanges, with the advantages and disadvantages that this brings. However, OTC options also exist for particular hedging needs. A key type of OTC option is an out-performance option, which provides a payment if the returns on one asset exceed those on another by more than a certain amount. These can be useful for pension schemes or insurance companies wishing to protect the returns on their investment portfolio relative to an interest rate-sensitive set of liabilities. CDSs are all traded OTC.

16.2.6 Hedging exposure to options

Whilst derivatives can be used to reduce risk, any institution writing a derivative might wish to hedge their exposure. For a future or a swap, the amount of the underlying asset that must be held is clear, since for every unit of futures exposure, a unit of the underlying asset must be held or sold short. However, for options the issue is more complex. The higher the price of the call option, the greater the sensitivity of the price to a change in the price of the underlying asset. This sensitivity is known as the delta of the option, Δ, and for an option with price C_t whose underlying asset has a price of X_t, both at time t, it is defined as:

$$\Delta = \frac{\partial C_0}{\partial X_0}. \tag{16.10}$$

This partial derivative can be calculated directly from an option pricing formula, or approximated by calculating the change in the price of the option for a small change in the price of the underlying asset from empirical data. If the Black–Scholes formula is used, the deltas for a call and put option, Δ_C and Δ_P respectively, are:

$$\Delta_C = e^{-r_D T} \Phi(d_1), \tag{16.11}$$

and:

$$\Delta_P = e^{-r_D T} [\Phi(d_1) - 1]. \tag{16.12}$$

The delta is important because it defines how much of an underlying asset is needed to hedge the exposure from an option based on that asset. In particular, if one unit of the option is held, then Δ units of the underlying asset must be held. However, the delta will change as the option price changes. This means that to remain *delta neutral* the amount of the underlying asset must be changed constantly. This process is known as dynamic hedging.

The amount by which a holding in the underlying asset should change is given by the gamma of the option, Γ. This is the second partial derivative of the option price with respect to the price of the underlying asset:

$$\Gamma = \frac{\partial^2 C_0}{\partial X_0^2}. \tag{16.13}$$

If the Black–Scholes formula is again used, the gammas for a call and put option, Γ_C and Γ_P respectively, are:

$$\Gamma_C = \frac{e^{-r_D T} \Phi(d_1)}{\sigma_X X_0 \sqrt{T}}, \tag{16.14}$$

and:

$$\Gamma_P = \frac{e^{-r_D T} [\Phi(d_1) - 1]}{\sigma_X X_0 \sqrt{T}}. \tag{16.15}$$

Two other measures of option price sensitivity are the theta, Θ, and vega, v. Although these are less important from a hedging perspective, it is useful to know what they represent.

Even if the price of the underlying asset stays the same, the price of an option will change as the option moves closer to its expiry date. The rate of change of an option with time is known as its theta. This is defined as:

$$\Theta = \frac{\partial C_0}{\partial t}. \tag{16.16}$$

The sensitivity of the price of an option to a change in the volatility is known as the vega. This is defined as:

$$\Theta = \frac{\partial C_0}{\partial \sigma_X}. \tag{16.17}$$

16.3 Interest rate risk

Interest rate risk arises from having assets and liabilities with different exposures to changes in interest rates. This suggests a particular type of risk management that addresses this type of risk specifically.

In terms of risk management, interest rate risk is dealt with slightly differently from other market risks. This partly because of time dimension, but also because unlike most other market risks, there is little reward for taking interest rate risk. Price inflation is included in this aspect of risk treatment, since the interest rates managed include nominal and real rates, the latter being the rate in excess of price inflation.

In terms of hedging, there are two broad categories of interest rate risk. The first relates to a need to pay or receive payments of interest at a particular level. This is referred to here as direct exposure (to interest rates). The second category relates to cash flows due at some point in the future, making their value sensitive to interest rates. This is referred to here as exposure to interest-sensitive liabilities

16.3.1 Direct exposure

This type of risk occurs when, for example, a financial institution which has interest rate-sensitive outgoings. For example, an insurance company might have designed a product paying a variable rate of interest.

Forward rate agreements

The easiest way to hedge such a risk is through the use of a forward rate agreement (FRA). This is an OTC contract that requires one counter-party to pay another a series of cash flows calculated as a particular rate of interest applied to a particular notional amount.

Interest rate caps and floors

Rather than simply locking into a particular interest rate, it is also possible to gain protection from rises in interest rates above or falls below particular levels. This can be done through the use of an interest rate cap or floor. An interest rate cap – which is made of individual interest rate caplets – is an option that makes a payment in any period that the interest rate rises above a predetermined level equal to the difference between the interest rate and that level. Conversely, an interest rate floor – which is made of individual interest rate floorlets – is an option that makes a payment in any period that the interest rate falls below a predetermined level equal to the difference between the interest rate and that level.

16.3.2 Indirect exposure

Indirect exposure to interest rates is most commonly experienced by pension schemes and life insurance companies, each of which might have an obligation to make fixed or inflation-linked payments long into the future.

Cash flow matching

The most basic way in which this type of interest rate risk can be controlled is by matching individual liability cash flows in order to neutralise the effect of interest rate changes. For example, consider a series of pension scheme cash flows that extend for the next fifty years. If these cash flows are discounted back to today to give a present value of liabilities, then this present value will change depending on the interest rate used – a rise in interest rates will cause the liabilities to fall, whilst a drop in interest rates will cause the liabilities to increase.

One way to reduce the risk is to invest in bonds whose coupon and redemption payments match the liability cash flows as closely as possible. For nominal liabilities, where the cash flows are known in absolute terms, conventional bonds can be used; for index-linked liabilities, where the cash flows are known only in real terms, index-linked bonds can be employed.

Whilst this can give a reasonable reduction in risk, it means that the investment strategy is also necessarily low risk. This might not be what is wanted – the desire might be to remove only the interest rate risk from the liabilities, whilst retaining market risk in the assets, for which a risk premium is expected.

A way of dealing only with the interest rate coming from the liabilities is to use interest rate swaps. Each series of payments is know as a leg. These are agreements between two parties where one side agrees to pay a fixed rate of interest in exchange for receiving a floating rate of interest from the other party. The fixed rate is based on the expected rate of interest over the term of the swap. This rate is agreed at the outset of the swap. The floating rate of interest is based on the actual short-term rate of interest as it develops over the lifetime of the swap.

A pension scheme wishing to hedge its cash flows could, therefore, enter into a series of interest rate swaps where it would pay floating and received fixed. In this case, the fixed payments it received would be set to exactly match the pensions that it needed to pay to members. In return, it would need to pay the short-term rate of interest. Since the net effect of changes in long-term liabilities would be cancelled out by their effect on the swap, the interest rate sensitivity of the liabilities would be neutralised.

A pension scheme might instead want to enter into this type of protection only if interest rates fell below a particular level. In this case, the scheme could buy an interest rate swaption. This would give the pension scheme the right – but not the obligation – to enter into an interest swap should rates reach a particular pre-arranged limit. This way, interest rate risk could be eliminated on the downside with the upside potential from a risk in interest rates – which would reduce the liabilities – being retained. Of course, this optionality is not free. Whilst a swap is an agreement with no initial cost, a swaption must be bought. This means that there is an initial outlay, and if the swaption is not exercised, the premium paid is lost.

Redington's immunisation

Cash flow matching is not the only way of managing long-term interest rate risk. In fact, given the range of additional risks faced by pension schemes and life insurance company annuity books, the cash flow matching approach is often viewed as having spurious accuracy. Longevity risk and investment risk can mean that a much less exact approach will often suffice. Furthermore, if the cash flows change, due to differences between actual and expected longevity for example, the swaps will also need to be changed.

The simplest way to limit interest rate risk is to ensure that when investing in a portfolio of bonds or interest rate swaps to hedge a set of liabilities:

- the present value of the bonds or the swaps' fixed legs is equal to the present value of the liabilities; and
- the modified duration of the bonds or swaps' fixed legs is equal to the modified duration of the liabilities.

If this is the case, then a very small change in interest rates will result in both the assets and the liabilities changing by the same amount. However, this approach can be improved by also allowing for the convexity of the assets and liabilities. The additional condition required is that:

- the convexity of the bonds or the swaps' fixed legs is greater than the convexity of the liabilities.

This means that for small change in interest rates the present value of the assets will always increase in value by more (or fall in value by less) than the present value of the liabilities.

This is known as Redington's Immunisation, named after Frank Redington (1952). This offers an elegant approach, but it relies on the change in interest rates being the same at each term. It also requires regular rebalancing of the assets to ensure that the conditions for immunisation are met. Practical difficulties can also exist. In particular, it might be difficult to obtain assets with a long enough duration and great enough convexity if the liabilities have a very long term.

Hedging using model points

An acceptable degree of hedging can be achieved for bonds or swaps at only some terms or model points. In this case, the amount of each position should be chosen such that the overall interest rate sensitivity of the liabilities and the bonds or swaps is as close as possible. For example, swaps with terms of five, ten, fifteen, twenty and thirty years could be chosen.

The notional value of each swap can be determined using stochastic interest rate modelling. For example, assume that a stochastic model produces N simulations of an instantaneous change in the full yield curve, so for each simulation gives T yields, covering terms 1 to T. These yields could be used to calculate a revised liability value for each simulation. They could also be used to calculate the value of the fixed leg of a portfolio of swaps.

Let W be an $N \times T$ matrix of present values based on the simulated yields, where N is the number of simulations and T is the term of the liabilities being hedged. In particular, let the element $w_{n,t}$ be the present value of a payment of one unit due at time t in simulation n. Then let X be a vector of length T containing a pension scheme's cash flows at each term t where $t = 1, 2, \ldots, T$. The N-length vector $L = WX$ contains the value of the liabilities under each interest rate simulation.

Let Y be an $N \times S$ matrix of present values based on simulated yields, where $S < T$, and each term s, where $s = 1, 2, \ldots, S$, represents a term at which a swap is to be used. Then let Z be a vector of length S, where each element is the fixed payment due from each swap.

If an N length vector, ϵ, is defined as the difference between the value of the liabilities and the swaps in each simulation, then these items can all be related as follows:

$$L = YZ + \epsilon. \tag{16.18}$$

If the criterion for optimisation is that the sum of squared differences be minimised, then this becomes an ordinary least square problem that must be solved for Z. The estimate of Z under these assumptions, $\hat{Z}$, is therefore given by:

$$\hat{Z} = (Y'Y)^{-1}Y'L. \tag{16.19}$$

16.4 Foreign exchange risk

Foreign exchange risk can also be mitigated using forwards, futures, options, swaps and other derivatives. On the face of it, this risk does not provide any systematic additional return, only an additional level of risk. For overseas bonds, this means that exposures are typically hedged, unless the investment position includes some view on relative currency movements.

However, the question of how much of this risk to hedge in relation to equity exposure is not straightforward. For example, if a UK pension scheme holds shares in a firm listed on the New York Stock Exchange, then it would appear that this holding exposes the UK firm to foreign exchange risk. However, if the firm derives profits from all over the world – profits that are not hedged – then efficient markets would reflect these foreign exchange exposures in the market price, meaning that any hedging should reflect the firm's own exposure to foreign markets and the extent to which these exposures themselves are hedged. But even this is not the whole picture. If the firm has to buy materials or labour from a range of markets, then these will affect the price of goods or services sold overseas, suggesting yet another layer of convexity. For this reason, overseas equities are often hedged either according to some rule of thumb or not at all.

Before any currency risk is treated, it is important to establish the net level of exposure to the currency in question. In particular, if amounts are owed to one party and due from another in a particular currency, then only the difference between these two amounts need to be hedged.

16.5 Credit risk

There are a range of ways in which credit risk can be managed, reflecting its importance for financial institutions. Some of these relate to the credit risk an institution poses by virtue of its structure, whilst others relate to the way in which credit risk is taken on and, once present, managed.

16.5.1 Capital structure

For a bank, raising or distributing capital, particularly debt capital, is a primary method of managing its own credit risk. A typical approach for an investment bank is to consider the volume of business that it believes it can carry out, consider the credit rating that it needs to target in order both to write this business and to maximise its risk-adjusted return on capital and then to raise the capital it needs to achieve this. Predominantly, retail banks are less likely to follow such an approach, being less well-able to change the volume of business written.

Whilst insurance companies might take the approach of investment banks, operational constraints faced by insurance companies for many lines of business mean that many insurers are less likely to change their level of capital on a tactical basis; however, like retail banks, strategic changes are possible if an insurer undertakes a review of its strategic business mix or finds itself systematically unable to profitably invest shareholders, funds.

Pension schemes frequently require additional capital injections from their equity shareholders (the sponsors, in other words), and determining the level of capital injection (or return of capital) is one of the key roles of the scheme actuary. However, this should ideally be carried out together with any review of investment strategy and the value of the sponsor covenant, all of which are inextricably linked. Considering each in turn is likely to lead to inertia.

A secondary question for pension schemes is whether alternative methods of contribution to cash payments (such as the securitisation of future sponsor earnings or letters of credit) would be appropriate. If such proposals are made, then their amounts should not be taken at face value; they should also be modelled consistently with the other assets and the liabilities, and should again reflect the credit risk of the sponsor.

Another option for a pension scheme, rather than raising equity capital is to reduce or cease the issue of debt capital – in other words, reduce or cease benefit accrual. This has only a gradual effect on the level of liabilities, in particular if a pension scheme is closed only to new entrants.

Rather than raising or distributing capital, an alternative approach might be to change the mix of capital, such as a debt-financed equity share buy back. Whilst there is no first order difference in the value of a firm from such a change, there are clear second-order advantages relating to tax, free cash flow, transaction costs and signalling. For pension schemes, the impact of the capital structure of the scheme on the capital structure of the sponsor should also be allowed for, and the two considered together.

16.5.2 The volume and mix of business

For banks and insurance companies, a simple way to reduce the level of own credit risk – particularly if the level of free capital is low – is to write less business, since capital is required to write business. This is an approach that is likely to be used by an insurance company where the level of capital available varies less over the short term. However, this is not necessarily always the best approach. For example, some risks are reduced if more business is written, for example on a particularly small book of annuity business. Similarly, if the mix of business within a particular class is improved – for example, by introducing geographical diversification (either directly or through reciprocal reinsurance agreements), then the level of risk can be reduced without the expected return being diluted by too much.

Similar results can be obtained through similar approaches by diversifying between types of businesses which have low correlations, for example different classes of insurance. An extreme example of this can occur within insurance companies, where the mortality risk borne by the life insurance book can be partially offset by the longevity risk borne by the pensions book. The degree to which this is possible depends on the natures and ages of the two books of business.

16.5.3 Underwriting

Before a bank issues a loan or approves a mortgage, it will usually carry out a process of underwriting to ensure that the amount being borrowed is likely to be repaid, or to ensure that the rate of interest charged reflects the risk that the bank is taking. This process will use the results of GLM analysis, or some more basic credit scoring approach. In particular, discriminant analysis has been used widely in the past.

All financial institutions will perform a similar – though perhaps more tailored – approach to determine the amount of collateral required from a counter-party to an OTC derivative. More broadly, obligations of both counter-parties in relation to collateral are outlined in the credit support annex (CSA) of the ISDA agreement, as described earlier.

16.5.4 Due diligence

Due diligence can be regarded as a non-standard type of underwriting used for some credit risks. This includes incidental credit risk – that is, credit risks taken on other than as part of a firm's core business – but also counter-party risk arising from the use of reinsurance and other similar exposures. Due diligence involves assessing the party that will be providing the goods or services. This means considering the financial strength of a firm, but also carrying out a more subjective assessment of the way a firm is run. In this sense, it is essentially the same approach that a credit rating agency would use when looking at a firm for the purposes of determining a rating.

The result of due diligence might be a decision not to use that particular counter-party, or to structure payment in such a way as to limit the exposure to credit risk.

16.5.5 Credit insurance

Credit insurance might be appropriate for limiting losses where there is incidental credit risk. This provides protection against the insolvency of a supplier of goods or services where payment has been made before delivery, for example in respect of an IT system. Unless the sums involved are large, such insurance has a negligible effect on the total amount of risk being carried – for small sums at risk, self-insurance is probably more appropriate. However, credit insurance can be important if advance payment is made in respect of significant projects.

16.5.6 Risk transfer

Any transfer of risk will affect the creditworthiness of the institution transferring that risk, but the ways of transferring non-credit risks are dealt with in relation to each of those risks individually.

In relation to credit risk, the most important examples relate to capital market risk transfer, or securitisation. One of the earliest examples of the securitisation of credit risk was the regulatory arbitrage performed by banks. They found that they were treated more favourably under the first Basel Accord if they converted some of their loan portfolios into securities, which were then sold in capital markets. This approach has been extended to instruments such as CDOs.

However, even under Basel II, securitisation offers a way of capitalising the profit – or crystallising the loss – on particular tranches of business. It also offers a way to fine-tune the aggregate exposure of a bank to its range of credit exposures.

Some pension schemes also in deficit can mitigate sponsor risk by buying a CDS, although the extent to which the CDS exposure will cover any deficit can only be approximate as the size of the deficit will change in response to movements in the interest rate and investments.

16.5.7 Credit default swaps

A CDS is similar in nature to insurance bought against the default of a bond issuer. However, unlike insurance there is no requirement to have any insurable interest – in this case, financial exposure to the default of the issuer – meaning that a CDS can also be used as an alternative to selling a bond short. CDSs are traded OTC rather than via an exchange. This means that the buyer of a CDS does not have the protection afforded by exchange trading – exchanges will typically pool trades meaning that there is no exposure to a single counterparty – but the buyer will also be exempt from the regulation that surrounds exchange trading.

These factors have led to criticism of the CDS market. In particular, it is possible for investors to drive down the price of a bond through the CDS market whilst remaining anonymous.

The fact that CDSs are OTC also means that they do not have a single, standardised structure; however, they usually share a number of common features. The buyer of CDS is known as the protection buyer, since protection is being bought in case of default by the bond issuer, known as the reference entity. This can be a single firm, a group of firms or a whole corporate bond index. The institution providing the cover – the protection seller – is usually a bank.

The protection is usually paid for through regular premiums, paid quarterly or semi-annually, based on notional value of bond. If a reference entity defaults, then the protection buyer receives a payment.

The definition of default is not fixed and must be agreed. Settlement of a CDS on default can also be achieved in more than one way, two of the most common being physical and cash settlement. With physical settlement, the protection seller pays the buyer the par value of bond – the value on which interest is charged, usually equal to the amount repaid when the bond is redeemed – and takes delivery of the bond. Under cash settlement, the protection seller pays the buyer a cash amount equal to the difference between par value and current market value. For this, the time at which the market value is calculated is crucial.

The structure of CDSs is shown graphically in Figure 16.4. Note that, whilst a CDS might be bought to give protection, the reference entity has no direct links with either the buyer or the seller.

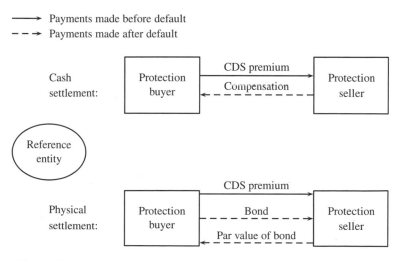

Figure 16.4 CDS structures

✕16.5.8　Collateralised debt obligations

CDOs have been mentioned several times as examples of the sort of complex credit derivatives constructed by banks. Their original purpose was to reduce the capital that banks needed to hold by converting loans sold by banks into securities, thus removing them from the balance sheets of banks. These types of CDOs – known as collateralised loan obligations (CLOs) are still used to transfer banks' risks from their balance sheets, but their use has expanded. In particular, a bank might put together a portfolio of bonds that it believes to be under-priced, frequently with the same low credit rating, and sell tranches of the resulting product to a range of investors with a range of risk appetites, thus allowing all of these investors to benefit from the mis-pricing. Both of these are examples of asset-based CDOs. However, it is also possible to create a synthetic CDO from CDSs instead.

A CDO is formed by setting up an investment entity known as a special purpose vehicle (SPV). This is used to purchase a portfolio of bonds, mortgages or credit derivatives. These investments can either be fixed or actively managed by an investment manager. The money used to purchase these securities or derivatives comes from external investors. These investors can purchase different classes of share in the SPV, each of which receives returns from the SPV. The riskiest tranche of shares – known as the equity tranche – suffers the full impact if any bonds default in the SPV. In other words, if a bond defaults, then only holders of the equity tranche suffer a reduction in their income stream. However, to compensate for this increased risk, these investors have the highest expected returns relative to their initial investment. At the other end of the scale, the safest tranche of shares does not suffer the impact of any defaults until all of the funds allocated to lower tranches have been exhausted through defaults. The high level of security means that investors in this tranche have the lowest expected return. For this reason, it is known as the senior or super-senior tranche. In the middle, with a moderate level of both risk and return, is the mezzanine tranche. This structure is shown in Figure 16.5.

The aggregate loss at which the payments on a particular tranche start to reduce are defined by attachment points. The returns for investors in a particular tranche can therefore be defined as follows:

- if the loss for the portfolio as a whole is less than the attachment point for this tranche, then the investor will receive the maximum possible investment;
- if the loss is greater than the attachment point for the next most senior tranche – which can also be regarded as the detachment point for the investor's own tranche – then the investor suffers a total loss; and

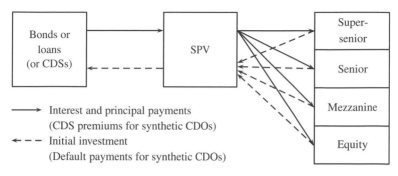

Figure 16.5 CDO structure

- if the loss is between these two points, then the return to the investor is the fund value less the detachment point.

The return received for each tranche is in return for an initial investment. The total investment over all tranches must equal the total initial value of the fund, but the greater the investment required for investment in a particular tranche, the lower the potential return for that tranche. The return for each tranche therefore depends on both the attachment points and the initial investment required from investors in each tranche.

The attachment points and levels of investment are determined using quantitative models that are frequently agreed with credit rating agencies. This means that the tranches themselves get credit ratings. However, it is important to note that, whilst it is often possible to recover some value from a defaulting bond, the loss on a defaulting tranche can be complete if it is defined in terms of portfolio loss.

CDOs can be priced using the credit portfolio models described in Chapter 14. The choice of model and parameters is crucial for determining the attachment points, since relatively small changes can have a major impact on the estimated return distributions for the different tranches. Some of the most important decisions relate to the degree of dependency between the underlying credits. This means not just the overall correlation, but the shape of that correlation, particularly in the tails – since senior and super-senior losses occur only when losses are in aggregate extreme, it is important that the degree of tail dependency is adequately allowed for.

16.5.9 Credit-linked notes

It is also worth mentioning credit-linked notes (CLNs). These are collateralised vehicles consisting of a bond and a credit derivative. As a result they are regarded as bonds for investment purposes, which can allow investors to gain exposure to credit derivatives even if systems or rules do not allow this to happen directly.

16.6 Liquidity risk

The main technique for managing all liquidity risk is to actively monitor liquidity needs. This should be done within and across legal entities allowing for legal, regulatory and operational limitations to the transfer of liquidity – just because there is sufficient liquidity in one part of an organisation does not mean that this liquidity can necessarily be transferred to another part if needed.

Ensuring that employees have an incentive to allow for liquidity risk is also important. This can be done by ensuring that liquidity management is included in employees' remuneration objectives.

Market liquidity risk can be managed through the investment strategy. This means that the maturity schedule of liabilities must be borne in mind when putting together a portfolio. Swaps can also be useful here in ensuring that fixed payments are received when they must be paid out to meet liabilities. Institutions should also maintain a cushion of high-quality, liquid assets.

It is also important to allow for liquidity risk in the design of any product where there is the opportunity to withdraw funds before the product's maturity date.

Funding liquidity risk is a bigger issue for banks given the nature of the business model, which involves long-term lending funded by short-term borrowing. To limit the risk of illiquidity, it is important to ensure diversification in the term and source of funding, that is the choice of equity and bond finance, the choice between short- and long-term bonds and so on. These decisions are linked to the management of credit risk, suggesting that both should be considered together.

Firms should constantly gauge their ability to raise capital from each source, whether or not they need to raise funds at that particular point in time. They should also have a contingency funding plan to provide liquidity in times of stress. This can include uncommitted bank lines of credit, other standby or back-up liquidity lines, and – for insurers – the ability to issue new products.

16.7 Systemic risk

The responses to systemic risk depend on the type of systemic risk.

The effect of exposure to a common counter-party can be limited by using a range of counter-parties. Unfortunately, the number of counter-parties can only be increased to the extent that an economically viable relationship remains with each one. A more extreme solution is to use exchange-traded instruments and derivatives, where the obligations of all counter-parties are essentially pooled. However, the exchange-traded route also has problems. Because exchanges deal only in standardised contracts, the level of tailoring that would otherwise be provided by OTC derivatives might not be available.

The impact of feedback risk for a particular security can be limited to an extent by holding a diversified portfolio. However, if the feedback is systemic, this is unlikely to help significantly. Indeed, the control of systemic feedback risk is more likely to be the responsibility of regulators than investors. One blunt instrument to reduce the problem of feedback risk is for stock exchanges to limit the extent to which a share price can change within a particular period. Many stock exchanges have such controls – known as circuit breakers – to limit excessive volatility for the market as a whole. For example, the New York Stock Exchange has the following limits:

- for a 10% fall in the Dow Jones Industrial Average (DJIA):
  (1) if the fall is before 14:00, the market closes for 1 hour;
  (2) if the fall is between 14:00 and 14:30, the market closes for 30 minutes;
  (3) if the fall is after 14:30, the market remains open;
- for a 20% fall in the DJIA:
  (1) if the fall is before 13:00, the market closes for 2 hours;
  (2) if the fall is between 13:00 and 14:00, the market closes for 1 hour;
  (3) if the fall is after 14:00, the market closes for the day; and
- for a 30% fall, the market closes for the day.

Regulators can also have an impact through the way in which solvency regulations are imposed. Feedback risk can be caused by solvency requirements, where a worsening financial position causes sales, which further reduces the price of those assets and so on. By reducing the extent to which immediate price changes feed through to the solvency position, this pro-cyclicality can be avoided. However, it is difficult to know the extent to which a price change is the result of forced selling and how much reflects a genuine change in sentiment. It is therefore important that any rules introduced to avoid feedback risk do not result in the value of some stocks being overstated for the purposes of statutory solvency.

Basel III uses another approach to try to avoid feedback risk. It requires firms to build up capital buffers when times are good so that additional reserves exist when times are bad. Which times are 'good' and which are 'bad' is not defined. The main constraint is that when the capital buffer is used, distributions to shareholders should be curtailed. This part of Basel III is discussed in more detail in Chapter 19.

In relation to systemic liquidity risk, the same principles apply as for less extreme liquidity risk. However, governments can also act to limit the impact of this risk by providing funding for banks directly. They can also seek to limit damage through relaxing monetary policy, for example by lowering interest rates.

However, it is difficult to limit the systemic risk arising from a number of organisations following the same strategies. One approach is to ensure that different activities are carried out by different firms. In the European Union, the First Life Directive of 1979 essentially does this by prohibiting the establishment of new composite life insurance companies. The 1933 Glass–Steagall Act in the United States performed a similar task for banks, requiring the separation of merchant and investment banking activity in that country, although separation was later allowed by the 1999 Gramm–Leach–Bliley Act. The separation described above protects certain customers or policyholders if a different type of business suffers catastrophic losses. However, this principle cannot be sensibly extended to, say, require different classes of insurance to be run by different firms, not least because of the positive diversifying effect that comes from having different classes of business in the same firm. However, some regulatory encouragement towards a degree of specialisation might ensure a healthier degree of variety in firms' strategies.

16.8 Demographic risk

There are two areas of demographic risk that can be considered: before risk is taken on, and once risk already exists.

16.8.1 Premium rating

Premium rating for individuals usually means using the results of GLM or other analysis to arrive at rating criteria which are used to calculate different premiums for different people.

For this aspect of underwriting in particular, it is important that the cost of underwriting does not exceed the benefit of improved differentiation. For example, carrying out a full medical examination on everyone applying for life insurance would give a good indication of risk classification, but would be very costly and would result in unrealistically high premiums. In reality, there are different levels of underwriting depending on the size of the policy, with full medical underwriting being used only where the sum assured is very high, or where a less expensive underwriting method – such as a medical questionnaire – has indicated that further investigation might be advisable.

Underwriting for a life insurance policy is focussed on trying to find factors that might lead to higher than average mortality. This means that the focus is on trying to find information that a policyholder might prefer not to share. However, if underwriting an annuity, potential policyholders are likely to be much more forthcoming about health issues, since a lower life expectancy leads to a larger annual annuity payment for a given premium. In this case, underwriting is less about trying to protect an insurance company from unexpectedly large claims, and more about trying to offer a lower premium where possible.

For groups of lives, premium rating might also include experience analysis, if it is thought that the mortality experience of that group can give a credible estimate of future survival probabilities.

16.8.2 Risk transfer

A method of risk transfer fundamental to insurance companies is reinsurance. This can be proportional (thus allowing an insurer to improve the mix of business written) or excess-of-loss (thus protecting an insurer from extreme events).

Pension schemes use an approach similar to proportional reinsurance when they buy annuities, either as a matter of course for retiring members or as part of a bulk buyout of a tranche of members (perhaps the entire membership). More recently, opportunities for deferred buyout have arisen from a number of specialist providers.

Reinsurance is typically 'with-asset' in nature, as is annuitisation, which means that a premium is paid and money is returned once there are claims. For annuitisation in particular, the long-term nature of the cash flows means that a significant amount of capital is tied up. However, it is also possible to structure this sort of protection in the form of a swap. For a pension scheme this would mean that it made fixed payments based on the expected longevity of its members, whilst receiving variable payments based on their actual survival. *Longevity Swaps*

For such a swap to be classed as risk transfer, the reference population upon which the swap payments were based would need to be the population of the pension scheme. However, swaps also exist that are based on the mortality experience of some other population, usually national. Hedging using such swaps is really risk reduction rather than transfer, but the effect is similar.

Life insurance companies also use securitisation to reduce their risk exposures. In particular, mortality catastrophe bonds have been issued which pay a generous level of interest that is reduced if aggregate claims rise above a certain level.

16.8.3 Diversification

It is important for life insurance companies to have large portfolios of business so that they are not overly exposed to losses from a single policy. However, diversification is also important, as this can help to avoid concentrations of risk. There should be geographic diversification, but also diversification by risk factors such as occupation. If diversification is difficult, then proportional reinsurance can be used to reduce the impact of losses from any one policy and, more importantly, allow an insurance company to take on more business to increase diversification.

In extreme cases, it is possible to go beyond diversification and into implicit hedging. This is the name given to the use of mortality risk in term assurance and similar products to hedge the longevity risk that arises from annuities. The implicit hedge is only approximate. For a start, annuities tend to be bought by older people, whilst term assurance is more important for those of working age. Furthermore, there may be different mixes of socio-economic or geographic groups in each portfolio. Whilst mortgage-holders are often required to have term assurance, large pensions are increasingly held by only the wealthiest groups.

14   16.9 Non-life insurance risk

Non-life insurance has many of the characteristics of life insurance, so the risk responses are similar. However, a key difference is that, whilst insured lives will only change state – from alive to dead – once, non-life insurance offers the possibility of a large number of claims over a number of years. This is particularly important in premium rating.

16.9.1 Premium rating

As with life insurance, underwriting is a key way of controlling risk. However the nature of non-life insurance means that an individual's claim experience can also be used to help determine a premium. The most obvious way in which this occurs is through the no claims discount (NCD) mechanism of motor insurance policies. This uses the number of claims in the past to scale the premium chargeable.

Experience and risk rating are combined to varying degrees in other classes of insurance, but when data are limited it is often necessary to use more subjective approaches to premium rating.

16.9.2 Risk transfer

Proportional and excess-of-loss reinsurance are commonly used to transfer non-life insurance risk, with the type being used depending on the class of insurance and the situation of the insurance company. For example, a class with high claim frequency, low claim amounts but a risk of geographic concentration – such as motor insurance – might be a candidate for quota share insurance, whilst a class with low claim rates but high claim amounts – such as product liability insurance – might have more used for excess or loss reinsurance.

Non-life insurance companies have used securitisation to reduce their risk exposures for many years. For example, catastrophe risk can be managed by the issuance of catastrophe bonds to limit the exposure of an insurer or reinsurer to catastrophes such as hurricanes, floods or earthquakes.

16.9.3 Diversification

Diversification offers an important way of controlling non-life insurance risk. Within each class, geographic diversification and diversification by risk factor together ensure that a portfolio of policies is not too greatly exposed to a concentration of risk. However, diversification between classes can also offer protection. Having said this, it is still important to consider the exposure to risk factors of all classes combined – for example, a flood could easily result in higher claims for not just buildings insurance but also motor insurance and consequential loss insurance from firms unable to carry on business.

16.10 Operational risks

Controlling operational losses is generally more important than quantifying them. Risks will generally be controlled through the use of appropriate systems and processes, although outsourcing some of the processes to external organisations can also be used to manage risk. However, whilst outsourcing might provide a benefit through the use of dedicated expertise, it requires additional resources to be spent on monitoring and results in less control over the outsourced function.

There are also particular approaches that can be used to mitigate many of the operational risks. These are detailed below.

16.10.1 Business continuity risk

Business continuity risk can cause a firm to lose a substantial amount of money – in many cases, the loss of profits will be even greater than the cost of the physical damage causing the disruption.

One way in which impact can be limited is to have contingency plans for an alternative business location. This can either be a property owned outright or an option to use a property at short notice. Many firms specialise in providing appropriate space, complete with computers, telephones and other office equipment.

It is also important to ensure that data are backed up regularly, preferably to a location away from the main site. This means that if the primary location is destroyed or damage, records remain safe. If servers are run in parallel at a secondary location, then this also means that the disruption to business can be kept to a minimum.

If this is thought to be too extravagant, it is important at least to ensure that key personnel can work from home. This can be helpful if there is widespread disruption to transport networks as a result of bad weather. If staff can access emails and a network server, then many organisations – particularly in the financial services sector – can still get a meaningful amount of business done through the period of disruption.

Whatever contingency plans are put in place, it is important that they are tested regularly to ensure that they do what they are supposed to.

If even these measures are impractical, then it is worth considering consequential loss insurance cover on top of other business insurance. This compensates a policyholder for profits lost as a result of business disruption.

16.10.2 Regulatory risk

It is important to keep abreast of regulatory changes since breaching regulations can have serious implications in terms of fines, reputation and even ongoing authorisation. Many firms will have in-house departments whose role is to learn about imminent changes and to disseminate them around the firm. If this is impractical, then subscribing to alert services can be helpful. Many consultants will also offer this sort of information to their clients for free.

However, as well as keeping track of changes to regulations, it is also important to take action if any proposed changes are likely to have an adverse effect.

This can be done by lobbying directly or by supporting an existing lobby group.

16.10.3 Technology risk

Managing technology risk requires a coherent strategy considering the risk on an organisation-wide basis. One of the key decisions in this respect relates to how much work relating to technology to carry out in-house and how much to outsource. Outsourcing can reduce the infrastructure that is required, but it also means relying on a resource over which there is only partial control.

To ensure that information technology (IT) is adequately managed, it is important to have a dedicated central IT resource. This itself can be internal or external (outsourced), but should provide a response to IT problems in a time scale appropriate to the nature of the issue.

The importance of backing up data and running secondary servers has already been discussed in the context of business continuity. However, these actions are also important to ensure that data corruption does not result in damage to a business.

It is also important to ensure that software is kept up to date and that security patches are implemented as soon as they are available. Failure to do so can leave a system vulnerable to hacking, which could also result in system failure or even data theft.

New software can also pose a problem in terms of its interaction with other systems. This aspect of new software should be considered at the earliest opportunity, as it can have a big impact on the eventual cost and effectiveness of software.

More mundane IT issues can still have an impact on the smooth running of a firm. For example, an inability to access email can result in hours of lost work.

Another issue that many firms face is the question of whether to develop bespoke IT systems or to use an off-the-shelf solution. This involves balancing the relative costs with the differences in functionality, but when considering the cost it is important to recognise both the ongoing burden of maintenance faced when a system is developed in-house and also the lack of recourse to an external provider if any issues are discovered.

16.10.4 Crime risk

One aspect of crime risk – hacking – has already been covered, but the full scope of this risk is huge, ranging from petty theft to solvency-threatening fraud. In all cases, the main principle is that the cost of mitigating the risk should reflect the size of that risk. A good example is in relation to employee expenses. Any additional monitoring that reduces the cost of fraudulently claimed expenses but costs more than that saving is not a good system. This does not mean that no action should be taken, but that the action to be taken should be carefully considered.

16.10.5 People risk

People risks are some of the most important in the financial services industry, where human capital is the main driver of profitability. This means that it is important to spend sufficient time considering how to to respond to particular risks.

Employment-related risks

The first employment-related risk to consider is the risk that the wrong people are employed. To ensure that this does not happen, it is important to use good recruitment procedures, and the starting point here is to employ a sufficiently skilled human resources team. How to recruit good recruiters in the first place is clearly not straightforward, but investing in the training of the existing team can help.

The first stage in recruitment is finding the right candidates. This means ensuring that direct advertisements appear in the right publications, and sometimes involves using recruitment consultants or 'head-hunters'. If recruitment consultants are used, then it is preferable that only one is put onto a particular assignment. This can reduce the risk of being inundated with candidates, and can also mean a lower fee being agreed. It is helpful if a good relationship can be built up with the recruitment consultants as this can lead to a better understanding of what is needed in particular roles, which itself can lead to more suitable candidates being put forward.

Once candidates have been put forward, the next part of the process – which can include a number of interviews and aptitude tests – should be rigorous enough to distinguish between candidates, but no more rigorous than that. Here again it is important that the money spent on the selection process is consistent with the value that each individual can add to the firm.

Once employees have been recruited, similar techniques should be employed when promotions or transfers are considered, although an element of 'experience rating' can be derived from the employee's performance to date in the firm.

It is important that the right employees are retained, which means that pay and conditions should be sufficient, but also that good employees are given sufficient responsibility to keep them interested. It is also important to support good employees by providing access to counselling, supporting flexible working and otherwise making it easier for them to stay with the firm.

Conversely, it is important that poorly performing employees are identified. Poor performance can be in terms of the quality of their work or in terms of absenteeism. Having identified these employees, it is important to offer support where possible, with any disciplinary action being a last resort.

All employees should also be supported with training for their roles. This should include CPD to ensure that skills remain up to date. One way of ensuring that relevant CPD is undertaken is to encourage employees to undertake appropriate professional qualifications. In financial services firms, these will often be required to carry out certain roles anyway. Once a qualification has been obtained, it is likely that there will be a CPD requirement to ensure that the qualification does not lapse.

As well as dealing directly with individual employees, it is also important to maintain good relationships with any collective bodies such as trade unions. This is particularly true if any changes affecting large groups of members are planned – clear and open communication before any such changes are finalised can reduce the risk of industrial action.

Adverse selection

A people risk that is particularly important in the context of underwriting is adverse selection. For example, if all life insurance companies but one charge a higher premium for people who smoke, then smokers will tend to use only the insurance company that does not differentiate. This is not necessarily an issue if the additional risk is reflected in the premium. In this example, one might expect the end result to be that all smokers would gravitate to the single insurer that did not ask them to disclose their smoking status, but that all would pay smoker rates; non smokers would still pay non-smoker rates at the other insurers.

However, until an equilibrium such as this had been reached – an equilibrium that assumes a far more efficient transfer of information than is likely to exist in practice – the single insurer that fails to distinguish between policyholders will be selected against.

This suggests that adverse selection can be dealt with by underwriting, with several conditions:

- underwriting should not cost more than it saves;
- the premium for each risk category should be no less than the average premium required for each individual in that category; and

- the heterogeneity within each risk category should not be so great that lower-risk members of a category choose not to belong to that group.

Moral hazard

Whilst adverse selection affects the decision of which insurance policy to buy or which loan to take out, moral hazard affects ones actions once cover or financing is in place. It can affect a range of actions, from the decision of whether to default on a debt to whether to falsely claim on a household contents policy.

There are a number of ways in which moral hazard can be limited. A key action is to make the consequences as unattractive as possible. For example, personal bankruptcy has a number of implications, including difficulties in obtaining future credit. Claiming for non-existent breakages is fraudulent and therefore a criminal offence.

However, whilst bankruptcy is self-evident, fraudulent insurance claims are more difficult to spot. However, a principle that has been mentioned already also applies here – namely that investigations should be made to the extent that they result in a net saving. This typically means that claims above a certain amount will need to be assessed by a loss adjuster. It might also be worth randomly investigating smaller claims – and advertising the fact that this happens – in order to discourage smaller frauds.

It is important to note that the levels of fraud do vary over time. In particular, they tend to rise when the economy suffers. This means that it is worth lowering the limit for which claims are assessed in times of economic stress.

Agency risk

Agency risk has led to some of the biggest financial disasters in recent years, so it is important that appropriate responses exist. As with many responses, there are two types of response that can be categorised broadly as 'sticks' and 'carrots'. The sticks are generally rules requiring certain actions to be taken. Many aspects of the corporate governance codes include these types of responses in an attempt to limit agency costs, covering areas such as remuneration, experience, education and board composition. These are important, but if only rules exist then agents will be tempted to find ways round them.

Carrots – in the form of incentives – can instead be used to encourage desirable behaviour. The best incentives are ones that align the interests of the agent with those of the principal. For example, if a firm's directors are encouraged to buy shares in that firm, then their interests are aligned with those of the shareholders. If director shareholdings are publicised, then directors are further incentivised to become shareholders.

16.10.6 Bias

Both deliberate and unintentional bias are important to guard against, but it is difficult to ensure that either is absent. One way of ensuring that reports, assessments and so on are balanced is by ensuring that they are checked by someone both competent and independent. If possible, the checking should use criteria that are as objective as possible.

Comparisons are also helpful. For example, to see whether an underwriter is charging premiums that are too high or too low, it is worth considering the spread of quotations with the spread seen with similar underwriters. If the quotations require significant subjective input and claims amounts are volatile, it may be difficult to do anything more.

If a board is being asked to assess a particular proposal, one way of ensuring that there is as little bias as possible is by making sure that the board has the skills to ask the right questions.

For unintentional biases, it can also be helpful simply to make people aware that these biases exist – as the saying goes, forewarned is forearmed.

16.10.7 Legal risk

In terms of responses, legal risk is similar to regulatory risk: the solution is to keep informed. However, legal risk can occur on a case-by-case basis. For example, there are often legal considerations when there is discretion over whether to pay a pension benefit, or when considering non-standard clauses in relation to an insurance policy. In every case, the safest solution must be that if there is any doubt over the legal status of a particular course of action, then legal advice should be sought.

16.10.8 Process risk

The large list of processes in any financial organisation means that there is ample scope for things to go wrong. This means that it is important to regularly review the processes and systems used.

If a new process is introduced, it is important to stress test not only that process but the way it fits into the broader structure. The structure as a whole can best be managed by using risk-focussed process analysis, described earlier.

16.10.9 Model risk

The risk that the model chosen has been incorrectly implemented is best min-imised by having a rigorous, documented process for model coding, together with a clear audit trail. This is merely a way of avoiding a type of process risk. However, the more interesting issues arise in relation to the choice of model.

It is important to ensure that all models are actually designed for the use to which they are put, or that there is a sound reason for putting such a model to another use. However, the biggest risk is that the results from a modelling exer-cise are driven disproportionally by the choice of model. This can be an issue if the choice of parameters has a negligible impact on the results compared with the choice of model. The presence of such a problem can be tested only by using a range of models to verify a set of results. Whilst this might sound daunting, crude models will often be sufficient to determine whether there is a problem.

16.10.10 Data risk

Data risk is in part another type of process risk, in that it is important that the processes are designed such that the possibility or incorrect data being entered is as low as possible. However, in relation to personal data there are particular issues. Errors can occur here if incorrect entries are made, or entries are omitted altogether

The first stage to limiting data risk is to limit the data that can be input. This is most easily done if data are entered electronically. For example, if a date of birth field allows entries only in two blocks of two and one block of four digits, with only numerical entries being accepted, then only dates can be entered. Similarly, for gender it is sensible to allow only entries of 'M' or 'F', or even to have a tick box instead. It is also important to check that as well as being valid, dates are reasonable. For example, dates of birth must be in the past and dates or retirement must follow dates of commencing employment. Finally, there can be requirements that a page of data cannot be accepted unless all fields are completed using valid entries.

As well as checking data at the point of entry, it is also important to re-check it if it is transferred, particularly if the system from which data are transferred does not employ strict checks. All of the above checks should be carried out, as well as checks on suspected default entries. For example, if a date is not known it might be entered as 1 January 1901, or '1/1/1'. This might also be the date that a system interpreting numbers as dates would derive from a blank entry. If there is an unusually large frequency of a single date, this might merit further investigation.

If personal data are being used for analysis, particularly mortality investigations, then it is important to combine duplicate entries. If any individuals have more than one entry in a set of data, as might be the case if more than one insurance policy has been bought, then any resulting analysis will be biased towards these individuals. De-duplication is a complicated process. Whilst a unique identifier such as a National Insurance or Social Security number might exist, this will not necessarily be the case. It is possible to create such an identifier by combining information such as the surname, gender and date of birth into a single field. However, an individuals surname will sometimes be written more than one way, particularly if it is normally written in a non-western script. It is important therefore to standardise surnames with the same or similar sound before constructing a unique identifier.

16.10.11 Reputational risk

Responding to reputational risk is difficult to do in advance, since damage to a reputation often arises from some other risk. It might be possible to scan every possible action for potential reputational damage, but such a strategy would make a firm slow to respond to challenges. Instituting a sound ERM framework offers the best way of realistically avoiding reputational risk.

Once an organisation's reputation has been damaged, it is important to rebuild it as quickly as possible. The public reputation can be rebuilt with the help of marketing specialists, but much of the damage will be to individual relationships. The stronger these are before any reputational damage, the easier they will be to rebuild afterwards.

16.10.12 Project risk

Since project risk covers the way in which risks are combined, it is best managed by having a comprehensive ERM framework. In particular, it should allow for the interaction between the different processes and the resulting concentrations of risk.

16.10.13 Strategic risk

Strategic risk is best limited by ensuring that a firm is correctly positioned in its market. This means not just that a strategy must be chosen, but that it must be constantly reviewed in the light of changes to the industry and changes to the strategies of competitors.

A key strategic decision that a firm needs to make is whether it will attempt to compete on price or through having a differentiated product. Cost leadership is the most obvious strategy for many financial products, where there is little discernable difference in quality. This fact has led to the development of a large number of price comparison websites. However, differentiation does exist. For savings products, a bank or insurance company might advertise its strong credit rating, highlighting the greater security of the product, and, with insurance and banking products, people are sometimes willing to pay more for better service. With some bank accounts and credit cards this has been extended to a full concièrge service with a range of additional benefits being on offer.

Firms also need to decide whether to focus on a single product or market or to offer a range of products. Whilst diversification has historically been more common, there have been an increasing number of specialists including mono-line insurers dealing only with pension scheme liabilities, and internet-only insurers who have no 'traditional' presence. Diversification has clear benefits in terms of the aggregate level of risk. However, some firms prefer to concentrate on products and markets where they have the greatest understanding. Furthermore, the economies of scale that might be available from offering a range of products could well turn into dis-economies of scale if the firm grows too large and complex.

16.11 Further reading

The references describing the assessment of the various risks also cover some of the responses. However, additional information on the operation derivatives markets is given in Hull (2009) and Wilmott (2000), whilst McNeil *et al.* (2005) discusses credit derivatives and their uses in detail. There is also useful information on securitisation, with particular reference to insurance risk, in Barrieu and Albertini (2009).

17

Continuous considerations

The previous chapters have outlined the stages that comprise a risk management process. However, as well as following these stages a number of activities should be carried out on a continuous basis. These can be summarised as:

- documentation;
- communication; and
- audit.

Documentation refers to the process by which all aspects of the risk management process are recorded, whilst communication refers to collation and circulation of information, both within an organisation and between that organisation and outside agencies. The final process, audit, covers the ongoing validation of the risk management process.

Whilst the scope of documentation and audit are relatively straightforward, communication covers a wide range of overlapping areas. The systems used to keep track of information could be described as monitoring, whilst the circulation of key items of risk information is also referred to as reporting. However, rather than trying to separate these items arbitrarily, they are included in the same section.

17.2 Documentation

Risk registers and their roles in the identification of risk have already been discussed. However, it is important to document the risk management process much more broadly. This means that the reasoning behind the process as a whole should be documented. However, there should also be adequate documentation of all decisions taken, and the reasons for those decisions.

457

The development of all systems should also be documented in detail, so that any future development can be carried out more easily. This is also true for financial models, the assumptions that they use and the data employed in the calculations. As well as recording this information, the reasons for the choices made should also be clearly set out.

Finally, information on risk management failures should also be recorded in a risk incident log. This should refer to the nature of the failure and the financial implication. Information on whether it was caused by a failure to follow process or despite the controls that were in place should also be recorded. This is partly to help assess the effectiveness of the risk management process, but also to inform future developments.

Not absolutely every detail can be recorded, but there should be sufficient information to understand the background to any decisions made.

17.3 Communication

Communication is a crucial aspect of risk management. This includes both formal and informal aspects of communication. Communication can also be either internal, involving only the firm's employees, or external between the firm's employees and outside agencies.

17.3.1 General issues

Information is needed on a range of areas. Many of these relate to what is happening within a firm, in terms of whether procedures are being followed, the cash flow position of the firm and of individual departments and so on. However, information on a firm's competitors is also important when issues such as pricing and marketing are being considered, and it is important that information on issues such as regulatory change are promptly sent to appropriate departments.

The way in which information is communicated should also allow for the fact that different departments – indeed, different individuals – will view the same risks in different ways. This means that the way in which information is framed is important.

The trade-off between the level of detail used in communication and the timeliness of the information should also be considered. Whilst it might be thought important to have as much information as possible about, say, an emerging risk, this is counter-productive if it means that communication concerning that risk is delayed unduly. However, the accuracy of information should not be compromised by a need for timely delivery.

Related to this is the issue of frequency. In relation to risk controls, it is important that information is received often enough for appropriate action to be taken, but that it is not received so often that it is not considered fully.

17.3.2 Informal communication

Informal communication can play an important role in the transfer of important information between departments and from the front line to the board room. For example, a potential risk may be discovered in one department that is also relevant for other departments. If this fact is discovered in a chat over coffee, action can be taken much more quickly than if it is reported to the CRF which at some point passes this information on in a risk bulletin.

The efficiency of informal communication is driven to a large extent by the culture of an organisation. The greater the perceived divisions between different levels, the less freely information is likely to flow between these levels. Also, the greater the extent to which an organisation operates in silos, with different business lines operating in isolation, the less likely information is to flow freely between them.

17.3.3 Formal communication

Much of the communication will be more formal, meaning it will be in the form of emails, newsletters and bespoke risk management systems. If the information shared is about new risks or changes to existing risks, it is important that there is clear two-way communication. This means that when those working in a particular department become aware of a change in the risk profile, they notify the CRF and *vice versa*.

A lot of the formal communications will relate to financial information such as investment positions, hedging strategies and solvency. These will typically be produced for the relevant managers, for the board and for external parties. It is therefore important that information is transferred manually as little as possible – reporting systems should be designed to draw information from front-line systems automatically. Other approaches – typing numbers from printouts, or copying and pasting numbers into a spreadsheet – leave far too much room for error.

Information on operational issues will also be transferred, including the extent to which procedures are followed and details of risk management breaches.

17.3.4 Internal communication

Internal communication relates to the transfer of information relating to individual risk and return measures as well as the overall strategy. In relation to existing risk responses and performance generally, this includes both financial and operational information. However, it is also includes the transfer of information on new risks.

It is important that reporting of risk management failures and near-misses is encouraged, and that a 'blame culture' is avoided – information on past failures can provide useful information that can help to strengthen risk management processes.

It is also important that performance indicators for risk management are developed for internal use. These can then be built into the performance indicators for individuals, preferably with a link to remuneration, so that staff have strong incentives to consider risk management.

17.3.5 External communication

Much of the risk management information that is collected is for use by external stakeholders. For example, shareholders need information on the risk management systems and processes, as do regulators. This information will often be provided in a set format, and may differ from the information produced for external use.

There is also communication that organisations receive from outside. A good example is regulatory change. It is important that there are clear responsibilities for finding information on such changes, and for distilling and disseminating such information to the appropriate people in an organisation.

Another form of external communication that can be more difficult to obtain is information on competitors.

4 17.4 Audit

Whilst the monitoring aspect of communication covers the outputs of the risk management processes, this section considers the processes themselves. It is important that an organisation constantly reviews the risk management process as a whole as well as the components of that process. This should include:

- the way in which risks are identified, both on a high-level basis and day to day;
- the way in which risks are communicated from the CRF to business units, and the other way;
- the methods of risk assessment, including the models and assumptions used to quantify risks;
- the choices and effectiveness of risk responses; and
- investigations of risk management failures.

As well as an internal review, external validation of the risk management process is also important. An external audit can provide insights with the benefit of being one step removed from the risk management process. Those providing the external audit can also apply the lessons learned from other clients, so ensuring that the risk management approaches are in line with best practice.

External audit might also be required as a condition of regulatory approval. In particular, the third pillars of both Basel II and Solvency II involve regulatory oversight and the validation of an organisation's risk management process by the regulator.

17.5 Further reading

Practical aspects of these continuous considerations are found in most of the advisory risk frameworks.

18

Economic capital

18.1 Introduction

The calculation of economic capital brings together many of the principles discussed throughout this book, covering risk measures and aggregation in particular detail. The issue of economic capital is also important to a number of departments within a financial organisation. One way to see the extent to which this is true is to consider why economic capital might be calculated. However, it is important first to understand exactly what economic capital is.

18.2 Definition of economic capital

There are a number of ways that economic capital can be defined, but most definitions contain three similar themes:

- they refer to additional assets or cash flows to cover unexpected events;
- they refer to an amount needed to cover these unexpected events to a specified measure of risk tolerance, with risk being measured in some way; and
- they consider the risk over a specified time horizon.

A common definition of economic capital is the additional value of funds needed to cover potential outgoings, falls in asset values and rises in liabilities at some given risk tolerance over a specified time horizon. It can also be defined as the funds needed to maintain a particular level of solvency (ratio of assets to liabilities) or the excess of assets over liabilities, again at some given risk tolerance over a specified time horizon

Risk tolerance can also have a number of meanings, referring to a percentile of the results, a value of loss or the result of some other key indicator.

462

18.3 Economic capital models

Economic capital is calculated using an economic capital model. This is used to create simulations of the future financial state of an institution so that the range of potential outcomes can be analysed. These outcomes are then used in the calculation of some measure of risk that allows for an assessment of the level of capital that should be held, given a pre-specified risk tolerance and time horizon.

Economic capital models can be internal or generic. Each type is discussed below.

18.3.1 Internal capital model

An internal capital model allows a firm to determine how much capital it should hold to protect it against adverse events. It not only gives a better understanding of the financial implications of the current strategy, but also allows the implications of any potential change in strategy to be assessed.

In relation to the products offered, this can mean helping to price new products. However, internal capital models can also be used to decide how to allocate capital across business lines. The economic capital allowance in a price should include an adequate margin for the additional risk being taken on, so that the reserves held are sufficient to cover claims variability, credit risk or some other source of uncertainty. The margin included in the price can depend on a number of factors, including:

- the uncertainty implicit in the product;
- the extent to which the product acts as a diversifier to other businesses – a product that offsets some of the other risks taken by a firm requires a smaller amount of economic capital, and can reduce the economic capital required by other parts of the business;
- the volume of the product sold – as more of a product is sold, the extent to which it can act as a diversifier reduces; and
- the experience that emerges from the product.

Together, these factors can be used to help determine the optimal mix of business.

Once business has been taken on, internal capital models can be used to assess the amount of economic capital that should be held to in respect of these products as they develop over time. This may change as views of risk also develop.

Internal capital models can also be used to assess the impact of changes in investment strategy and the capital structure of an organisation. Furthermore, they can be used to help to determine the optimal mixes of assets and funding sources.

As well as helping make decisions based on average outcomes, these models can also be used to look at how an organisation copes in the face of extreme events. This too can have an impact on the decision a firm takes about various aspects of its strategy.

This aspect of internal models can be useful in determining risk limits, in terms of business taken on, but also in terms of investment and funding strategies.

Models can also be used to help measure performance. In particular, it can be used when calculating the return on capital when either the return or capital is risk-adjusted. This means that the true economic cost of business is allowed for when performance is compared across an organisation. This gives a better indication of where capital should be used with the results being fed into decisions on optimal capital allocation. It also means that compensation can be linked to the outputs of these models, by ensuring that risk-adjusted performance is rewarded.

Internal capital models are also useful when carrying out due diligence for corporate transactions, as they give an idea of the strengths and weaknesses of an organisation. They can also be used to provide information on the financial state of an organisation to a regulator.

Internal capital models should allow for all risks faced by an institution. These risks should be modelled in a consistent, holistic way. In particular, this means modelling the dependencies between the various components of an institution's assets and liabilities.

18.3.2 Generic capital model

Whilst internal capital models give a firm-specific view of the economic capital needed, they require thorough investigation by regulators if they are being used to calculate regulatory capital. For this reason, generic capital models are sometimes used instead to give a consistent assessment of the capital required across a range of firms.

18.4 Designing an economic capital model

The first stage in designing an economic capital model is to agree what the model is for. A number of potential uses are described above, and whilst some will allow for a model that takes some time to set up and run, others will require near real-time results.

There should also be agreement over the risks that will be modelled. For example, an insurance company might model market and mortality risk stochastically, but allow for the risk of reinsurer failure only in separate scenario analysis.

There are also a number of different approaches that can be used to determine economic capital. These can be summarised as factor tables, deterministic approaches and stochastic approaches.

The factor table approach requires a certain amount of economic capital to be held in respect of each unit of a particular type of activity. This is a simplistic approach used by many regulators. The deterministic approach is essentially a stress test, which considers the amount that a firm would lose under different scenarios, with the amount of economic capital required being related to the losses under the various scenarios. The stochastic approach involves the use of a model, although the approach could be genuinely stochastic, parametric or empirical. The genuinely stochastic approach involves the construction of a full economic model capable of producing a large number of simulated results. However, a parametric approach might instead be used, where results are calculated based on an assumed statistical distribution, or an empirical approach might be used where results are based on past data on a firm's own losses or rating agency data. The choice of approach depends on the cost of the approach and complexity of the institution involved. It also depends on the purpose of a model. For example, if extreme events are being considered, a deterministic approach might be preferred over a stochastic one.

Another decision that must be taken is whether a model will be run on an enterprise-wide basis, or whether individual models will be run for each business line with the results being subsequently combined. If the latter case is true, then it is important to decide how the results are combined. In particular, if combining results after the calculation of a risk metric, it is important that a coherent risk measure is used.

A final decision that must be made is the nature of the output required, in particular the output needed in the calculation of the risk metric and thus to determine the capital requirement.

18.5 Running an economic capital model

Once a model has been built, decisions still need to be taken. First, there should be agreement on the risk metric to be used. Next, the time horizon for calculation must be considered. The level of risk that is acceptable – the risk tolerance – must be determined.

Decisions are also needed on management actions. These are changes made at future points in time in response to particular scenarios. For example:

- changes to investment strategy in response to performance;
- sources and amounts of capital;
- decisions on the withdrawal of particular products;
- levels of reinsurance;
- premium rates;
- dividends payable; and
- bonuses payable on with-profits policies.

Finally, a decision must be taken on whether the model is to be implemented on a run-off basis – assuming that no new business is won – or allowing for new business, in which case new business volumes must be estimated.

18.6 Calculating economic capital

There are a number of facets to the calculation of economic capital. A theoretical approach is discussed first, followed by some practical factors that must be considered.

18.6.1 Theoretical approach

Let the capital required at time zero be $K_0 = A_0 - L_0$, where A_0 is the value of the assets and L_0 is the value of the liabilities. These will typically be marked-to-market values, meaning that they will have been derived from traded instruments for which a market price is available. Where no such instrument is available, a marked-to-model approach may instead be used – with caution. K_0 must be set such that at any point in the future, t, the probability that $K_t = A_t - L_t$ will be less than some critical value, kL_t, must be below a particular level of risk tolerance, α. In other words, $\Pr(K_t \geq kL_t) \geq 1 - \alpha$, where $k \geq 0$, for all $t \geq 0$.

K_0 increases as the volatility of the assets relative to the liabilities increases, and falls as the expected return on the assets relative to the liabilities increases. An increased level of correlation between the assets and liabilities also results in a reduction in K_0.

If the assets and liabilities are projected stochastically, then the time interval used is important. Because solvency is measured only at discrete intervals, there is the possibility that, whilst a firm is solvent at two adjacent observations, it might have been insolvent at some point between them. Whilst this can be explained by the volatility present in the models, there are also practical reasons why this might occur. In particular, if cash outflows occur before inflows over the course of a year, then a potentially solvent position at the end of the year is irrelevant if insufficient assets were available to meet the outflows at the start of the year.

Another issue with this approach is that the range of parameters required can be enormous. In particular, the number of correlations required increases exponentially with the number of variables. Parameters such as correlation and variance are also unstable over time, meaning that it can be dangerous to place too much reliance on the results of a model such as this one.

18.6.2 Practical approaches

For the above approach to be implemented, measures of assets and liabilities must be chosen and projected, and measures of risk must be chosen.

The most obvious approach is to consider the probability of ruin based on the market or market-equivalent value of the assets and liabilities. The probability of ruin is the probability that the value of assets will fall below the value of liabilities. If a maximum acceptable probability of ruin is defined, then this calculation returns the additional value of assets that must be held to achieve this level of security. Along with all of the other assumptions, a decision must be taken on how these additional assets will be invested, since this will have an impact on the value of assets that must be held.

The concept of the probability of ruin can be extended to the cost of ruin for policyholders (in an insurance company) or account holders (in a bank). This is the amount lost by policyholders in the event of a firm's insolvency. As this is a value rather than a ratio, it makes sense to standardise it somehow, perhaps in relation to the total value of policyholder benefits. However, whilst this approach gives a more relevant measure of risk – for the policyholders, at least – the calculations required are more involved.

18.7 Economic capital and risk optimisation

Economic capital can be used as a way of optimising the way in which a firm carries out its business. In particular, it can be used to ensure that the limited amount of capital that an institution has is put to the best use.

Optimisation means that the highest return is achieved for the level of risk that is taken. However, there are a number of ways in which this criterion can be defined in practice.

18.7.1 Return on capital

Returns as a proportion of capital have already been discussed, and it is helpful to recognise that the measure of capital considered is economic capital. This means that the return that a financial firm makes should be considered in the context of the capital it needs to hold – the excess of assets over liabilities – for the business it has written. However, since the return on capital could simply be augmented by reducing the amount of capital held, it is more instructive to consider a measure where either the return or the economic capital is adjusted for risk. The most common example of this is the risk-adjusted return on capital (r_A). This is the ratio of the risk-adjusted return to the economic capital held. This means that if a firm reduces the amount of capital it holds and this results in the return being more risky, the return will be reduced correspondingly.

The return used can be an actual or an expected return, and the measure can be calculated for an entire firm or for an individual department. In fact, the nature of this measure means that it is well suited for comparing different lines of business within a firm as well as different firms.

18.7.2 Economic income created

Returns on capital consider the standardised rate of return. In contrast, the economic income created (EIC) returns the amount of return generated. It is calculated as:

$$EIC = (r_A - r_H)EC, \qquad (18.1)$$

where r_H is the hurdle rate of return and EC is the economic capital. This is rate of return that each unit of a product sold must earn to cover the additional amount of risk it generates. This is important, as the hurdle rate takes into account not just the riskiness of a product on a stand-alone basis, but also the extent to which it diversifies other products sold.

18.7.3 Shareholder value

Both the return on capital and the EIC are single-period measures. This means that the term of any opportunity is ignored. One way of allowing for this is by considering the present value of a business. This is known as the shareholder value (SV), defined as:

$$SV = \frac{r_A - r_G}{r_H - r_G}EC, \qquad (18.2)$$

where r_G is the rate of growth of the cash flows. This expression represents the discounted present value of all future cash flows. A related measure is the shareholder value added (SVA), which represents the present value of future cash flows in excess of the economic capital invested in a product:

$$SVA = \left(\frac{r_A - r_G}{r_H - r_G} - 1\right) EC. \tag{18.3}$$

18.8 Capital allocation

As important as calculating the total capital requirement of a financial institution is the need to allocate the capital that an institution has at its disposal between business lines. This has an impact on the amount of business that different departments can write, but also on the performance of each department in terms of the return on that capital.

The allocation of capital depends on the level of risk inherent in each department, but also on the extent to which each line of business acts as a diversifier to the rest of an organisation.

18.8.1 Allocating the benefits of diversification

If a new business line is launched, then the total capital requirement for a firm in unlikely to rise by as much as the stand-alone capital requirement for the new firm. Of course, a firm may choose not to allocate capital to individual business lines at all, holding all capital centrally and allocating business arbitrarily. However, this means that products might be sold without a full understanding of their impact on the capital requirements for the business as a whole. If capital is allocated to business lines – as would usually be the case – then a decision must therefore be taken on how to allow for the difference between the stand-alone requirement and the marginal addition.

A first thought might be for the company to retain the difference centrally. This is a simple approach, and means that the capital allocated to each business line is not subject to a potentially arbitrary allocation formula. However, this is not a particularly efficient use of capital, and could make lines of business uncompetitive if other firms are able to set their prices with an allowance for the diversifying effect of that business.

Another easy approach would be to leave the capital requirements for all existing business lines unchanged, giving the full benefit of diversification to the new business line, on the grounds that it is only the existence of the new business line that created this benefit. However, this is to give the benefit based on an accident of timing – had this new business line been in place with an existing line being the new one, the diversification benefit would rest elsewhere. Such an approach is therefore arbitrary.

A fairer approach is to perhaps start with the stand-alone capital requirements, but to allocate the diversification between the business lines somehow. There are a number of ways in which this could be done. For example, the reduction in capital could be divided in proportion to the undiversified reserves held. This is a simple approach, which is easy to justify, but might be perceived as unfair – a business line that provides more diversification might believe that it is due a higher proportion of the diversification benefit.

An approach that makes such an allowance is one that considers the marginal contribution of each additional unit of business to the overall capital required by the firm. This approach – known as the Euler capital allocation principal – gives the fairest allocation of capital between business lines, but it is also the most complicated approach.

18.8.2 Euler capital allocation principle

The Euler capital allocation principal can be used if a risk measure displays positive homogeneity. Positive homogeneity is one of the axioms that must be satisfied for a risk measure to be coherent. In this context, it was defined in terms of a risk measure $F(L)$ based on a loss function L that satisfied the expression $F(kL) = kF(L)$, where k was a constant. In fact, Euler's homogeneous function theorem is more general than this, and can be applied to any function exhibiting positive homogeneity of order q, so where $F(kL) = k^q F(L)$.

Consider a firm with N business lines. Let the loss in each line of business be L_n for the current volumes of business, such that:

$$L = \sum_{n=1}^{N} L_n. \tag{18.4}$$

Let k_n be some multiple of each business line n, with each $k_n = k$. If $F(L)$ is a risk measure based on the loss function L that exhibits positive homogeneity of order q, then Euler's theorem states that:

$$q k^{q-1} F(L) = \sum_{n=1}^{N} k_n \frac{\partial F(kL)}{\partial k_n}. \tag{18.5}$$

If $q = 1$, then this reduces to:

$$F(L) = \sum_{n=1}^{N} k_n \frac{\partial F(kL)}{\partial k_n}. \tag{18.6}$$

Equation (18.6) can be used to give the allocation of capital for a particular risk measure.

Standard deviation of losses

For example, let $F(L) = \sigma_L$, the standard deviation of the loss, and let the losses be linked by a covariance matrix, Σ. The total risk is therefore:

$$F(L) = \sigma_L = (k' \Sigma k)^{1/2}, \tag{18.7}$$

where k is a column vector of weights k_n. Partially differentiating this expression with respect to k_n gives:

$$\frac{\partial F(L)}{\partial k_n} = \frac{\sum_{m=1}^{N} \sigma_{L_m, L_n} k_m}{(k' \Sigma k)^{1/2}} = \frac{\sigma_{L_n, L}}{\sigma_L}, \tag{18.8}$$

where $\sigma_{L_n, L}$ is the covariance between the loss in line n and the total loss. Setting each $k_n = 1$, this means that the marginal contribution to the total risk of the risk in line n is $\sigma_{L_n, L}/\sigma_L$. If the economic capital required is proportional to the standard deviation of the losses, then this expression gives the multiple of the risk capital needed for line n.

Value at Risk and tail Value at Risk

If the capital required is instead proportional to the VAR of the loss function, L, at some level of confidence, α, $VaR_\alpha(L)$, then it can be shown that the marginal contribution of risk is given by $E[L_n \mid L = VaR_\alpha(L)]$. Similarly, if the capital required is instead proportional to the tail Value at Risk of the loss function, L, at some level of confidence, α, $TVaR_\alpha(L)$, then it can be shown that the marginal contribution of risk is given by $E[L_n \mid L \geq VaR_\alpha(L)]$.

18.9 Further reading

McNeil *et al.* (2005) gives further technical information on the calculation of risk capital, whilst Society of Actuaries [2004] provides a detailed practical assessment of this topic. Whilst this document is mainly concerned with insurance companies, a banking perspective is available in Matten (2000).

19

Risk frameworks

Whilst looking at the various parties that have an opinion on risk in financial institutions, it is clear that many rules are in place to control these risks. However, in many cases these rules consider only one aspect of a financial institution. In contrast, risk frameworks look at financial institutions, or even systems, as a whole and try to manage all of these risks in a consistent manner. There are three broad types of risk framework:

- mandatory;
- advisory;
- proprietary.

Mandatory risk frameworks must be followed in order for an organisation to carry out some types of business. However, they often have features that are useful to a wider range of institutions. Advisory risk frameworks offer guidelines for firms wishing to set up their own risk management framework. These are usually generic, which means that they can be used for a many different types of organisation, but also that a considerable amount of work must be carried out to tailor them to specific institutions. Finally, there are proprietary risk frameworks. These are frameworks used by firms for some specific purpose, the most common of which is credit rating.

All of the risk frameworks covered here are comprehensive, covering a range of risk types for an organisation. This is what differentiates a framework from a more narrowly focussed code.

19.1 Mandatory risk frameworks

Mandatory risk frameworks are those that must be complied to by firms working in particular industries. Two of the most important – and relevant – are

Basel II and Solvency II. The former is concerned with solvency in the banking sector, whilst the latter deals with the insurance industry.

19.1.1 Basel II

Basel II is the global risk framework designed to promote stability in the banking sector. It is published and updated by the Basel Committee on Banking Supervision (BCBS), which was established in 1974 by the governors of the central banks of the Group of Ten (G10) countries under the auspices of the Bank for International Settlements (BIS). It has no formal supranational authority and merely recommends statements of best practice. However, these recommendations are taken up not just by regulators in the G10 countries, but also by those in other countries, although the exact implementation can differ substantially from country to country.

Background to Basel II

The BCBS was founded in response to an evident lack of cross-border co-ordination in financial transactions. It issued guidance throughout the 1970s and 1980s, culminating in 1988, with a set of minimum capital requirements for banks as given by the Basel Committee for Banking Supervision (1988). This was originally known as the 1988 Basel Accord, although when it was superseded it became known as the First Basel Accord, or Basel I. Basel I focuses mainly – and, originally, focussed exclusively – on credit risk. However, as it became clear that the changing nature of banks meant that they were increasingly exposed to market risk, Basel I was updated to allow for this with an amendment as described by the Basel Committee for Banking Supervision (1996). In this context, credit risk can be defined as the risk that funds owed are not paid, and market risk as the risk that the value of assets will move in such a way as to cause a financial loss.

At a high level, the methodology behind Basel I was straightforward. First, credit-related assets and liabilities that were off-balance sheet were converted to on-balance sheet equivalents. These were then risk weighted, together with the existing on-balance sheet credit exposures. Very low risk assets, such as AAA-rated government bonds, had a risk weight of zero, whilst assets which were more risky could have a risk weight of up to 100% of their face value. Exposures to market risk were then adjusted using either a risk-weighting approach (as for credit risk) or a firm's agreed internal model. The model generally involved calculating the risk weight based on a 99% ten-day VaR. These risk-weighted assets were then summed, the total representing the level of risk to which the institution was exposed, and multiplied by a minimum capital requirement of 8%. This meant that firms had to hold additional capital worth at least 8% of risk-weighted assets.

The available capital was originally classed as either core (tier 1) or supplementary (tier 2) capital. Tier 1 capital consisted of a bank's equity capital and disclosed reserves. Tier 2 capital was made up of undisclosed reserves, revaluation reserves, general loss reserves, hybrid debt instruments and subordinated debt. This capital was generally subject to discounting, upper limits or both. There was a limit on the total amount of tier 2 capital of the amount of tier 1 capital. Essentially, this meant that tier 1 capital of at least 4% of risk-weighted assets was needed. When market risk was added to Basel I, the concept of tier 3 capital also arrived. This made additional allowance for certain types of shorter-dated capital to cover market risk, although again there was a limit on the amount of tier 3 capital that could be used, in this case 250% of the tier 1 capital used to support market risk. Goodwill and unconsolidated banking subsidies were deducted from the sum of these sources of capital, and unconsolidated non-banking subsidies were risk weighted.

As mentioned above, Basel I was simple – but crude. The scope for regulatory arbitrage through methods such as securitisation led to excessive risk being maintained. This process involved packaging loans into instruments such as CDOs, whilst also buying similar products in the market. This could leave a firm's economic exposure to credit risk unchanged, but the credit risk under Basel I would be lower. The risk would instead have been converted to market risk, which could be allowed for in an internal model. This could significantly reduce the amount of capital required.

However, a more important issue was that even with the addition of market risk to the original credit-based formulation, the range of risks considered is still narrow. In particular, some banks ran into difficulties despite appearing healthy from a Basel I point of view. Operational risk – not covered by Basel I – often featured heavily in these cases.

Basel II, the Second Basel Accord, was introduced in 2004. It seeks to address many of the issues with Basel I. Basel II is based on a concept of 'three pillars':

- minimum capital requirements;
- supervisory review process; and
- market discipline.

Minimum capital requirements

The first pillar is similar to the minimum capital requirement under Basel I in that it uses tiers 1, 2 and 3 capital with only minor changes. It also allows for market and credit risk, with market risk being unchanged. In terms of valuation, liquid assets are marked to market (so the market value of assets is used), whereas illiquid assets are marked to model, meaning that the values are 'benchmarked, extrapolated or otherwise calculated from a market input'. As with Basel I, market risk can be calculated using either the risk-weighting approach or an internal model.

Credit risk changes in Basel II. First, the standardised model for credit risk from Basel I is updated to allow for a greater range of creditors. This increased granularity seeks to treat the different credits more equitably. However, Basel II also allows for the use of an internal model in the same way as for market risk, in this case known as the internal ratings based (IRB) approach. This means that, in theory, market and credit risks can be treated consistently. Basel II also makes explicit allowance for securitisation in an effort to limit this aspect of regulatory arbitrage.

Perhaps the greatest change from the first to the second Basel Accord is that an allowance is made for operational risk. In Basel II, operational risk is defined as 'the risk of loss resulting from inadequate or failed internal processes, people and systems or from external events'. Three approaches can be used to calculate the reserves required. The simplest method is the basic indicator approach. This involves applying a fixed multiple, α, to the gross income. The measure of gross income used under Basel II is the average over last three years, with any years for which the gross income is less than or equal to zero being excluded from the calculation. This means that if the gross income is positive for only two of the last three years, the income in these two years should be added together and divided by two rather than three. If the capital required under this approach is K_{BIA}, the approach can be described as:

$$K_{BIA} = \frac{\sum_{t=1}^{3} \max(GI_t, 0)\alpha}{I(GI_t \geq 0)}, \tag{19.1}$$

where GI_t is the gross income in year t and $I(GI_t \geq 0)$ is an indicator function that is equal to one if $GI_t \geq 0$ and zero otherwise.

A more advanced approach is to take a multiple of gross income across each business line. This is the standardised approach. This is similar to the basic indicator approach except that the firm is divided into eight separate business lines with the capital being calculated separately for each. Also, negative gross incomes are not excluded from the calculation. This should make the standardised approach more attractive to banks than the basic indicator approach. If

Table 19.1. *Values of β_n for standardised approach to operational risk under Basel II*

n	Business line	β_n
1	Corporate finance	18%
2	Trading and sales	18%
3	Retail banking	12%
4	Commercial banking	15%
5	Payment and settlement	18%
6	Agency services	15%
7	Asset management	12%
8	Retail brokerage	12%

Source: Basel Committee on Banking Supervision: *International Convergence of Capital Measurement and Capital Standards – A Revised Framework* (2004).

the capital required under this approach is K_{SA}, the approach can be described as:

$$K_{SA} = \frac{\sum_{n=1}^{8} \sum_{t=1}^{3} GI_{n,t}\beta_n}{3},$$
(19.2)

The values of β_n for each business line are given in Table 19.1.

Finally, with the agreement of the regulator, a firm can use internal models and scenario analysis to calculate a bespoke reserve requirement. This is the advanced measurement approach. This involves using internal and external data to determine the probability of loss events and the expected size of loss, given that an event has occurred in each business line. For this approach, each loss event type must be allowed for. The different event types are:

- internal fraud;
- external fraud;
- employment practices and workplace safety;
- clients, products and business practices;
- damage to physical assets;
- business disruption and system failures; and
- execution, delivery and process management.

This approach gives banks the opportunity to model operational losses consistently with market and credit risks.

Supervisory review

The second pillar of Basel II is supervisory review. This is important as it recognises explicitly that holding capital is not a substitute for inadequate risk management, although the result of the review might be a requirement to hold additional capital against risks not covered in the first pillar. There are a number of aspects to supervisory review. The first is that firms have an internal process for monitoring capital adequacy. This forms the basis for review by the regulator, who needs to make sure that the process is sound. The regulator also needs to ensure that the firm is operating above minimum level, and has an obligation to intervene quickly if there is a risk of capital falling below minimum levels. There are a number of aspects of the process that the regulator must pay particular attention to. Interest rate risk should be considered, as should various aspects of credit risk, including concentration and counter-party risk. The regulator should also verify that the approach used to quantify operational risk is consistent with the business, and whether market risk is correctly measured.

Market discipline

Market discipline is the third pillar of Basel II. This is essentially a case of promoting transparency by requiring firms to publish details of their risks, their capital and the ways in which they manage risk. The aim is to make sure that sufficient information about a firm is disclosed for the market to assess the risks faced by the firm, and for the cost of capital – the price of equity and debt – to be adjusted accordingly. The rationale behind this is that a firm will seek to manage risks in order to manage the cost of capital.

√ Criticisms of Basel II

The Second Basel Accord is a major improvement on the First both in scope and in process. However, it still has a number of flaws.

Both Basel Accords focus on a single number as a measure of all risks. In one sense, this is helpful as it allows a variety of different firms to be compared on a consistent basis. However, the range of firms described by this figure and the range of risks aggregated into it mean that is dangerous to place too much emphasis on a single measure of risk such as this. In fact, the headline figure is arguably even less informative for Basel II than for Basel I, since the range of risks covered by the former is much greater than that covered by the latter.

A major improvement is the allowance for operational risk. However, the list or risks addressed remains incomplete. Importantly, liquidity risk is given only cursory treatment, although in 2009 the Basel Committee on Banking Supervision did issue proposals to allow for this risk more fully.

It is worth noting the difficulty in arriving at a firm estimate for the risk number. First, it is difficult to quantify many operational risks. However, quantification is not straightforward even for market risks: the level of confidence required is one in two hundred years, but some of the asset classes considered have existed for only a decade. In this context, the levels of confidence could be regarded as spurious.

However, despite the potentially reduced reliability of this number, there is a risk that the more risks included and the more complex the calculations become, the greater the confidence that will be put in the risk number. This equation of complexity with reliability is dangerous.

Basel II is also blamed for elements of pro-cyclicality in market cycles, leading to feedback risk. In particular, marking assets to market or even to model (which requires valuation using market inputs) requires risky assets to be sold if their market value has fallen. This forced sale can force asset prices down further. Pro-cyclicality is cited as a reason to avoid marking assets to market. However, it is not clear what alternative could be used – anything that does not reflect market values risks over-valuing assets held by a bank. Having said this, market values can seriously under-value certain fixed interest instruments where the risk of loss is slight – the question is, by how much.

Linked to the issue of pro-cyclicality is the failure of Basel II to deal with systemic risks. Basel II aims to control the risk of insolvency of each bank. However, this only serves to control the overall risk of the banking system if the risks of insolvency are reasonably uncorrelated across banks. Because banks are often similarly exposed to many risks, this means that an adverse event affecting one bank could actually affect many.

A final, practical issue with Basel II is that the added complications relative to Basel I require added expenditure on appropriate systems. It is not clear that this expenditure will improve outcomes for all companies.

Basel III

In response to the global financial crisis, certain aspects of the first pillar of Basel II are to be strengthened under what has become known as Basel III introduced by the Basel Committee for Banking Supervision (2010). From 1 January 2015, tier 1 capital will need to constitute at least 6% of risk-weighted assets. As part of this ratio, the amount of common equity will need to be equal to at least 4.5% of risk-weighted assets. The balance of the 8% minimum capital requirement can be met by tier 2 capital. Transitional arrangements will phase these changes in from 1 January 2013.

In addition, Basel III defines a conservation buffer of 2.5% of risk weighted assets that must also come from common equity. This is intended to provide banks with some breathing room in times of financial stress – in other words, a bank can draw on this buffer, but to do so will bring limits on the extent to which earnings can be distributed. The buffer will be phased in from 1 January 2014 to 1 January 2019.

Finally, further deductions from common equity – covering items such as investments in financial institutions, mortgage servicing rights and certain deferred tax assets – will be phased in over the period from 1 January 2014 to 1 January 2018.

Apart from the strengthening of capital requirements, the main advance in Basel III is that it attempts to limit pro-cyclicality by allowing for capital requirements to fall in times of financial stress. It also attempts to deal with the systemic risk by limiting the extent to which cross-holdings in other banks are allowed for in a bank's equity. It still does not deal with the risk that banks with similar strategies will be similarly exposed in financial crises.

19.1.2 Solvency II

Solvency II is the risk framework for insurance companies operating in EU member states that is due to be implemented on 1 November 2012. It is modelled on Basel II, but sponsored not by the G10 but by the EU. Unlike the BCBS, the EU can require adherence by member states to its directive.

Background to Solvency II

As with Basel II, Solvency II has a heritage stemming from the 1970s with the First Non-Life Directive and the First Life Directive published in 1973 and 1979 respectively. A number of additional directives followed. The 2002 Non-Life Directive further amended its predecessors, whereas the 2002 Life Directive was so substantial as to not only amend previous directives but to recast the remaining elements that were in force as a single document. Together, these were known as Solvency I.

Solvency I covered issues as diverse as the rights of policyholders, the good character of employees and insurance advertising. However, the bulk of the legislation referred to insurance company solvency. Under Solvency I, the EU delegated insurance company supervision to the regulators in each member state. The directives were clear that supervision included verification of the solvency of insurance companies, and that regulators must require these companies to have robust accounting, administration systems and internal controls.

Solvency I also placed duties on auditors to ensure that they scrutinised accounts to a sufficient degree.

The First Life Directive forbade the establishment of new composite insurers, but it also required that existing ones be treated as separate life and non-life insurance companies for the purposes of solvency calculations. Solvency I also set out the requirement to calculate technical provisions and to include expenses. There was also guidance on the choice of discount rate for valuation, as this is particularly important for the valuation of long-term business.

Solvency I then described the assets that could be used to back these technical provisions, and the maximum extent to which various assets can be allowed for. These were expressed in terms of a diversification requirement.

Finally, Solvency I outlined the calculation of the minimum solvency margin, and detailed what could be used to back it – specifically free reserves, share capital and retained profits. Intangible assets were specifically excluded. Part of the minimum solvency fund was the guarantee fund, the absolute minimum level of solvency that must be achieved below which regulatory intervention was required.

As with Basel I, Solvency I was a good attempt to provide a consistent and robust basis for solvency amongst insurers, but it suffered from many of the same issues: it was inflexible, different assets were treated as though they were identical and it concentrated on too few risks. Solvency II is an attempt to address this.

Solvency II is, like Basel II, designed around three pillars. These are similar in nature to those of Basel I, but are instead defined as:

- quantitative requirements;
- qualitative requirements; and
- disclosure.

Quantitative requirements

The detail for Solvency II is included in the Solvency II Framework Directive from the European Commission (2009). The quantitative requirements for Solvency II are in two parts: the Solvency Capital Requirement (SCR) and the Minimum Capital Requirement (MCR), the latter being a specific requirement rather than a pillar as in Basel II. The SCR is the main standard by which solvency is measured, below which regulatory action is taken; the MCR is a lower capital requirement, but any firm falling below the MCR would lose its authorisation.

The SCR must be achievable with a 99.5% level of confidence over a one-year time horizon. There are two approaches to calculation. The standard formula involves calculating the sum of basic solvency capital items, adding the requirement for operational risk, and making several adjustments. The components of basic solvency capital are:

- non-life underwriting risk;
- life underwriting risk;
- special health underwriting risk;
- market risk (including interest rate mismatch);
- counter-party default risk; and
- operational risk.

Under this standardised approach, required reserves are calculated according to a specified deterministic basis, although stochastic methodology is often required, particularly in the valuation of with-profits guarantees. Market risk is dealt with by limiting the assets which are admissible and, for firms writing long-term business, requiring stress testing in response to a small number of deterministic scenario tests; credit risk is dealt with by limiting exposure to reinsurance and other counter-parties. An additional solvency margin is calculated which differs by business: for non-life insurance business, the margin is the greater of various amounts determined from premiums written or earned, claims incurred and the previous year's reserves on a deterministic, formulaic basis; for firms writing long-term business, proportions of the liabilities, assets and sums assured are calculated, again in a deterministic and formulaic manner. However, instead of using this standard approach, a quantitative model designed by the firm can be used. This internal model-based calculation is somewhat different, and to be accepted by a regulator it must fulfil six criteria:

- the 'use test';
- statistical quality standards;
- calibration standards;
- profit and loss attribution;
- validation standards; and
- documentation standards.

The use test is intended to ensure that the model used is not simply constructed for regulatory purposes but is more widely employed. In particular, a firm should use the model as part of:

- the risk management systems;
- the decision-making processes;
- the economic capital assessment and allocation processes; and
- the solvency capital assessment and allocation processes.

The statistical quality standards are intended to ensure that realistic assumptions are used, based on accurate, appropriate and up-to-date information. It is permissible to allow for diversification, potential management actions and responses to risk, but credit risk and the impact of guarantees should also be adequately allowed for.

The calibration standards are in place to ensure that the outputs from the internal model can be used to calculate the SCR in such a way that it reflects the 99.5% level of confidence, whilst the profit and loss attribution requires that the causes and sources of profits and losses for each major business unit are analysed at least annually. This is closely related to the validation standards which require the performance of the internal model to be monitored. Specifically, the appropriateness of the model specification should be under constant review with model results being compared with actual experience.

Finally there must be minimum standards of documentation, with a detailed outline of the theory, assumptions and reasoning underlying the internal model being recorded.

In contrast to the SCR, the MCR is a simple, objective calculation. As discussed above, it is intended to be an absolute minimum level of solvency. The firm should be able to meet the MCR with a probability of 80% to 90% over a one-year period.

Under Solvency II, all assets and liabilities are taken at 'fair value', implying market consistency wherever possible. The excess of assets over liabilities forms part of what is known as a firm's 'basic own funds', an item which also includes subordinated liabilities. Another item known as 'ancillary own funds' includes unpaid share capital, letters of credit and insurance payments due. These funds are used to constitute the capital required to meet the quantitative requirements. Broadly speaking, basic own fund items count as tier 1 capital if they can be called upon to meet losses in ordinary circumstances, whereas basic and ancillary own fund items that can only be called upon in wind-up are tier 2. Anything less easily available is tier 3 capital. When trying to meet the SCR, tier 1 capital can make up no less than one-third of the capital, and tier 3 capital no more; for the MCR, tier 1 capital must be more than half of the basic own funds.

Qualitative requirements

The second, qualitative pillar of Solvency II is as much a message to regulators on how to treat firms as it is to firms on how they should behave. For example, regulators should analyse the strategies of firms, suggesting that business models are under scrutiny. This is an important check on the sustainability of a business. Processes also need to be investigated, which is important given their role in operational risk. The governance requirements in the second pillar are aimed at bringing consistency across the insurance, securities and banking industries.

Disclosure

Finally the third pillar – disclosure – is intended to improve risk management by encouraging firms to control risk in order to reduce the cost of capital. As with Basel II, this is an important way of encouraging firms to embrace good risk management, supplementing the requirements centred around capital adequacy.

Comparison of Basel II and Solvency II

Both Solvency II and Basel II have been designed with multi-national firms in mind. Both are also risk-based three-pillar frameworks, meaning not only that more capital is allocated to firms that run higher risks, but also that capital is not the only answer to risk management.

There are, though, major difference between the two frameworks. Solvency II is less prescriptive than Basel II. The former concentrates on broad principles with the expectation that the regulators in each country will provide the detailed rule for firms under their supervision. Basel II also has a greater focus on systemic risk than Solvency II. This is appropriate, since the operations of insurers are less intertwined than those of banks, so systemic risk is less of an issue. However, despite these differences, many of the criticisms levelled at Basel II have also been levelled at Solvency II.

19.2 Advisory risk frameworks

Advisory risk frameworks are not required for compliance with legislation, but can be helpful for all organisations – financial and non-financial – in defining their ERM frameworks. A number of these risk advisory frameworks exist. Those covered in this section are:

- Risk Analysis and Management for Projects (RAMP)
- the COSO ERM Integrated Framework;

- the IRM/AIRMIC/Alarm Risk Management Standard;
- the Treasury Board of Canada Integrated Risk Management Framework;
- the Orange Book;
- AS/NZS 4360:2004; and
- ISO 31000:2009.

19.2.1 RAMP

The methodology developed by the Institute and Faculty of Actuaries together with the Institution of Civil Engineers for the management of the risks in any kind of project is known as RAMP (Risk Analysis and Management for Projects). This approach, which was first published in 1998, is now well established and consists of eight stages:

- RAMP launch;
- risk identification;
- risk analysis;
- financial evaluation;
- risk mitigation;
- go/no-go decision;
- risk control; and
- RAMP closedown.

Once stakeholders and their viewpoints have been identified (part of the RAMP launch), many of the remaining sections are consistent with the analysis in the preceding sections. Because RAMP is intended for use with capital projects rather than in an ongoing business, a decision on whether to proceed at all is required, and post-project 'closedown' analysis forms part of the process. However, both could and perhaps should also be incorporated into broader risk management processes relating to the financial and non-financial operations of banks, insurance companies and pension schemes.

The methodology enables risks to be expressed in financial terms through the use of an investment model and facilitates decisions on whether projects should go ahead or not, and in what form. Because risks, including the eventual operational risks, are fully thought through at the outset of the project, costly mistakes should be minimised.

19.2.2 The COSO ERM integrated framework

COSO is the Committee of Sponsoring Organizations of the Treadway Commission, the commission itself being sponsored by the American Accounting Association (AAA) , the American Institution of Certified Public Accountants (AICPA), the Financial Executives International (FEI), the Institute of

Management Accountants (IMA) and the Institute of Internal Auditors (IIA). The COSO framework is set out in a detailed document, which sets out a detailed generic risk management process.

The document is keen to point out that events can have both a positive and a negative impact, and that these correspond to the opportunity and risk associated with the events. The document is also clear that ERM is an ongoing process rather than an event. This is an important concept, as it emphasises the importance of integrating ERM into everyday workings of an organisation. The document also notes that this process should be applied at all levels of an organisation and as part of the organisation's strategy.

According to the COSO approach, ERM encompasses:

- aligning risk appetite and strategy;
- enhancing risk response decisions;
- reducing operational surprises and losses;
- identifying and managing multiple and cross-enterprise risks;
- seizing opportunities; and
- improving the deployment of capital.

The COSO document addresses the context and scope of the risk management exercise by defining them in terms of a three-dimensional matrix, with each dimension being inextricably linked to the others. The first of these dimensions is the range of areas that the risk framework should cover. The framework divides these into four categories:

- operational;
- compliance;
- reporting; and
- strategic.

The COSO ERM framework emphasises that ERM should cover the operations of any firm, since failure here can lead at best to inefficiency and at worst to catastrophic losses. Compliance with rules and regulations, as well as internal procedures, falls within the remit of the COSO framework. It also covers the reporting area, which is important since it is impossible to control risks if accurate information about the state of the business is not available. The COSO document states that ERM should be implemented at a strategic level, being integrated into the way an enterprise seeks to meet its high level goals.

The second dimension described in the framework cover eight components of ERM:

- internal environment;
- objective setting;

- event identification;
- risk assessment;
- risk response;
- control activities;
- information and communication; and
- monitoring.

The internal environment of an organisation defines the context in which ERM is carried out. The COSO document highlights the risk management philosophy, risk appetite and broad values as being important components of this environment. Setting the objectives is also a key feature here, since objectives must be set before the risks of not meeting these objectives can be considered. The first stage here is to identify the risks and opportunities, both internal and external, that might have an impact on the ability of an enterprise to meet its objectives. The incidence and potential intensity of these risks are then analysed, and risks responses considered. The four standard responses to risk are discussed, defined here as avoidance, reduction, sharing and acceptance.

The final dimension in this framework is the level of application. This emphasises that risk management applies to all levels of an organisation, from the entity as a whole through divisions, business units and subsidiaries.

The interaction of these dimensions is shown graphically in Figure 19.1. This demonstrates the combinations that can be considered, but not every cell will necessarily be populated.

The COSO framework emphasises that ERM is not a serial process, but is multi-directional and iterative. It places ultimate responsibility for the framework with the CEO, but points out that everyone has some role in risk management. This is true for the board, which is responsible for overseeing the framework and liaison with senior management, for senior management and staff, but also for regulators, professional organisations and educators. Of course, these external parties are not responsible for implementing ERM within a particular firm, but they can provide useful information to such a firm.

The framework also explicitly describes the limitations of ERM, which are important to recognise. Even in a risk management framework, management processes might be inadequate and human error can occur. Also, the benefit of ERM, and any component of it, needs to be considered against the cost of implementation. Finally, it is impossible to guard against all opportunities for deliberate circumvention of risk controls, particularly if two or more individuals collude.

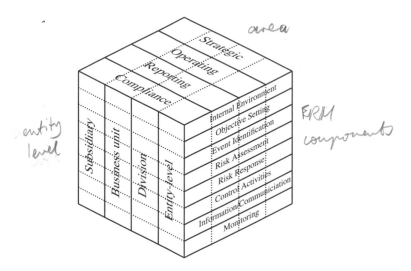

Figure 19.1 COSO Risk Framework, adapted from PricewaterhouseCoopers: *Enterprise Risk Management – Integrated Framework* (2004)

19.2.3 IRM/AIRMIC/Alarm risk management standard

The next risk framework discussed is provided by the Institute of Risk Management (IRM), the Association of Insurance and Risk Managers (AIRMIC) and the National Forum for Risk Management in the Public Sector (Alarm). Whilst AIRMIC has withdrawn its support from this framework in favour of the new global standard ISO 31000:2009, it remains useful not least because it is free to download. The IRM/AIRMIC/Alarm Risk standard has a number of similarities with the COSO framework. For example, risk is defined here in terms of a combination of the probability of an event and its consequences. It also recognises that risks can have upsides as well as downsides, but the focus in this framework is firmly on downside risks. Risks are classified in this framework as internally and externally driven, and are divided into financial, strategic, operational and hazard risks, as shown in Figure 19.2. These risks form part of the process described in Figure 19.3.

The document recommends that the risk identification process should be approached methodically, suggesting that a pro-forma approach is used to consider all aspects of risk. It also suggests that an in-house approach is likely to be more effective than an approach that uses external consultancies, since people already working for a firm are likely to have greater familiarity with the

Figure 19.2 IRM/AIRMIC/Alarm Risk Categories, adapted from IRM/AIRMIC/
Alarm: A Risk Management Standard (2002)

relevant issues. The document also emphasises that the reporting of risks is
part of good corporate governance, and makes reference to both internal and
external reporting. Internal audit is also seen as an important control.

There is no detailed discussion of the types of risk treatment, but there is
more focus on what risk treatments should lead to at a minimum: effective and
efficient operation, effective internal controls and compliance with laws and
regulation. The document also covers the monitoring of risk as well as the roles
of various interested parties: the board is responsible for taking a strategic view
on risk, and for considering the various costs and benefits, as well as the overall
review process; business units are responsible for the day-to-day management
of risk, and for including risk management in the fabric of their work; but the
greatest responsibility is saved for the risk management function. This function
is regarded as the primary champion for risk management in the organisation,
and is responsible for:

- setting risk management policy and strategy;
- building a risk management culture;
- educating employees;

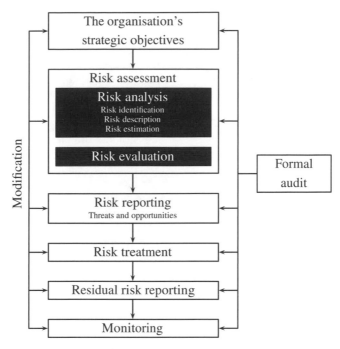

Figure 19.3 IRM/AIRMIC/Alarm Framework, adapted from IRM/AIRMIC/ Alarm: A Risk Management Standard (2002)

- establishing structures and policies within business units;
- designing processes;
- coordinating functions;
- developing responses to risks; and
- reporting risk.

The framework recommends that risks are described in a structured format, as shown in Table 19.2. This ensures that all aspects of a risk are considered.

For estimation, the framework points out that a quantitative, semi-quantitative or qualitative approach can be used, and that both threats and opportunities should be considered. It suggests attaching probabilities to low, medium and high estimations of risk, and estimating low, medium and high consequences. The result of this work is a risk profile for the enterprise. The risks can then be mapped to the business areas affected, with the primary control procedures in place being detailed. This makes it easier to discuss areas where the level of risk control investment might be increased, decreased or reapportioned.

Table 19.2. *IRM/AIRMIC/Alarm Risk Descriptions*

Name of risk	
Scope of risk	Qualitative description of the events, their size, type, number and dependencies
Nature of risk	For example, strategic, operational, financial, knowledge or compliance
Stakeholders	Stakeholders and their expectations
Quantification of risks	Significance and probability
Risk tolerance/appetite	Loss potential and financial impact of risk
	Value at Risk
	Probability and size of potential losses/gains
	Objective(s) for control of the risk and desired level of performance
Risk treatment and	Primary means by which the risk is currently managed
control mechanisms	
	Levels of confidence in existing control
	Identification of protocols for monitoring and review
Potential action for improvement	Recommendations to reduce risk
Strategy and policy developments	Identification of function responsible for developing strategy and policy

Source: IRM/AIRMIC/Alarm: *A Risk Management Standard* (2002).

The final area discussed, which is also covered in the COSO document, is reporting. An important distinction is made between internal reporting, for the benefit of the board of directors, business units and individuals around an organisation, and external reporting for investors and regulators. Both must be borne in mind.

19.2.4 The Treasury Board of Canada risk management frameworks

The Treasury Board of Canada has issued two risk management frameworks. The first – the Integrated Risk Management Framework – was issued in 2001.

Whilst this was superseded by the Framework for the Management of Risk in 2010, it still offers useful insights.

Integrated risk management framework
This document was created to help public-sector workers in Canada with their decision making. The framework has four stages:

- developing the corporate risk profile;
- establishing an integrated risk management function;
- practising integrated risk management; and
- ensuring continuous risk management learning.

The corporate risk profile sets out the ways in which an institution is exposed to various risks. The framework lists a number of relevant factors in the development of this risk profile that are internal to an organisation:

- the overall management framework;
- governance and accountability structures;
- values and ethics;
- the operational work environment;
- individual and corporate risk management culture and tolerances;
- existing risk management expertise and practices;
- human resources capacity;
- the level of transparency required; and
- local and corporate policies, procedures and processes.

The external factors are defined as arising from political, economic, social and technological sources. For risks that arise from both internal and external sources, the risk environment includes the type of risk, source of risk, exposure and ability to control the risk.

Having considered the risk profile, the framework then turns to risk capacity. An important aspect of this is risk tolerance, which the framework suggests can be determined through consultation or through observing responses to different levels of risk exposure.

The framework recommends using an integrated risk management function, and that risk management should be built into the decision-making process. It also recommends that a common risk management process is used throughout the enterprise, with the process shown in Figure 19.4.

The importance of adequate reporting is emphasised, and it is recommended that communication and consultation be employed at all levels.

Figure 19.4 The Treasury Board of Canada Integrated Risk Management Frame-work, from the Treasury Board of Canada: Integrated Risk Management Frame-work (2000)

There is also considerable comment on the importance of CPD and the importance of creating a supportive work environment with continuous learning being built into employees' development plans.

Framework for the management of risk

This revised framework is designed to help government officials in the management of risk. In particular, it aims to help officials

- identify and explain different risk types;
- make decisions and provide guidance on levels of risk tolerance and the treatment of risks;
- support continuing professional development;
- embed risk management principles and practices in their organisations;
- align their approach to risk management practices and policies of the Treasury Board of Canada;
- support wider government policies;
- manage risk in a way that recognises external and internal risk management contexts;
- add value through risk management;

- balance responses to risk with innovation;
- be transparent, inclusive, integrated and systematic; and
- improve the culture, capacity and capability of risk management in organizations.

At the time of writing, the Guide to Integrated Risk Management that supplements this new framework was not available.

19.2.5 The Orange Book

The Orange Book is designed to give general guidance on risk management in both the public and private sector. It is intended to operate at a higher level than other risk management standards, so that the other standards can operate within the framework set out in the Orange Book.

The broad risk management model is shown in Figure 19.5. It is noted that this is not a linear process, and that the various factors interact in a number of ways.

Starting with risk identification, a distinction is made between initial and continuous identification. In both cases, the importance of linking the risk to an objective is highlighted – if an adverse event does not affect an objective, then it is not a risk. A distinction is also made between the impact of a risk and

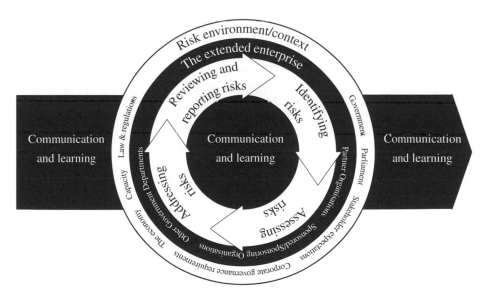

Figure 19.5 Orange Book Risk Management Model, from HM Treasury: *The Orange Book: Management of Risk – Principles and Concepts* (2004)

the risk itself. A number of risk identification approaches are discussed, with internal and external reviews considered.

When considering risk assessment, three principles are highlighted

- that there should be a structured process considering both likelihood and impact for each risk;
- that assessments should be recorded in such a way that risk can be monitored and prioritised easily; and
- that the distinction between inherent and residual risks is clear.

Prioritisation rather than quantification is the focus here, with classification into high, medium and low risk categories being considered.

Next, risk appetite is discussed. This is considered in terms of a series of boundaries, limiting the risk that can be taken in particular departments. A distinction is made between corporate risk appetite, which applies to the organisation as a whole; delegated risk appetite, which is the appetite allocated to a particular business lines or departments; and project risk appetite, which is allocated to stand-alone enterprises outside the day-to-day business of an organisation.

In terms of risk response, the four approaches are referred to as tolerate, treat, transfer and terminate. Also discussed is the option of taking the opportunity. This is regarded as being carried out in addition to any of the responses other than terminate. Risk treatments are classified as:

- preventive, which limit the possibility of an undesirable outcome;
- corrective, which correct undesirable outcomes that have been realised;
- directive, which seek to ensure a particular outcome is achieved; and
- detective, which provide notification after a loss event has occurred.

Regular reviewing and reporting is recommended, with the ultimate responsibility resting with the audit committee. The possibility of setting up a dedicated risk committee is also discussed. The importance of communication and learning are emphasised, as is the importance of recognising the context in which an organisation sits.

$\mathscr{E}$ 19.2.6 AS/NZS 4360:2004

The joint Australian Standard and New Zealand Standard 4360:2004 is the third revision of this standard. The original was produced in 1995, a second edition was published in 1999 and the latest standard appeared in 2004. It was produced by a joint committee from the two organisations, and is a relatively high-level document, but can be seen as the predecessor to many of the other

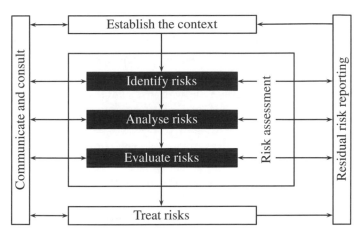

Figure 19.6 AS/NZS 4360:2004 Risk Management Process, from Standards Australia and Standards New Zealand: *adapted from Risk Management Guidelines – AS/NZS4360:2004*, (2004)

frameworks that exist today. In particular, it was used as the first draft for the global risk management standard ISO 31000:2009, which has been adopted in Australia and New Zealand as the successor to AS/NZS 4360:2004. The broad process for AS/NZS 4360:2004, shown in Figure 19.6, has fewer stages than some comparable frameworks, but covers many of the same points.

The first stage in the process here is to establish the context within which the risk management will be carried out. 'Context' refers to a range of factors such as the objectives of the exercise, the main stakeholders and the criteria by which the process will be measured. The internal context relates to aspects such as the culture of an organisation, its structure and goals, its stakeholders and their abilities. This is an important aspect of risk management as it defines the frame of reference for the risk management process. The external context refers to the various environments in which an organisation exists – business, social, regulatory, cultural, competitive, financial and political are all mentioned. It also covers 'SWOT' factors – the strengths, weaknesses, opportunities and threats to which an enterprise is exposed. There are also external stakeholders who need to be considered, whose influence on an organisation should not be underestimated. Finally, there is the risk management context, which defines the scope of the risk management framework and the enterprise's tolerance for risk.

The next stage is risk assessment, which encompasses identification, analysis and evaluation. Identification covers what, where, when and how risks can occur, and various techniques for risk identification are discussed. The section on risk analysis is one of the largest in the document. The issues covered appear to be more relevant for non-financial organisations, but this makes them particularly useful when considering operational risk for financial enterprises. Quantitative, semi-quantitative and qualitative approaches are considered. Evaluation of risks is dealt with in a short section in this framework, and simply recommends the comparison of the assessment of a risk with the appetite for that risk.

After the assessment of risks, their treatment is discussed at a high level. Potentially positive and negative outcomes are considered separately, although the actions are along the same lines. The first treatment is to seek an opportunity (for a positive outcome) or avoid it (for a negative outcome). Changing the likelihood and consequence of an opportunity to enhance positive and reduce negative outcomes is then suggested, followed by the option of sharing an opportunity to the same effect. Finally, the choice of retaining the residual opportunity is discussed.

All the while, communication and consultation with stakeholders – both internal and external – should take place. The key point is also made that this is not a linear process. The process should be monitored and kept under continual review.

The standard also considers how the risk management process can be made more effective. Most importantly, it recommends that the process be formalised into a risk management plan. The importance of getting senior management buy-in is also emphasised, as are communication and the establishment of accountability. For this to be accomplished, adequate resources in terms of money and people are needed, and these must be committed if a risk management framework is to be implemented successfully.

19.2.7 ISO 31000:2009

As mentioned above, AS/NZS 4360:2004 is important as it forms the basis for the international standard ISO 31000:2009. This has been used as the basis for a number of country-specific standards, not only in Australia and New Zealand but around the world. AIRMIC has also withdrawn its support from the IRM/AIRMIC/Alarm Risk Management Standard in favour of the new international standard; however, whilst the documentation for ISO 31000:2009 must be purchased, the IRM/AIRMIC/Alarm documentation has the attraction of being free.

The ISO 31000:2009 has three broad sections covering principles and guidelines, the risk management framework and the risk management process.

It also includes sections on risk management techniques and the vocabulary of risk management.

This standard defines risk as the effect of uncertainty on objectives. It notes that the effect may be positive, negative or just a deviation from the expected outcome. It also points out that risk is often described by an event, a change in circumstances or a consequence of some other change. This can be regarded as an important shift of emphasis from possibility of an *event* to the possibility of an *effect* and, in particular, an effect on an objective.

Unlike other frameworks, ISO 31000:2009 holds organisations, as well as managers, accountable for risk management. It also emphasises the fact that risk management is intended to create value. In other words, the cost of managing a risk should not exceed the saving arising from that risk, the saving being measured in a reduction in cost, an increased availability of capital or some other metric.

The risk management principles in ISO 31000:2009 can be summarised as follows:

- risk management should both create and protect value;
- it should be an integral part of all processes in an organisation;
- as such, it should also form a part in decision-making processes;
- it should address uncertainty explicitly;
- the processes of risk management should be carried out in a systematic, structured and timely manner;
- decisions taken should be based on the best available information;
- the approach to risk management should be tailored to the specific nature of the organisation;
- this means it should take into account all human and cultural factors;
- the approach should also be transparent, inclusive and relevant;
- it should not be static – the process should be dynamic, iterative and should respond to changing needs; and
- it should facilitate the improvement of an organisation on a continuous basis.

The risk management framework is driven by the mandate and commitment given by the board of an organisation, as shown in Figure 19.7. As this diagram also shows, such an undertaking allows the design of a framework which, once implemented, can be monitored, reviewed and improved on a continual basis.

The process for managing risk is clearly very similar to that described in AS/NZS 4360:2004, as shown in Figure 19.8. A notable difference is

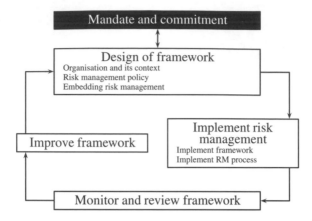

Figure 19.7 Framework for Managing Risk, adapted from IRM/AIRMIC/Alarm: *A Structured Approach to Enterprise Risk Management (ERM) and the Requirements of ISO 31000*, (2009)

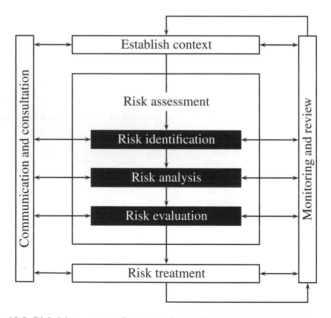

Figure 19.8 Risk Management Process, adapted from IRM/AIRMIC/Alarm: *A Structured Approach to Enterprise Risk Management (ERM) and the Requirements of ISO 31000*, (2009)

that monitoring and review replaces residual risk reporting, and replaces it as a more integral part of the process. In fact, reporting and disclosure are mentioned only briefly in ISO 31000.

One aspect of this process that merits further discussion is the area of risk treatment. The usual list of four approaches is expanded to seven:

- avoiding the risk by not starting or continuing with the activity that gives rise to the risk;
- taking or increasing the risk in order to pursue an opportunity;
- removing the source of the risk;
- changing the likelihood of the event;
- changing the consequences of the event;
- sharing the risk with another party or parties; and
- retaining the risk by informed decision.

19.3 Proprietary risk frameworks

Credit rating agencies have been mentioned a number of times already. They are stakeholders in a risk management context to the extent that the provide information to an institution's investors and bondholders. The quantitative approaches used by some rating agencies and the way in which credit ratings can be used in other approaches have also been discussed. However, qualitative methods used by most credit rating agencies to rate bond issuers constitute risk management frameworks in themselves. It is therefore worth looking at their approaches in this light to see what insights can be derived. The various long-term credit rating scales used are summarised in Table 19.3.

19.3.1 Fitch

Fitch Ratings was established as the Fitch Publishing Company in 1913, introducing its rating scale in the 1920s. It became a global firm when it merged with the UK-based IBCA in 1997.

When Fitch are rating companies, they consider both qualitative and quantitative factors, gaining information from public sources and through private meetings with issuers. Analysts' research is assessed by a ratings committee, which gives a rating based on a consensus decision. The exact criteria for each firm are determined by each rating team, but the factors below are considered to a greater or lesser extent.

Industry risk and operating environment
Ratings are determined in the context of industry fundamentals. For example, an industry may be in decline or thriving, may have high levels of competition

or significant barriers to entry, may be capital or labour intensive, may be inherently risky or safe. Overlaid onto this aspect of the analysis are differences in financial, management and country risk profiles affecting each industry.

Company profile

As well as considering the nature of industries, the firms position in an industry is clearly important. This includes its ability to prosper in the face of competition through product innovation and diversity. Size can help, since larger firms are more likely to benefit from economies of scale and greater financial flexibility, but these positive attributes must be displayed – large firms are not given credit simply for being large.

Management strategy and corporate governance

Analysts focus on areas such as corporate strategy, risk tolerance and funding policies. Capital structure is regarded as important due to the impact that this can have on cash flows. There is also a focus on the corporate governance of firms, since this has an impact on the effectiveness with which strategies are implemented.

Ownership and group structure

Fitch also take into account the relationship between issuer of a bond and its parent company. This is important as the extent to which a subsidiary can rely on the financial report of its parent – and the extent to which a parent might draw on funds from its subsidiary – can have an impact on the creditworthiness of the subsidiary.

Financial profile

Much of the focus on a firm's financial profile is on its cash flow. This is regarded as important as it has an influence over likelihood of a firm being able to raise further funds. A firm's cash flow is linked to its capital structure, so this too is considered, especially in relation to industry norms. In this context, preferred stock is treated as quasi-debt. Contingent liabilities and pensions also receive special attention.

Finally, accounting policies are considered. This does not mean that accounts are re-audited, but there is an analysis of the methods used to construct the accounts.

19.3.2 Moody's

Moody's Investors Service was incorporated in 1914 and by 1924 was producing Moody's Ratings covering almost all of the US bond market.

Table 19.3. *Comparison of credit rating scales*

	Fitch		Moodys		Standard & Poors	
Investment grade	AAA	Highest credit quality	Aaa	Minimal credit risk	AAA	Extremely strong
	AA	Very high credit quality	Aa	Very low credit risk	AA	Very strong
	A	High credit quality	A	Low credit risk	A	Strong
	BBB	Good credit quality	Baa	Moderate credit risk	BBB	Adequate
Speculative grade	BB	Speculative	Ba	Substantial credit risk	BB	Less vulnerable
	B	Highly speculative	B	High credit risk	B	More vulnerable
	CCC, CC,C	High default risk	Caa	Very high credit risk	CCC	Vulnerable
	DDD, DD,D	Default	Ca	Highly speculative	CC	Highly vulnerable
	C		C	Default	R	Regulatory supervision
					SD	Selective default
					D	Default
Modifiers	+,-		1, 2, 3 (1 high, 3 low)		+,-	

Source: www.fitch.com, www.moodys.com and www.standardandpoors.com

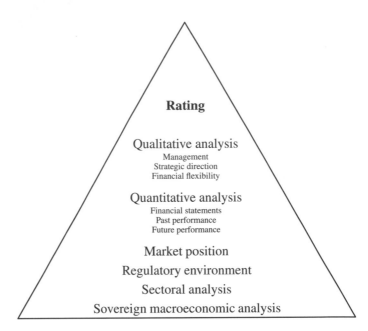

Figure 19.9 Moody's Rating Analysis Pyramid, from Moody's Investor Services: *Moody's Approach to Rating the Petroleum Industry*, (2003)

Analysts use both top–down and bottom–up approaches when analysing firms. The starting point is the macro-economic picture, where political, economic and industry considerations are covered. Analysts then move onto an assessment of a company's operating and competitive position.

The process can be illustrated in the form of a pyramid, as shown in Figure 19.9.

The exact approach used differs from sector to sector, but in all cases qualitative and quantitative aspects are combined with views on market-wide and industry trends.

19.3.3 Standard & Poors

Standard and Poor's has its origins in 1860, but it was formed by the merger of Standard Statistics and Poor's Publishing Company in 1941.

Firms are assessed by an analyst, with each firm having a particular individual assigned to that firm. The exact rating system used varies from analyst to analyst, but the principals are broadly the same. All of the factors discussed below are combined into ratings using a subjective approach, but a rating is given only if it is believed that there is sufficient information for credible analysis to be carried out.

The Standard and Poor's rating framework is divided into business analysis and financial analysis. However, the credit rating for a corporate entity can also depend on the creditworthiness of the country in which a firm is based.

Sovereign risk

Countries can demand cash flows before they are distributed to overseas creditors, often through the mechanism of taxation. They can also put in place currency controls, or impose other regulations that make it hard for firms to pay creditors. These can either be direct restrictions or general restraints on the extent to which a firm can carry out its trade. Governments also have a broader impact on the environment in which a firm must operate and, therefore, its profitability.

Business risk

The analysis of business risk starts with a rating of each company's environment, particularly in relation to the industry in which the firm operates. The prospects for the industry determine the baseline prospects for a firm. The industry considerations take into account long-term prospects for growth, stability or decline, but also cyclical factors. Having said this, the ratings remain long-term outlooks.

The rating process also allows for the benefit of diversification that might be enjoyed by a firm that is involved in a number of industries. Larger firms are also given credit for the fact that their size can give them a degree of resilience.

This leads to the wider issue of a firm's position in the industries in which it operates, and the competitive position is also taken into account by the rating agency. This is itself driven by aspects of the firm's operation, such as marketing, the use of technology and overall efficiency.

These factors are, of course, in the control of the management, and an assessment of management quality is an important factor in the rating process. Much of this analysis is based on impressions formed in meetings between analysts and the management, but the organisational nature of the management structure is also considered. This means that factors such as over-reliance on key individuals and the relationship between various operational and financial areas of the firm are taken into account.

Financial risk

The starting point when analysing financial risk is the consideration of a range of financial ratios. The value of a ratio that is consistent with a particular credit rating depends on the industry in which a firm is based. The ratios are based on audited data and, whilst accounting issues are reviewed, the audited data are taken to be correct.

At least as important to Standard and Poor's is the attitude that management takes to financial policy. This includes aspects such as leverage targets, but also more in-depth consideration of issues such as liquidity management.

Much of the ratio analysis concerns profitability. The level of profitability measured by the coverage of various outgoings is important, as is the volatility of this profit. Furthermore, analysts are also interested in how the firm plans to achieve profit growth. However, even a profitable firm can fail if it does not have adequate cash flow, so a range of cash flow ratios are also considered.

Capital structure is a particular focus of analysts, and the measures of leverage used consider both long- and short-term debt financing. The treatment of hybrid asset classes, such as preferred equity, varies between types and between firms. Measures also take into account off-balance sheet financing, such as operating leases on machinery and even guarantees. On the other side of the coin, guarantees to creditors, in the form of covenants applying to particular issues of debt, are also allowed for.

The degree to which a firm has financial flexibility is also considered. This relates to the extent to which a firm can alter its sources of funding in times of financial stress.

19.4 Further reading

All of the risk frameworks described above are worthy of further inspection. Information of Basel II (and III) is available at www.bis.org/bcbs, whilst details of Solvency II can be found at http://ec.europa.eu/internal_market/ solvency. Many consultants also provide free commentary on these frameworks. There are also a number of books on both banking and insurance standards frameworks, such as Sandström (2006) and Cruz (2009).

The advisory frameworks are also worth reading in their original forms. Some, such as the Orange Book are available free (www.hm-treasury.gov.uk/d/orange_book.pdf), whilst others are only available to purchase.

Finally, the credit rating agencies often make significant detail on their methodologies available through their websites, although registration may be required.

20

Case studies

20.1 Introduction

One way to help understand ERM is to use case studies. These can illustrate the issues faced in real organisations, and the causes of a range of risk management failures. It is, unfortunately, the failures that make up the majority of case studies. This is mainly because no-one ever hears about many successful risk management initiatives. If an investment banker fails to make increasingly desperate trades because it is impossible to hide any resulting losses in a hidden trading account, then the good design of the risk management protocols will attract little attention; however, the absence of such protocols and the bankruptcy of the banker's employer will make the news and can give valuable insights into how things should not be done.

The majority of the case studies here relate to financial institutions, since these are the ones that can be related most closely to the principles in this book. However, some non-financial examples are also included, since they highlight risk management issues that face all organisations, not just those in the financial services sector.

The information for this chapter is distilled from a number of books on the various episodes described. I recommend that you read these books, not only to understand the risk management more fully but also because the stories are often compelling in themselves.

20.2 The 2007–2011 global financial crisis

The global financial crisis that began in 2007 has continued into 2011, and may well persist into the future. It has been characterised by a lack of liquidity – particularly funding liquidity – and a corresponding fall in the creditworthiness of firms and governments. Whilst the popular view is that the crisis is the fault

505

of 'the bankers', it is important to understand both the background to the crisis and the particular risk management failures that caused it.

20.2.1 Causes of the crisis

The role of China

A key role in the build-up to the crisis was played by China. Over the last few decades, the Chinese economy has grown very quickly. Much of this growth has been driven by exports to the West. The capacity for growth in China has meant that the rise in production has occurred in parallel with relatively low rates of inflation. More importantly, the large flows of wealth into China – to pay for Chinese exports – did not prompt a strengthening of the Chinese renminbi. Such flows of funds normally cause a currency to appreciate significantly over time, but the People's Bank of China has deliberately maintained a reasonably stable exchange rate with the US dollar. This maintained demand for Chinese exports, but also meant that increasing demand in the West did not lead to price inflation.

The increasing demand was stoked by persistently low rates of interest in the United States and elsewhere, and China played a part here too. The Chinese government needed to invest the income that it was receiving from exports, and a significant proportion of that income was used to buy US treasury bonds. The demand for these bonds was high enough that the price was inflated, and as the price of a bond increases, its yield falls.

The role of housing markets

The low rates of interest on government borrowing meant that banks were also able to maintain low interest rates. This meant that mortgages became cheaper, causing house prices to rise.

The role of regulation

The Gramm–Leach–Bliley Act of 1999 allowed the commercial and retail banks to carry out investment banking activities, something that had been forbidden by the Glass–Steagall Act of 1933. In normal markets, this could be seen as desirable since it allows banks to benefit from the diversification of carrying out both types of business. However, in stressed markets, it means that catastrophic losses in the investment banking arm can adversely affect retail and commercial account holders. This is what happened in the recent financial crisis.

The Basel I system of banking regulation also played a part. In particular, it gave banks an incentive to convert credit risk in respect of mortgages to market risk by securitising loans. The resulting vehicles are known as mortgage-backed securities (MBSs). Whilst this was a form of regulatory arbitrage, it was not discouraged, except through removing the need for arbitrage by improving the allowance for credit risk under Basel II. Indeed, the ability to package risk and spread it around the market was seen as an important diversification tool. Furthermore, the tranching approach used in collateralised mortgage obligations (CMOs) – the MBS version of a CDO – was seen as a good way to allow investors with different risk appetites to gain exposure to a single pool of risk. However, since the main buyers of these securities were other banks, they were linked to each other through exposure to the housing market in a way they had not been before. Furthermore, the holder of an MBS does not have the same level of information on the borrowers underlying the security as the bank that sold the mortgages does.

The role of incentives

Securitisation also had an adverse effect on incentives. Until the advent of securitisation, a bank making a loan would hold the risk for that loan on its balance sheet. However, securitisation meant that once a loan had been taken on, the profit could be capitalised by packaging the loan – and much of the risk – into an MBS. This resulted in a reduced incentive to ensure the creditworthiness of borrowers and an overall decline in credit quality.

The pricing of CMO tranches was determined by credit rating agencies. Although there may have been concerns over the models used, as discussed below, there was an incentive for banks to exploit this mis-pricing by retaining particular (under-priced) CMO tranches (usually the equity tranche) and selling the (over-priced) remaining tranches.

There are also incentive issuers in terms of those structuring and trading MBSs and CMOs:

- a significant proportion of earnings is paid as bonus;
- bonuses are based on results over a relatively short period (often one year); and
- bonuses are based to a significant extent on team rather than individual performance.

The fact that a significant proportion of earnings are paid as bonus gives employees an incentive to take significant risks to earn these bonuses – if bets yield a positive result, a high bonus is paid; if they yield a negative result, then the result is no bonus, but the minimum bonus is zero rather than a negative number.

Because bonuses are often based on short-term results, there is an incentive to make short-term profits without considering the long-term impact of a trade. This is particularly true if the profit used to calculate bonuses requires assumptions in respect of future outcomes. In particular, if an MBS or CMO is bought or sold, then it is possible to calculate the profit earned on this trade at the point it is created; however, since the ultimate profit depends on the extent to which the borrowers underlying the security are able to make payments, the final result may be very different.

Finally, if bonuses are based on team performance, then there is little incentive to do anything differently from the rest of the team – a good (but different) individual result will not have a huge impact on an individual's bonus, but a bad (and different)individual result could result in being fired. If an individual copies the rest of the team, then that individual will share fully in the good times and will have protection in the bad – it is rare that an entire team is sacked.

The role of models

One common accusation in relation to the crisis is that the models used to price CDOs and CMOs were not sufficiently accurate. In particular, the Gaussian copula has been singled out. This is perhaps true for some market participants – particularly credit rating agencies – but many individuals recognised the short-comings of this copula when used in time-until-default models. Many also improved upon this assumption – but the bonus structure described above might have meant that many did not, if using a Gaussian copula gave higher expected profits and thus bonuses.

However, it is in any case important to recognise that even improved models cannot exactly replicate the world. As such, they should be treated only as guides to what might happen. Excessive reliance on the output of models by senior management, and a lack of understanding of the models' limitations, was at least as important as any shortcomings in the models themselves and perverse incentives.

Organisational issues

Many banks were exposed to housing market risks both directly through mortgages and indirectly through MBSs. However, it seems that senior managers did not recognise this concentration of risks. Furthermore, whilst some departments of banks were selling MBSs and CMOs, others were buying them. This concentration was similarly missed.

20.2.2 Evolution of the crisis

Many of the mortgages sold in the housing boom were to sub-prime borrowers. This meant that they had a higher-than-average risk of defaulting on those mortgages. If such a risk had been correctly priced through a sufficiently high interest rate, then this would not necessarily have caused a problem. However, as discussed above, securitisation led to a reduced incentive to check creditworthiness and indeed to correctly price risk. The situation was made more precarious by the sale of mortgages with low initial rates that rose sharply after a couple of years. Since again the incentive was to sell as many mortgages as possible, risks were not properly explained to borrowers. This led, inevitably, to a sharp rise in mortgage defaults.

The widespread exposure to mortgages meant that many banks made large losses. However, the complexity and opacity of some of the products used meant that many banks could not easily quantify their exposure to future losses. This made banks reluctant to lend to each other. Because banks rely on short-term funding to remain solvent, such a fall in funding liquidity left many banks on the brink of insolvency. Some were actually wound up and many others required government funding to remain solvent. The government assistance came in two forms. The first was to purchase certain illiquid assets from banks in order that they might exchange illiquid assets for liquid ones. However, some governments also provided cash in exchange for equity stakes in banks, in some circumstances going as far as complete nationalisation.

The reduced solvency of the banks meant that their own ability to lend was compromised. As a result, the liquidity crisis spread from the financial sector to the wider economy, as firms and individuals found it harder to borrow. When loans and mortgages were made available, the interest rates charged were higher, to compensate for the higher perceived credit risk, and also to help banks to rebuild their depleted risk capital.

The resulting slow down in economic growth combined with the cost to governments of stabilising financial institutions has resulted in large budget deficits. It has also exposed the structural differences between various Eurozone governments, these differences being reflected in large differences – hundreds of basis points – in the cost of borrowing between governments. Thus the crisis has extended into 2010, and may extend further.

20.2.3 Lessons from the crisis

There are a number of important lessons that might be learned from this crisis, although commentators and stakeholders will have differing views over the extent to which each is valid. However key ones in respect of risk management relate to:

- the organisational structure of banks;
- the capital structure of banks;
- the structure of bank bonuses; and
- the use of models.

Organisational structure

The Gramm–Leach–Bliley Act meant that banks profits could be smoothed over time. However, it also meant that commercial and retail account holders – who generally have a low risk tolerance – were exposed to excessive risks. Should this continue?

Within banks, ERM should have a much higher status, and CROs should have much more power. The CRF should have be able to see all similar risks across a bank and should have the authority to stop undue risk being taken.

Those who design and work with complex models should be given a greater say in the use of those models and should be encouraged to make their limitations known.

Capital structure

Banks will continue to borrow on a short-term basis to fund risks with longer terms. However, they should hold more capital in case the economic outlook changes adversely. The type of capital is also important – it should be liquid. Banks should also ensure that they have contingency plans in case normal sources of liquidity dry up.

Having said this, capital – even liquid capital – is not an alternative to good risk management and should not be regarded as such.

Bank bonuses

Bank bonuses should reflect the term of the instruments being traded. In particular, full bonuses should not be awarded before the risk inherent in any deal has fully run its course.

This principal should also extend to securitisations. If a bank transfers risk to the financial markets, then adequate risk should be retained to give the bank sufficient incentive to ensure the credit quality of the loans the bank sells in the first place. In particular, this risk should be reflected in the term and structure of the bonuses earned by bank employees.

Finally, good risk management – to the extent it avoids large losses – should be as well rewarded as the ability to generate large profits.

Models

Models are essential in risk management. However:

- models should be used as tools – their output should be used to help make decisions, but no more;
- those making decisions using the output from models should understand the model's capabilities and limitations;
- if models are used for a purpose other than that for which they were designed, they should be used with caution; and
- there are some risks for which simple models are better than complex ones – more complex models are not necessarily better.

20.3 Barings Bank

The collapse of Barings Bank in 1995 is the classic example of how organisation failure can allow a single individual to cause catastrophic losses. However, the collapse followed a series of failures by a number of individuals, and it is important to understand the range of factors that allowed these failures to occur.

20.3.1 Background

Nick Leeson was general manager and head trader of Barings Bank's futures trading subsidiary in Singapore. His role as head trader meant that he carried out both proprietary trading on behalf of the bank and trading as a broker on behalf of external clients. However, part of his role as general manger involved oversight of back office activities. This included accounting for trades and reporting to the London head office.

Leeson had two reporting lines. One, in respect of his management and proprietary trading roles was to London, whilst in respect of his broking activities he reported to Barings' Tokyo office.

The proprietary trading carried out by Leeson was supposed to be confined to futures arbitrage. This strategy involved trading Nikkei 225 futures on both the Singapore International Monetary Exchange (SIMEX) and Osaka Securities Exchange (OSE). In particular, short futures positions on SIMEX were supposed to be offset by long positions on the OSE, which long and short trades being put on simultaneously. This would mean that the net exposure to changes in the Nikkei 225 was zero at all times, and the only profit would be arbitrage profit made from temporary mis-pricing. As a further risk control, no positions were supposed to remain in place at the close of business each day.

20.3.2 Development of the collapse

Shortly after moving to Singapore in 1992, Leeson started to try to enhance the returns on the arbitrage strategy by holding net long and short positions in the futures contract – in other words, rather than putting long and short positions on simultaneously, he would put only a single trade on. This exposed Barings to movements in the Nikkei 225, contrary to the objective of Leeson's arbitrage strategy.

The strategy appeared to be very profitable, and as far as Barings in London was concerned, these profits were coming from the arbitrage strategy. Indeed, Leeson was earning several times his basic salary in bonuses each year. However, any losses made by the directional strategy he was actually carrying out were hidden in a secret account, 'Error Account 88888'. Leeson did not report on the results of the account to London and falsified other submissions.

On 17 January 1995, an earthquake struck Kobe, Japan. At the time of the earthquake, Leeson had a significant long position in Nikkei 225 futures. Following the earthquake, the Nikkei 225 fell sharply, causing large losses on Leeson's future positions. To try to recover his losses, Leeson started buying Nikkei 225 futures as the market fell, but the losses continued to accumulate. All were hidden in Error Account 88888.

In order to meet margin calls on the futures, Leeson requested funds from London which were transferred, despite – or perhaps because of – Leeson's apparent profitability. However, despite the continued trading, the losses mounted.

By this point, Leeson had added already another strategy. Despite being authorised to trade options only on behalf of clients in his role as a broker, he had started to use a straddle strategy with Nikkei 225 options. The purpose of this strategy was to accumulate premiums from the sale of put and call options, the premiums being paid into Error Account 88888 in order to offset losses. However, as shown in Figure 20.1, the straddle strategy provides a positive return only if index levels are stable; if the index either falls or rises significantly, then the writer of the options must make payments.

By the end of February 1995, the losses from these strategies had become so large that Leeson could no longer hide them – he fled Singapore, and Barings Bank was declared insolvent shortly afterwards.

20.3.3 Reasons for the collapse

A key reason for the crisis was the fact that Leeson was responsible both for trading and for accounting for those trades. This meant that a whole level of

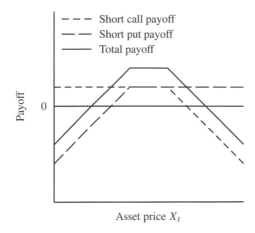

Figure 20.1 Payoff for an option straddle strategy

internal scrutiny was missing. However, internal audits failed to uncover Error Account 88888.

Internal audits did uncover weaknesses in the internal controls in place in Barings' Singapore operation, and there was a recommendation that the general manager – in other words, Leeson – should no longer be responsible for back office operations. However, the appointment of another individual to a part-time role to carry out this oversight did not change the situation in Singapore significantly, and the hidden account was still not uncovered.

There was also insufficient scrutiny from London – either from Barings Bank or the Securities and Futures Association – in what was happening in Singapore. In particular, key warning signs were missed, such as the size of the profits that Leeson was making from a low-risk arbitrage strategy. Because price differentials between futures based on the same underlying index should be small, the scope for profit is also small. There was also insufficient analysis of the size of transfers requested by Leeson to fund his trading.

Another key factor underlying the collapse was the reporting structure. The fact that Leeson had two separate reporting lines – one to London and one to Tokyo – meant that he was better able to hide his trades in Error Account 88888.

It is also worth noting the fact that Leeson was earning six-figure bonuses for his trades, bonuses that were paid out based on performance over a very short period. The bonus structure could well have influenced his desire to move beyond his arbitrage strategy.

20.3.4 Lessons from the collapse

Some of the clearest lessons are that internal and external auditors should carry out their jobs more rigorously. However, even if this had occurred, there were structural issues that could have been addressed.

A clearer and more direct reporting line, with clear responsibility for all aspects of Leeson's work, should have been in place. There should also have been separate parties responsible for trading and for back office work from the outset – once Error Account 88888 had been set up, it was difficult to trace.

There should also have been more robust analysis of the consistency of the profits being made with the strategy being undertaken, and of the reasons for the transfers of funds being required. Arbitrage strategies should be low risk and low return, and the profits being reported were not consistent with such an approach. It appears that the management in London were too pleased by the size of the profits to question too closely where it had come from.

The bonus strategy at Barings might also have played a part in the situation that developed. In particular, a bonus closer in size to the basic salary based on profits over a longer period might have dissuaded Leeson from undertaking such risky trades in the first place.

20.4 Equitable Life

The consequences of the collapse of Equitable Life, the oldest mutual insurance company in the world, are still being felt today. Investors continue to fight for compensation, and many feel that their retirement income was severely damaged by the actions of the firm. However, the operational failures at Equitable Life also led to improvements in the way in which members of the UK Actuarial Profession work.

20.4.1 Background

The Equitable Life Assurance Company was a mutual insurance company that had followed a successful strategy. This was to adopt a bonus strategy that involved distributing as much of its reserves to its with-profits policyholders as possible. As well as holding a limited cushion of free assets, Equitable Life also aimed to ensure that each generation of policyholders received its own asset share – that is, its accumulated contributions plus investment returns less costs – rather than there being any of smoothing of returns between generations.

This approach differed from most other insurers. As well as accepting a degree of inter-generational transfer, most insurers held a significant free estate to provide security against adverse financial conditions.

The strategy followed by Equitable Life ensured that it wrote a large volume of business – but it did also mean that it was more exposed to adverse market conditions. A significant proportion of the business written was in the form of with-profits pension policies that came with guaranteed annuity rates (GARs). The annuity rate is the rate of income received in exchange for a fixed amount, often quoted in the UK as a rate per £100,000. Therefore a GAR is intended to offer a rate of conversion below which the annuity rate will never fall, whatever the market price of annuities is at retirement. It was, therefore, a call option on an annuity

The policies written were open-ended. This meant that once a policy had been taken out, additional contributions could be invested on the same terms indefinitely. In particular, the GAR in place when a policy was sold would also apply to funds accumulating from these additional contributions.

A large number of these policies were sold in the 1970s and 1980s, when interest rates high. As a result, annuity rates were low – much lower than the rate provided by the GAR. This meant that the option could be regarded as being deeply out of the money – the chances of an option being exercised would have appeared small.

In 1988, when personal pensions introduced in the UK, Equitable Life's terminal bonus system was redesigned and GARs were dropped for new contracts. This meant essentially that the policies being sold constituted a new class of with-profits contract. However, policyholders were not told this, and no separate bonus series was started. This was partly because it was thought that the guarantees were unlikely to apply. Even if they did, the directors thought that they would have full discretion over bonuses, and that they would be able to adjust terminal bonuses to recoup the cost of any guarantees.

In 1991, Roy Ranson, Equitable Life's appointed actuary, was also appointed to the role of chief executive. In a proprietary insurance company, this would clearly be an issue – the chief executive acts for shareholders, whilst the appointed actuary – when role existed – had at least some responsibility to policyholders. However, since Equitable Life was a mutual, meaning that policyholders were also essentially shareholders, the issue was less clear-cut. Combining both roles can help to ensure that a consistent view is taken, but a chief executive might be more concerned with new business, whilst an appointed actuary should be looking after the position of existing policyholders. There is also the issue that combining both roles results in too great a concentration of power in a single individual.

20.4.2 Development of the collapse

As interest rates fell during the 1990s, GARs started to appear more attractive, and in 1993, GARs exceeded the annuity rates available in the market. Since Equitable Life had not properly quantified, let alone hedged, the options against it, it was faced with a significant cost.

It sought to deal with this through the terminal bonus which was paid at point of annuitisation. In particular, it awarded different rates of terminal bonus depending on whether a policy had a GAR or not. This meant, in effect, deducting the cost of the guarantee from the terminal bonus.

This was not unprecendented, and Equitable Life had taken similar action – albeit in reverse – in the 1970s, when GARs were lower than market rates. At that time, it increased the terminal bonus payable to those who took annuities at the guaranteed rate. However, when it tried to reduce terminal bonuses for some members, Equitable Life faced a legal challenge from policyholders.

The board of Equitable Life apparently expected the High Court to confirm legality of action on bonuses, which it did. However, the Court of Appeal found in favour of policyholders, as did the House of Lords. The underlying reason given – which was simplified in the final judgement – was that the directors should not be allowed to use their discretion to negate a benefit which a policyholder might reasonably be entitled to expect. The final ruling also left Equitable Life unable to ring-fence its GAR policies. The resulting cost meant that in the first half of 2000 no reversionary bonus was paid. Shortly afterwards, the Equitable Life Assurance Society closed to new business.

20.4.3 Reasons for the collapse

On the face of it, the reasons for the collapse are clear – Equitable Life held insufficient capital; it failed to model accurately the extent of options against it; and failed to take action at the appropriate time to separate different types of policies. However, these more obvious conclusions hide important cultural aspects that allowed mistakes to be made.

Sir Howard Davies, Chairman of the FSA, described Equitable Life as having 'arrogant superiority' in the way it dealt with the regulator, and the Corley Inquiry (Corley, 2001) set up by the Institute and Faculty of Actuaries found that Equitable Life had held an isolated position in the insurance industry. These factors meant that Equitable Life took insufficient notice of changes in the financial landscape that affected all insurance companies. It left the company unwilling to learn from the practice adopted by other firms, and meant that there was insufficient scrutiny of its own business model.

Admittedly, the FSA could have been more robust in the way it dealt with Equitable Life, but culture is determined by senior management. The concentration of the key roles of appointed actuary and chief executive in a single individual would make this even more true. In particular, if a position of isolation was adopted at the top, it is likely that this would be reflected throughout the organisation.

20.4.4 Consequences of the collapse

In the wake of the collapse of Equitable Life, a number of reviews appeared making a range of recommendations. At a basic level, the Corley Inquiry recommended that appointed actuaries be subject to external peer review, but other reports went further.

Not long after the collapse, the FSA replaced the role of appointed actuary with two roles: actuarial function holder and with-profits actuary. This was intended to give much greater protection to with-profits policyholders by removing some of the conflicts that existed when these two roles were combined.

The Morris Review (Morris, 2005) described a degree of insularity in the whole actuarial profession, and the review led to independent regulation for UK actuaries, a new body for the setting of actuarial standards and, importantly, more robust requirements for CPD. In particular, a proportion of CPD must be external – that is, from outside of an employee's firm. This is intended to ensure that actuaries are exposed to a wider range of views. There is also an increasing focus on interdisciplinary research.

Whilst none of these measures can absolutely change the culture of an organisation, they are intended to reduce the extent to which a firm can carry on in isolation from other similar organisations.

20.5 Korean Air

In 1999, the safety record for Korean Air was so bad that Delta Air Lines and Air France suspended their flying partnership with the firm. Whilst the loss rate for United Airlines in the ten years to 1998 was 0.27 aircraft per million departures, for Korean Air it was 4.79 per million departures. The 1997 crash of Korean Air flight 801 in Guam – which killed 228 of the 254 passengers on board – provides a classic example of what went wrong (Gladwell, 2008).

20.5.1 Korean Air flight 801

In August 1997, Korean Air flight 801 was trying to land in bad weather. The glide scope – used by pilots to guide them along the correct trajectory to the runway – was out of action. The captain had therefore decided to follow a radio beacon until the runway was in sight, and then carry out a visual approach.

The captain spotted the airfield from some distance, so decided to carry out a visual approach from some distance instead of locking onto the beacon. However, a localised storm then moved in and the weather worsened dramatically. Worried about the conditions, the flight engineer sought to warn the captain, but only indirectly saying, 'the weather radar has helped us a lot'. This was intended to be a warning, but was made so politely that it was ignored. Similarly, the first officer remarked to the captain, 'don't you think it rains more? Here?' This warning too was ignored.

20.5.2 Reasons for the accident

The roots of the crash can be traced to the hierarchical nature of Korean society. This meant that members of the flight crew were reluctant to challenge openly decisions made by the captain. In other words, the culture resulted in a lack of challenge, even though challenge was obviously needed.

20.5.3 Lessons from the accident

Since 1999, Korean Air has not had a single accident. This is due to a number of changes made by David Greenberg in response to the Guam crash and a number of similar incidents. Greenberg was taken on from Delta Airlines as the director of flight operations for Korean Air in 2000. A key change that Greenberg made was that the only language allowed in the cockpit of a Korean Airlines jet became English. Improving the proficiency of flight crew in English, which is the international language of aviation, has an obvious role in improving safety: it reduces the risk of communication problems between flight crew and air traffic control. However, it also helps in a more subtle way. In particular, by communicating in English, Korean flight crew are able to remove themselves more easily from the roles they might otherwise have in a more traditional Korean setting. This makes them better able to challenge decisions made by other members of the crew.

20.5.4 Implications for financial organisations

The decisions made in financial services firms are rarely a matter of life and death, but the Korean Air example shows the importance of recognising the external context in which business is carried out, the problems the context can cause and at least one way of dealing with the issue.

20.6 Long Term Capital Management

The near-collapse of Long Term Capital Management (LTCM) almost led to a liquidity crisis ten years before the one we currently find ourselves in. However, despite the market events that led to the collapse, this is again a story of operational failure.[1]

20.6.1 Background

LTCM was a US hedge fund set up in 1993. Its partners included Myron Scholes and Robert Merton, each of whom would later win the Nobel Prize for Economics.

The strategy used by the hedge fund was fixed income arbitrage between on-the-run and off-the-run treasury bonds within the US, Japanese and European government bond markets. This involves taking long and short positions in government bonds of a similar term, some of which have been recently issued – described as being 'on-the-run' – and some of which have not been issues for some time – described as 'off-the-run'. On-the-run bonds are generally more liquid than off-the-run bonds, and so are more expensive. This means that if an investor buys an off-the-run bond, whilst simultaneously taking a short position in an on-the-run bond, then an arbitrage profit will be realised if both positions are held to maturity.

However, because this is an arbitrage strategy, the potential profits are small. In order to increase the profits, trades are therefore often levered. This means that money is borrowed and added to the equity capital invested in the trade. The proceeds of the trade are then used to pay off the borrowing, with the excess being profit on the initial equity capital invested. By 1998, LTCM was borrowing $25 for every $1 of capital invested. In other words, they needed to make enough profit on each trade to pay interest on the $25 borrowed as well as making a reasonable return on the $1 of equity.

As LTCM's asset base grew, it became clear that there were not enough potential arbitrage trades to maintain levels of profitability. The partners therefore started exploring a wider range of trades. These included expanding trades from arbitrage *within* bond markets to arbitrage *between* markets. In particular, they used models to describe the long-term relationship between

---

[1] A fuller account of these events is given in Lowenstein (2002).

different government bond markets, and placed trades to exploit any perceived mis-pricing in relation to expected long-term relationships.

However, this was not true arbitrage. Whilst taking positions in on- and off-the-run bonds in a single market can lock an investor into a profit if the position is held until the bonds are redeemed, attempted arbitrage between markets relies on convergence of markets to some long-term norm.

20.6.2 Market events

In 1997, after trying to support its currency against speculative attacks, Thailand decided to allow Thai baht to float freely and break its link with the US dollar. However, under the weight of speculation the currency collapsed.

The same speculators then suspected that the currencies of Thailand's neighbours were exposed to similar weaknesses. To exploit this view, these currencies were sold short, leading to dramatic falls in other South-East Asian currencies with the contagion ultimately spreading as far as Japan. This resulted in Japanese and US bond prices diverging as investors sold Japanese government bonds for the safety of US bonds.

The currency crisis led to a wider financial crisis in Asia, and the fall in economic activity resulted in sharp falls in commodity prices. This ultimately triggered a crisis in Russia, for whom commodities account for a significant amount of national income. As a result of the rapid deterioration in its financial position, Russia defaulted on its government debt in 1998. Fearing that the contagion would spread from Russia across Europe, investors sold European government bonds to buy US treasuries.

All of these movements caused large losses for LTCM. As a result of the losses, LTCM was forced to close out many of its trades. This included arbitrages that would have ultimately converged, but that had to be sold. The closure of positions on this large scale caused prices to move further against LTCM, leading to fears that LTCM might collapse – by this time its leverage ratio had increased from 25–1 to 250–1. Because LTCM had such large positions with so many counter-parties, such a failure could well have caused a wider liquidity crisis. The Federal Reserve Bank of New York therefore arranged a bail-out, funded by a number of large investment banks. LTCM's positions were eventually liquidated, making a small profit for these banks and LTCM was itself liquidated in 2000.

20.6.3 Lessons from LTCM

The obvious issue here seems to be that LTCM was caught out by unexpected market conditions. However, the factors that allowed this to happen are instructive.

LTCM's strategy relied heavily on models. In particular, models were used to make decisions rather than simply inform them. These models underestimated the likelihood of large, adverse movements. LTCM's failure shows that over-reliance on models at the expense of good judgement can be damaging. In particular, the models assumed convergence within a reasonable time period. Whilst many of the strategies did ultimately prove to be profitable, they remained unprofitable for too long – as the economist John Maynard Keynes noted, 'the market can remain irrational longer than you can remain solvent'.

As with Barings, the large profits from what were supposedly arbitrage strategies should have concerned investors, although unlike Barings the change from arbitrage strategies to riskier bond trading was carried out with the full knowledge of partners. However, the presence of such eminent partners encouraged people to invest – investors were star-struck. It it important to objectively assess the returns available from strategies that are being advertised, and it is important to understand what risks are being taken when investing in any fund.

20.7 Bernard Madoff

The case of Bernard Madoff appears at first sight to be a clear-cut case of investors taken in by a fraudster. However, it is important to recognise why investors were taken in and how the fraud was allowed to continue.

20.7.1 Background

Bernard Madoff founded and ran Bernard Madoff Investment Securities LLC. One arm of the company was devoted to asset management for a range of charities, trusts and high net worth individuals. Madoff provided high, stable returns to his investors. Demand for his asset management services was therefore high, and investment was essentially by invitation only.

The strategy that Madoff claimed to be using was a collar strategy. This would have involved buying out-of-the-money put options and out-of-the-money call options on a stock, whilst holding the stock itself, as shown in Figure 20.2. The put options would limit the losses below the strike price of that option, whilst the premiums received on the call options would pay for the puts. Selling the call options would have involved foregoing equity returns above the call option strike price.

However, it ultimately became clear that rather than investing assets, Madoff had in fact been running a Ponzi or pyramid scheme. This

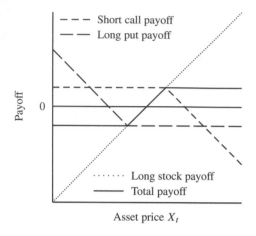

Figure 20.2 Payoff for an option collar strategy

involved funding payments to existing investors from contributions made by new investors. When the fraud was uncovered, many investors were bankrupted.

20.7.2 How the fraud happened

The fraud was allowed to happen partly because Madoff was working in a small organisation, with little or no internal oversight. Furthermore, the only external auditing was carried out by a personal friend who was the only auditor in his three-person firm. At best, this meant that the auditor lacked the expertise to carry out a thorough audit; at worst, the auditor was complicit in the fraud.

However, the fraud also exposes the reluctance of potential investors to analyse to any extent Madoff's investment strategy. Those that did found such inconsistencies in the trades carried out that they not only avoided investing but also notified the regulator. However, it took Madoff declaring the fraud himself for action to be taken.

20.8 Robert Maxwell

The theft of assets by Robert Maxwell from the pension schemes of companies in his group caused a number of changes in the law. However, it is important to recognise the cultural and organisational issues underlying the episode since these can still occur today.

20.8.1 Background

Robert Maxwell, owner of the Mirror Group died suddenly in 1991. Following his death, it became apparent that assets had been misappropriated from a number of pension schemes in Maxwell's group of companies in an attempt to support the group. These transfers had not been authorised by the trustees, and would have been illegal anyway. The transfers had happened despite the oversight of the scheme's auditors, and it was many years before pension scheme members' benefits were finally secured.

20.8.2 How the fraud happened

The fraud was possible because a number of key positions in the group and on the trustee boards were held by the same small group of individuals. It was made easier because all assets were held by an in-house manager, Bishopsgate Investment Management.

Maxwell himself was an imposing man, and his dominance of the group could easily have ensured that assets were transferred when this would not have been allowed elsewhere.

20.8.3 Consequences of the fraud

The UK government commissioned a report by Professor Roy Goode and many of the recommendations in this report found their way into the 1995 Pensions Act. Many of the recommendations were aimed at improving the independence of trustees, such as the requirement to have a third of all trustees nominated by scheme members. The act also clarified that the only power of investment that could be delegated was the management of assets, which could be given to one or more investment managers. However, it required that other powers be delegated. Trustees were no longer allowed to act as either auditor or actuary to the scheme, and two new statutory roles were created: scheme actuary and scheme auditor.

Despite this, the most important step that can be taken is to avoid the concentration of risk at the head of a company. When a company is run in such a dominant way, the risk that other directors and managers will cave in to unwarranted requests is severely increased.

20.9 Space Shuttle Challenger

Following the loss of the Space Shuttle Challenger in 1986, a presidential commission was set up under the chairmanship of William Rogers. One of the twelve members of the commission was the Nobel Prize-winning physicist Richard Feynman. Feynman insisted on a separate appendix being added to the report containing his personal observations. These give some interesting insights into organisational issues and perceptions of risk that affected the National Aeronautics and Space Administration (NASA), but that could affect any organisation[2].

20.9.1 Background

On 28 January 1986, the Space Shuttle Challenger broke apart, 73 seconds after take-off. The cause of the accident was determined to be the failure of a rubber seal – an 'o-ring' – in one of the booster rockets. The night before launch had been particularly cold, and the cold weather had made the o-ring less elastic and so less able to provide a good seal. Fuel had escaped from the booster rocket, ignited and destroyed the shuttle.

20.9.2 Organisational failures

The key underlying failure at NASA related to management's perception of risk. For example, NASA engineers estimated the risk of failure of a shuttle flight as around 1%, whilst management estimated it as around 0.001%. The latter figure was arrived at by adding together the assumed rates of failure for each component. However, there was too little experience to determine with any accuracy the failure rates of individual components, so assumptions as low as 0.0000001 were used, with little evidence.

The definition of failure was also flawed. In particular, if a component did not behave as it was designed to, but the behaviour did not cause an accident, then this behaviour was not deemed a failure. The o-ring leaks fell into this category, since leaks had occurred on a number of previous flights. This failure was compounded by the fact that each launch that did not result in the loss of a craft was regarded as an argument in favour of the safety of components.

More direct management failures also occurred. For example despite the manufacturers of the o-rings cautioning against a launch after cold weather due to the risk of leaks, NASA countered that, since leaks had occurred when the weather had been warmer, there was no reason to suspend a launch on the grounds of low temperature.

Feynman believed that the problem stemmed from NASA managers wanting to deliver, and to be seen to deliver. Whilst this was a fine motive, it may

---

[2] The appendix together with Feynman's thoughts on his time with the commission are included in Feynman (2002).

have led to managers making promises that the engineers did not think they could keep.

20.9.3 Implications for financial organisations

The loss of a space shuttle might seem some way from the challenges faced by financial services firms, but there are a number of lessons that could be learned. How often do those working more directly with financial models understand the risks better than their managers? How often do managers choose to ignore warnings that conflict with their aims to increase profits? Financial risk management is often compared with rocket science, and the risk management failures that occurred with the loss of the Space Shuttle Challenger show just how close this analogy is. However, by learning from past failures, it is possible to limit the risk of future losses in both industries.

20.10 Conclusion

Despite these case studies covering a wide range of failures, a number of factors occur over and over again.

First, people seem willing to invest in high-returning investment opportunities without determining with any rigour the source of these returns. This is even more true if the returns are accompanied by a big name, an element of exclusivity or some other factor that makes investors feel lucky to have the opportunity to invest. The maxim that is being forgotten here is that if something looks too good to be true, it probably is.

A second feature seen more than once is the importance of culture. Too strict a hierarchy can limit useful communication; too strong a leadership can stifle dissent; and too isolated a position can restrict the flow of useful external ideas. There is also a link to the earlier point about thorough investigation – any culture that does not encourage a thorough questioning of new ideas and an open discussion of the risks involved can easily find itself in dire financial straits.

Finally, the importance of incentives is important. It is important not to have any remuneration structure in place that encourages excessive risk taking, short-termism or even fraud. Although competitive pressures can make it difficult, the importance of designing a remuneration strategy that creates the right incentives cannot be overstated.

20.11 Further reading

Some of the case studies in this chapter can be found as parts of other books. The Korean Air story is described in more detail by Gladwell (2008), whilst the first-hand account of the Rogers Commission following the Challenger disaster can be found in Feynman (1988). Dennett (2004) includes some discussion of the Maxwell scandals and the problems with Equitable Life.

There are also many books devoted to other financial disasters. One of the best is the account of LTCM given in Lowenstein (2002), which provides a number of important insights. Patterson (2010), on the other hand, gives an entertaining account of the role of hedge funds in the recent liquidity crisis together with a history of the hedge fund industry itself.

Moving away from the topics in this chapter, the definitive book on the Enron scandal is by Elkind and McLean (2004). There is a useful book written by Jorion (1995) on the failures that led to the Orange County disaster, the largest municipal failure in US history. However, good books on these topics go all the way back to Galbraith (2009) and the updated version of his 1954 book on the 1929 financial crash.

References

Akerlof, G.A. 1970. The market for 'lemons': Quality uncertainty and the market mechanism. *Quarterly Journal of Economics*, **84**(3), 488–500.

Altman, E.I. 1968. Financial ratios, discriminant analysis and the prediction of corporate bankruptcy. *Journal of Finance*, **23**(4), 589–609.

Andersen, L.B.G. and Piterbarg, V.V. 2010a. *Interest Rate Modeling. Volume 1: Foundations and Vanilla Models*. London: Atlantic Financial Press.

Andersen, L.B.G. and Piterbarg, V.V. 2010b. *Interest Rate Modeling. Volume 2: Term Structure Models*. London: Atlantic Financial Press.

Andersen, L.B.G. and Piterbarg, V.V. 2010c. *Interest Rate Modeling. Volume 3: Products and Risk Management*. London: Atlantic Financial Press.

Arcot, S.R. and Bruno, V. 2006. *In Letter but not in Spirit – An Analysis of Corporate Governance in the UK*. Social Science Research Network working paper.

Artzner, P., Delbaen, F., Eber, J.-M. and Heath, D. 1999. Coherent measures of risk. *Mathematical Finance*, **9**(3), 203–228.

Auditing Practices Board. 2008. *Ethical Standards*. London: Auditing Practices Board.

Azzalini, A. and Capitanio, A. 2003. Distributions generated by perturbation of symmetry with emphasis on a multivariate skew *t*-distribution. *Journal of the Royal Statistical Society Series B*, **65**, 367–389.

Bagehot, W. 1972. Risk and reward in corporate pension funds. *Financial Analysts Journal*, **28**(1), 80–84.

Bahar, R. and Brand, L. 1998. *Recoveries on Defaulted Bonds Tied to Seniority Rankings*. Standard and Poor's.

Bailey, J.V. 1992a. Are manager universes acceptable performance benchmarks? *Journal of Performance Management*, **18**(3), 9–13.

Bailey, J.V. 1992b. Evaluating benchmark quality. *Financial Analysts Journal*, **48**(3), 33–39.

Barrieu, P. and Albertini, L. (eds). 2009. *The Handbook of Insurance-Linked Securities*. Chichester: John Wiley & Sons Ltd.

Basel Committee on Banking Supervision. 1988. *International Convergence of Captial Measurement and Captial Standards*, Bank for International Settlements, Switzerland.

Basel Committee on Banking Supervision. 1996. *Amendment to the Capital Accord to Incorporate Market Risks*, Bank for International Settlements, Switzerland.

Basel Committee on Banking Supervision. 2001. *Working Paper on the Regulatory Treatment of Operational Risk*, Bank for International Settlements, Switzerland.

Basel Committee on Banking Supervision. 2003. *Sound Practices for the Management and Supervision of Operational Risk*, Bank for International Settlements, Switzerland.

Basel Committee on Banking Supervision. 2004. *International Convergence of Capital Measurement and Capital Standards – A Revised Framework*, Bank for International Settlements, Switzerland.

Basel Committee on Banking Supervision. 2008. *Principles for Sound Liquidity Risk Management and Supervision*, Bank for International Settlements, Switzerland.

Basel Committee on Banking Supervision. 2010. *Basel III: A Global Regulatory Framework for more Resilient Banks and Banking Systems*, Bank for International Settlements, Switzerland.

Baxter, N.D. 1967. Leverage, risk of ruin and the cost of capital. *Journal of Finance*, **22**(3), 395–403.

Besar, D., Booth, P., Chan, K.K., Milne, A.K.L. and Pickles, J. 2009. Systemic risk in financial services. Paper presented to the Institute of Actuaries.

Bhattacharya, U. and Daouk, H. 2002. The world price of insider trading. *Journal of Finance*, **57**(1), 75–108.

Birla, K.M. 2002. *Report of the Kumar Mangalam Birla Committee on Corporate Governance*. Securities and Exchange Board of India.

Black, F. 1980. The tax consequences of long-run pension policy. *Financial Analysts Journal*, **36**(4), 21–28.

Black, F. and Karasinski, P. 1991. Bond and option prices when short rates are lognormal. *Financial Analysts Journal*, **47**(4), 52–59.

Black, F. and Litterman, R. 1992. Global portfolio optimization. *Financial Analysts Journal*, **48**(5), 28–43.

Black, F. and Scholes, M. 1973. The pricing of options and corporate liabilities. *Journal of Political Economy*, **81**(3), 637–654.

Blackburn, R. 2002. *Banking on Death, or Investing in Life – the History and Future of Pensions*. London: Verso.

Blake, D. 2003. *Pension Schemes and Pension Funds in the United Kingdom*. Oxford: Oxford University Press.

Blake, D. 2005. Pension funds need three pillars too – What the pension protection fund can learn from the financial regulation of banks and insurance companies. Paper presented to the Savings, Pension Provision and Retirement Programme at the University of Exeter.

Blake, D. and Burrows, W. 2001. Survivor bonds – Helping to hedge mortality risk. *Journal of Risk and Insurance*, **68**(2), 339–348.

Blomqvist, N. 1950. On a measure of dependence between two random variables. *Annals of Mathematical Statistics*, **21**(4), 593–600.

Board of Banking Supervision. 1995. *Report of the Board of Banking Supervision Inquiry into the Circumstances of the Collapse of Barings.*

Bodie, Z., Light, J.O., Mørck, R. and Taggart, R.A. 1987. Funding and asset allocation in corporate pension plans – an empirical investigation. In Z. Bodie, J.B. Shoven and D.A. Wise (eds), *Issues in Pension Economics.* Chicago, Illinois: University of Chicago Press.

Box, G. and Jenkins, G. 1970. *Time Series Analysis: Forecasting and Control.* San Francisco, California: Holden-Day.

Branco, M.D. and Dey, D.K. 2001. A general class of multivariate skew elliptical distributions. *Journal of Multivariate Analysis*, **79**(1), 99–113.

Brennan, M.J. and Schwartz, E.S. 1982. An equilibrium model of bond pricing and a test of market efficiency. *Journal of Financial and Quantitative Analysis*, **17**(3), 301–329.

Brouhns, N., Denuit, M. and Vermunt, J.K. 2002. A Poisson log-bilinear regression approach to the construction of projected life-tables. *Insurance: Mathematics and Economics*, **31**(3), 373–393.

BSi British Standards. 2008. *BS 31100:2008 – Risk Management Code of Practice.*

Buckmaster, D. and Saniga, E. 1990. Distributional forms of financial accounting ratios – Pearson's and Johnson's taxonomies. *Journal of Economic and Social Measurement*, **16**(3), 149–166.

Bühlmann, H. and Gisler, A. 2005. *A Course in Credibility Theory and Its Applications.* Berlin, Germany: Springer.

Burtschell, X., Gregory, J. and Laurent, J.-P. 2009. A comparative analysis of CDO pricing models under the factor copula framework. *Journal of Derivatives*, **16**(4), 9–37.

Cadbury, A. 1992. *Report of the Committee on the Financial Aspects of Corporate Governance.* London: Committee on the Financial Aspects of Corporate Governance.

Cairns, A.J.G. 2000. A discussion of parameter and model uncertainty in insurance. *Insurance: Mathematics and Economics*, **27**(3), 313–330.

Cairns, A.J.G. 2004. *Interest Rate Models: An Introduction.* Princeton, New Jersey: Princeton University Press.

Cairns, A.J.G., Blake, D. and Dowd, K. 2006. A two-factor model for stochastic mortality with parameter uncertainty: theory and calibration. *Journal of Risk and Insurance*, **73**(4), 687–718.

Cairns, A.J.G., Blake, D., Coughlan, G., Epstein, D., Ong, A. and Balevich, I. 2009. A quantitative comparison of stochastic mortality models using data from England and Wales and the United States. *North American Actuarial Journal*, **13**(1), 1–35.

CFA Institute. 2005. *Code of Ethics and Standards of Professional Conduct.*

Chapman, R.J. 2006. *Simple Tools and Techniques for Enterprise Risk Management.* Chichester: John Wiley & Sons Ltd.

Chernow, R. 2010. *The House of Morgan: An American Banking Dynasty and the Rise of Modern Finance.* New York: Grove Press.

Cherubini, U., Luciano, E. and Vecchiato, W. 2004. *Copula Methods in Finance*. Chichester: John Wiley & Sons Ltd.

Chief Risk Officer Forum. 2008. *Liquidity Risk Management – Best Risk Management Practices*, 29 October.

Chow, G.C. 1960. Tests of equality between sets of coefficients in two linear regressions. *Econometrica*, **28**, 591–605.

Committee of Sponsoring Organizations of the Treadway Commission. 2004. *Enterprise Risk Management – Integrated Framework*.

Copeland, T.E., Weston, J.F. and Shastri, K. 2004. *Financial Theory and Corporate Policy*. Reading, Massachusetts: Addison-Wesley.

Corley, R.D. 2001. *Report of the Corley Committee of Inquiry Regarding the Equitable Life Assurance Society*. London: Institute and Faculty of Actuaries.

Cortes, C. and Vapnik, V. 1995. Support vector networks. *Machine Learning*, **20**(3), 273–297.

Cox, J.C., Ingersoll, J.E. and Ross, S.A. 1985. A theory of the term structure of interest rates. *Econometrica*, **53**(2), 385–407.

Crosbie, P. and Bohn, J. 2003. *Modeling Default Risk*. Moody's KMV.

Crouzet, F. 1982. *The Victorian Economy*. London: Methuen.

Cruz, M. (ed.). 2009. *The Solvency 2 Handbook: Developing Enterprise Risk Management Frameworks in Insurance and Reinsurance Companies*. London: Risk Books.

de Haan, L. and Ferreira, A. 2006. *Extreme Value Theory – An Introduction*. New York: Springer Science & Business Media.

de Jong, P. and Heller, G.Z. 2008. *Generalized Linear Models for Insurance Data*. Cambridge: Cambridge University Press.

de Servigny, A. and Renault, O. 2004. *Measuring and Managing Credit Risk*. New York: McGraw Hill.

DeAngelo, H. and Masulis, R.W. 1980. Optimal capital structure under corporate and personal taxation. *Journal of Financial Economics*, **8**(1), 3–29.

Dennett, L. 2004. *Mind over Data – an Actuarial History*. Cambridge: Granta Editions.

Dey, P. 1994. *Where Were the Directors? Guidelines for Improved Corporate Governance in Canada*. Toronto Stock Exchange.

Dickey, D.A. and Fuller, W.A. 1979. Distribution of the estimators for autoregressive time series with a unit root. *Journal of the American Statistical Association*, **74**, 427–431.

Dickey, D.A. and Fuller, W.A. 1981. Likelihood ratio statistics for autoregressive time series with a unit root. *Econometrica*, **49**(4), 1057–1072.

Dickson, D.C.M., Hardy, M.R. and Waters, H.R. 2009. *Actuarial Mathematics for Life Contingent Risks*. Cambridge: Cambridge University Press.

Dimson, E., Marsh, P. and Staunton, M. 2002. *Triumph of the Optimists – 101 Years of Global Investment Returns*. Princeton, New Jersey: Princeton University Press.

Dowd, K. 2003. Survivor bonds – A comment on Blake and Burrows. *Journal of Risk and Insurance*, **70**(2), 339–348.

Dowd, K. 2005. *Measuring Market Risk*. Chichester: John Wiley & Sons Ltd.

Durbán, M., Currie, I. and Eilers, P. 2002. *Using P-splines to Smooth Two-Dimensional Poisson Data*. Proceedings of 17th International Workshop on Statistical Modelling, Crete, 207–214.

Durbin, J. 1970. Testing for serial correlation in least squares regression, when some of the regressors are lagged dependent variables. *Econometrica*, **38**, 410–421.

Durbin, J. and Watson, G.S. 1950. Testing for serial correlation in least squares regression, I. *Biometrika*, **37**, 409–428.

Durbin, J. and Watson, G.S. 1951. Testing for serial correlation in least squares regression, II. *Biometrika*, **38**, 159–179.

Easterbrook, F.H. and Fischel, D.R. 1985. Optimal damages in securities cases. *University of Chicago Law Review*, **52**(3), 611–641.

Eeckhoudt, L., Gollier, C. and Schlesinger, H. 2005. *Economic and Financial Decisions under Risk*. Princeton, New Jersey: Princeton University Press.

Eilers, P.H.C. and Marx, B.D. 1996. Flexible smoothing using splines and penalties. *Statistical Science*, **11**(2), 89–121.

Eilers, P.H.C. and Marx, B.D. 2010. Splines, knots and penalties. *Wiley Interdisciplinary Reviews – Computational Statistics*, **2**(6), 637–653.

Elkind, P. and McLean, B. 2004. *The Smartest Guys in the Room*. London: Penguin.

Elton, E.J., Gruber, M.J., Brown, S.J. and Goetzmann, W.N. 2003. *Modern Portfolio Theory and Investment Analysis*. Hoboken, New Jersey: John Wiley & Sons Ltd.

Embrechts, P., Lidskog, F. and McNeil, A.J. 2001. Modelling dependence with copulas and applications to risk management. Unpublished working paper.

Embrechts, P., McNeil, A.J. and Strauman, D. 2002. Correlation and dependency in risk management: Properties and pitfalls. In M. Dempster (ed.), *Risk Management: Value at Risk and Beyond*. Cambridge: Cambridge University Press.

European Commission. 1973. *First Non-Life Directive*.

European Commission. 1979. *First Life Directive*.

European Commission. 2003a. *Market Abuse Directive*.

European Commission. 2003b. *Pensions Directive*.

European Commission. 2004. *Market and Financial Instruments Directive (MIFID)*.

European Commission. 2006. *Capital Requirements Directive*.

Exley, C.J., Mehta, S.J.B. and Smith, A.D. 1999. Pension funds – A company manager's view. Paper presented to the Joint Institute and Faculty of Actuaries Investment Conference.

Fama, E.F. and MacBeth, J.D. 1973. Risk, return and equilibrium – Empirical tests. *Journal of Political Economy*, **81**(3), 607–636.

Feynman, R.P. 1988. *What do You Care what Other People Think?* London: Penguin.

Financial Services Authority. 2000. *The Combined Code – Principles of Good Governance and Code of Best Practice*, London.

Financial Reporting Council. 2005. *Internal Control – Revised Guidance for Directors on the Combined Code*.

Financial Reporting Council. 2008. *Combined Code on Corporate Governance*, London.

Financial Reporting Council. 2010. *The UK Corporate Governance Code*, London.

Fisher, J.B. 2010. *When Money Was In Fashion: Henry Goldman, Goldman Sachs, and the Founding of Wall Street*. New York: Palgrave Macmillan.

Fisher, R.A. 1936. The use of multiple measurements in taxonomic problems. *Annals of Eugenics*, **7**, 179–188.

Fitch Ratings. 2009. *Corporate Rating Methodology*.

Fréchet, M. 1951. Sur les tableaux de corrélation dont les marges sont données. *Annales de l'Université de Lyon Serie 3*, **14**, 53–77.

Fréchet, M. 1957. Sur les tableaux de corrélation dont les marges et des bornes sont données. *Annales de l'Université de Lyon Serie 3*, **20**, 13–31.

Frees, E.W. 2010. *Regression Modeling with Actuarial and Financial Applications*. Cambridge: Cambridge University Press.

Frees, E.W. and Valdez, E.A. 1998. Understanding relationships using copulas. *North American Actuarial Journal*, **2**(1), 1–25.

Galbraith, J.K. 1929. *The Great Crash 1929*. London: Penguin.

Giesecke, K. 2003. A simple exponential model for dependent defaults. *Journal of Fixed Income*, **13**(3), 74–83.

Gilbart, J.W. 1834. *The History and Principles of Banking*. London: Longman, Rees, Orme, Brown, Green and Longman.

Gladwell, M. 2008. *Outliers – the Story of Success*. London: Allen Lane.

Gompertz, B. 1825. On the nature of the function expressive of the law of human mortality, and on a new mode of determining the value of life contingencies. *Philosophical Transactions of the Royal Society of London*, **115**, 513–585.

Good, I.J. 1969. Some applications of the singular decomposition of a matrix. *Technometrics*, **11**(4), 823–831.

Goode, R. 1993. *Pension Law Reform: The Report of the Pension Law Review Committee*. Pension Law Review Committee.

Gorsky, M. 1998. The growth and distribution of English friendly societies in the early nineteenth century. *Economic History Review*, **51**(3), 489–511.

Graham, J.R. 1996. Proxies for the corporate marginal tax rate. *Journal of Financial Economics*, **42**(2), 187–221.

Green, E. 1989. *Banking – an Illustrated History*. Oxford, England: Phaedon.

Greenbury, R. 1995. *Directors' Remuneration*. Greenbury Committee.

Greene, W.H. 2003. *Econometric Analysis*. Upper Saddle River, New Jersey: Prentice Hall.

Gupton, G.M., Finger, C.C. and Bhatia, M. 1997. *CreditMetrics Technical Document*. J.P. Morgan.

Hamilton, J.D. 1994. *Time Series Analysis*. Princeton, New Jersey: Princeton University Press.

Hampel, R. 1998. *Corporate Governance*. Hampel Committee, European Corporate Governance Institute.

Hannah, L. 1986. *Inventing Retirement – the Development of Occupational Pensions in Britain*. Cambridge: Cambridge University Press.

Harris, M. and Raviv, A. 1990. Capital structure and the informational role of debt. *Journal of Finance*, **45**(2), 321–349.

Haugen, R.A. and Senbet L.W. 1978. The insignificance of bankruptcy costs to the theory of optimal capital structure. *Journal of Finance*, **33**(2), 383–393.

Haugen, R.A. and Senbet, L.W. 1988. Bankruptcy and agency costs – Their significance to the theory of optimal capital structure. *Journal of Finance and Quantitative Analysis*, **23**(1), 27–38.

He, G. and Litterman, R. 1999. *The Intuition Behind Black–Litterman Model Portfolios*. Goldman Sachs Quantitative Resources Group.

Higgs, D. 2003. *Review of the Role and Effectiveness of Non-Executive Directors.* Department of Trade and Industry.

HM Government. 1973. *Social Security Act.*

HM Government. 1975. *Policyholders Protection Act.*

HM Government. 1980. *Companies Act.*

HM Government. 1984. *Health and Social Security Act.*

HM Government. 1985. *Company Securities (Insider Dealing) Act.*

HM Government. 1986a. *Finance Act.*

HM Government. 1986b. *Financial Services Act.*

HM Government. 1986c. *Social Security Act.*

HM Government. 1987a. *Banking Act.*

HM Government. 1987b. *Building Societies Act.*

HM Government. 1988. *Income and Corporation Taxes Act (ICTA).*

HM Government. 1995. *Pensions Act.*

HM Government. 1996. *Employment Rights Act.*

HM Government. 2000a. *Financial Services and Markets Act.*

HM Government. 2000b. *Limited Liability Partnerships Act.*

HM Government. 2000c. *Trustee Act.*

HM Government. 2004a. *Finance Act.*

HM Government. 2004b. *Pensions Act.*

HM Government. 2006a. *Companies Act.*

HM Government. 2006b. *Pension Protection Act.*

HM Government (Canada). 1985. *Canada Business Corporation Act.*

HM Treasury. 2004. *The Orange Book – Management of Risk, Principles and Concepts.*

Ho, T.S.Y. and Lee, S.B. 1986. Term structure movements and pricing interest rate contingent claims. *Journal of Finance*, **41**(5), 1011–1029.

Höffding, W. 1940. Massstabinvariante Korrelationstheorie. *Schriften des Mathematischen Seminars und des Instituts für Angewandte Mathematik der Universität Berlin*, **5**(3), 181–233.

Holland, R.G. and Sutton, N.A. 1988. The liability nature of unfunded pension obligations since ERISA. *Journal of Risk and Insurance*, **55**(1), 32–58.

Hull, J.C. 2009. *Options, Futures and Other Derivatives*. Upper Saddle River, New Jersey: Pearson.

Hull, J.C. and White, A. 1994a. Numerical procedures for implementing term structure models I. *Journal of Derivatives*, **2**(1), 7–16.

Hull, J.C. and White, A. 1994b. Numerical procedures for implementing term structure models II. *Journal of Derivatives*, **2**(2), 37–48.

Institute of Directors. 2008a. *The Duties, Responsibilities and Liabilities of Directors*.

Institute of Directors. 2008b. *The Key Differences between Directors and Managers*.

Institute of Risk Management, Association of Insurance and Risk Managers and ALARM – National Forum for Risk Management in the Public Sector. 2002. *A Risk Management Standard*.

Institute of Risk Management, Association of Insurance and Risk Managers and ALARM – National Forum for Risk Management in the Public Sector. 2009. *A Structured Approach to Enterprise Risk Management (ERM) and the Requirements of ISO 31000*, London: Airmic.

International Actuarial Association. 2004. *A Global Framework for Insurer Solvency Assessment: Report of the Insurer Solvency Assessment Working Party*.

International Actuarial Association. 2008. *Practice Note on Enterprise Risk Management for Capital and Solvency Purposes in the Insurance Industry*.

International Organization for Standardization. 2009. *Risk Management: Principles and Guidelines – ISO 31000:2009*.

Jarque, C.M. and Bera, A.K. 1980a. Efficient tests for normality, homoscedasticity and serial independence of regression residuals. *Economics Letters*, **6**(3), 255–259.

Jarque, C.M. and Bera, A.K. 1980b. Efficient tests for normality, homoscedasticity and serial independence of regression residuals: Monte Carlo evidence. *Economics Letters*, **7**(4), 313–318.

Jensen, M.C. 1986. Agency costs of free cash flow, corporate finance, and takeovers. *American Economic Review*, **76**(2), 323–329.

Jensen, M.C. and Meckling, W.H. 1976. Theory of the firm – managerial behavior, agency costs and ownership structure. *Journal of Financial Economics*, **3**(4), 305–360.

Johnson, R.A. and Bhattacharyya, G.K. 2010. *Statistics: Principles and Methods*. Chichester: John Wiley & Sons Ltd.

Johnston, J. and Dinardo, J. 1997. *Econometric Methods*. New York: McGraw-Hill.

Jones, A.R., Copeman, P.J., Gibson, E.R., Line, N.J.S., Lowe, J.A., Martin, P., Matthews, P.N. and Powell, D.S. 2006. A change agenda for reserving. *British Actuarial Journal*, **12**(3), 435–619.

Jorion, P. 1995. *Big Bets Gone Bad: Derivatives and Bankruptcy in Orange County*. Burlington, Massachusetts: Academic Press.

Kealhofer, S. 2003a. Quantifying credit risk I: Default prediction. *Financial Analysts Journal*, **59**(1), 30–44.

Kealhofer, S. 2003b. Quantifying credit risk II: Default valuation. *Financial Analysts Journal*, **59**(3), 78–92.

Keyworth, J. 1994. Our debt to the Goldsmiths. *Goldsmiths Review*, London: Worshipful Company of Goldsmiths, pp. 34–36.

Kim, E.H. 1978. A mean-variance theory of optimal capital structure and corporate debt capacity. *Journal of Finance*, **33**(1), 45–63.

Kim, E.H. 1982. Miller's equilibrium, shareholder leverage clienteles and optimal capital structure. *Journal of Finance*, **37**(2), 301–319.

King, M.E. 1994. *The King Report on Corporate Governance*. Institute of Directors in Southern Africa.

King, M.E. 2002. *The King Report on Corporate Governance for South Africa*. Institute of Directors in Southern Africa.

King, M.E. 2009. *The King Report on Governance for South Africa*. Institute of Directors in Southern Africa.

Kirby, M. 1998. *The Governance Practices of Institutional Investors: Report of the Standing Senate Committee on Banking, Trade and Commerce*. Senate of Canada.

Kohn, M. 1999. *Merchant Banking in the Medieval and Early Modern Economy*. Dartmouth College Department of Economics Working Paper 99-05.

Kolari, J., McInish, T.H. and Saniga, E. 1989. A note on the distribution types of financial ratios in the commercial banking industry. *Journal of Banking and Finance*, **13**(3), 463–471.

Mardia, K.V. 1970. Measures of multivariate skewness and kurtosis with applications. *Biometrika*, **57**(3), 519–530.

Lam, J. 2003. *Enterprise Risk Management – from Incentives to Controls*. Hoboken, New Jersey: John Wiley & Sons Ltd.

Lee, P.J. and Wilkie, A.D. 2000. A comparison of stochastic asset models. Paper presented to the AFIR Colloquium at Tromsø, Norway.

Lee, R.D. and Carter, L.R. 1992. Modeling and forecasting US mortality. *Journal of the American Statistical Association*, **87**(419), 659–671.

Leitch, M. 2010. ISO 31000:2009 – the new international standard on risk management. *Risk Analysis*, **30**(6), 887–892.

Leland, H.E. and Pyle, D.H. 1977. Informational asymmetries, Financial structure, and financial intermediation. *Journal of Finance*, **32**(2), 371–387.

Lewin, C.G. 2003. *Pensions and Insurance Before 1800 – a Social History*. East Linton, Scotland: Tuckwell Press.

Li, D.X. 2000. *On Default Correlation – a Copula Function Approach*. Riskmetrics Working Paper 99-07.

Linstone, H.A. and Turoff, M. 2002. *The Delphi Method – Techniques and Applications*. Available online at http://is.njit.edu/pubs/delphibook/delphibook.pdf.

Lintner, J. 1965. The valuation of risk assets and the selection of risky investments in stock portfolios and capital budgets. *Review of Economics and Statistics*, **47**(1), 13–37.

Lloyds of London. 2006. *Famously Providing Insurance*.

Lowenstein, R. 2002. *When Genius Failed – the Rise and Fall of Long Term Capital Management*. London: Fourth Estate.

Mahalanobis, P.C. 1936. On the generalised distance in statistics. *Proceedings of the National Institute of Sciences of India*, **2**(1), 49–55.

Malevergne, Y. and Sornette, D. 2006. *Extreme Financial Risks: From Dependence to Risk Management*. Berlin, Germany: Springer-Verlag.

Marcus, A.J. 1987. Corporate pension policy and the value of PBGC insurance. In Z. Bodie, J.B. Shoven and D.A. Wise (eds), *Issues in Pension Economics*. Chicago, Illinois: University of Chicago Press.

Mardia, K.V. 1970. Measures of multivariate skewness and kurtosis with applications. *Biometrika*, **36**, 519–530.

Marshall, A. and Olkin, I. 1967. A multivariate exponential distribution. *Journal of the American Statistical Association*, **62**(317), 30–44.

Marx, B.D. and Eilers, P.H.C. 1999. Generalized linear regression on sampled signals and curves – a P-spline approach. *Technometrics*, **41**(1), 1–13.

Matsumoto, M. and Nishimura, T. 1998. Mersenne twister – a 623-dimensionally equidistributed uniform pseudo-random number generator. *ACM Transactions on Modeling and Computer Simulation*, **8**(1), 3–30.

Matten, C. 2000. *Managing Bank Capital*. Chichester: John Wiley & Sons Ltd.

McCarthy, D.G. and Neuberger, A. 2002. Pricing pension insurance: The proposed Levy structure for the pension protection fund. *Fiscal Studies*, **26**(4), 471–489.

McNeil, A.J., Frey, R. and Embrechts, P. 2005. *Quantitative Risk Management – Concepts, Techniques and Tools*. Princeton, New Jersey: Princeton University Press.

McWilliam, E. (ed.). 2011. *Longevity Risk*. London: Risk Books.

Merton, R.C. 1973. The theory of rational option pricing. *Bell Journal of Economics and Management Science*, **4**(1), 141–183.

Merton, R.C. 1974. On the pricing of corporate debt – the risk structure of interest rates. *Journal of Finance*, **29**(2), 449–470.

Merton, R.C. 1976. Option pricing when underlying stock returns are discontinuous. *Journal of Financial Economics*, **3**, 125–144.

Merton, R.C. 1992. *Continuous-time Finance*. Malden, Massachusetts: Blackwell.

Meucci, A. 2009. *Risk and Asset Allocation*. New York: Springer.

Michaud, R.O. 1998. *Efficient Asset Allocation*. Boston, Massachusetts: HBS Press.

Miller, M.H. 1977. Debt and taxes. *Journal of Finance*, **32**(2), 261–275.

Miller, M.H. and Modigliani, F. 1961. Dividend policy, growth and the valuation of shares. *Journal of Business*, **34**(4), 411–433.

Miller, M.H. and Scholes, M. 1972. Rates of return in relation to risk – A re-examination of some recent findings. In M.C. Jenson (ed.), *Studies in the Theory of Capital Markets*. New York: Praeger.

Mitchell, D. 1995. Innovation and the transfer of skill in the Goldsmiths' trade in restoration London. In D. Mitchell (ed.), *Goldsmiths, Silversmiths and Bankers – Innovation and the Transfer of Skill, 1550–1750*. London: Alan Sutton Publishing Ltd/Centre of Metropolitan History.

Modigliani, F. 1982. Debt, dividend policy, taxes, inflation and market valuation. *Journal of Finance*, **37**(2), 255–273.

Modigliani, F. and Miller, M.H. 1958. The cost of capital, corporation finance and the theory of investment. *American Economic Review*, **48**(3), 261–297.

Modigliani, F. and Miller, M.H. 1959. The cost of capital, corporation finance and the theory of investment – Reply. *American Economic Review*, **49**(4), 655–669.

Modigliani, F. and Miller, M.H. 1963. Corporate income taxes and the cost of capital – a correction. *American Economic Review*, **53**(3), 433–443.

Modigliani, F. and Miller, M.H. 1965. Corporate income taxes and the cost of capital – Reply. *American Economic Review*, **55**(3), 524–527.

Moody's Investor Services. 2010. *Corporate Default and Recovery Rates, 1920–2009*.

Morris, D. 2005. *Morris Review of the Actuarial Profession – Final Report*. HM Government.

Morrison, A.D. and Wilhelm, W.J. 2007. Investment banking – past, present, and future. *Journal of Applied Corporate Finance*, **19**(1), 42–54.

Morrison, A.D. and Wilhelm, W.J. 2008. The demise of investment banking partnerships – theory and evidence. *Journal of Finance*, **63**(1), 311–350.

Mossin, J. 1966. Equilibrium in a capital asset market. *Econometrica*, **34**(4), 768–783.

Myers, S.C. 1977. Determinants of corporate borrowing. *Journal of Financial Economics*, **5**(2), 147–175.

Myers, S.C. 1984. The capital structure puzzle. *Journal of Finance*, **39**(3), 575–592.

Myers, S.C. and Majluf, N.S. 1984. Corporate financing and investment decisions when firms have information that investors do not have. *Journal of Financial Economics*, **13**(2), 187–221.

Myners, P. 2001. *Institutional Investment in the United Kingdom – a Review*. HM Treasury.

Neal, L. and Quinn, S. 2001. Networks of information, markets, and institutions in the rise of London as a financial centre, 1660–1720. *Financial History Review*, **8**(1), 7–26.

Nelsen, R.B. 2006. *An Introduction to Copulas*. New York: Springer.

OECD. 1999. *Principles of Corporate Governance*.

OECD. 2004. *Principles of Corporate Governance*.

Patterson, S. 2010. *The Quants*. London: Random House.

Penrose, G.W. 2004. *Report of the Equitable Life Inquiry*. Penrose Inquiry.

Pensions Investment Research Company Limited. 2007. *Review of the Impact of the Combined Code – PIRC's Response to the FRC Consultation Paper*.

Porter, M.E. 1980. *Competitive Strategy*. New York: Free Press.

Purdy, G. 2010. ISO 31000:2009 – Setting a new standard for risk management. *Risk Analysis*, **30**(6), 881–886.

Quinn, S. 1995. Balances and Goldsmith-banking – the co-ordination and control of inter-banker debt clearing in seventeenth century London. In D. Mitchell (ed.), *Goldsmiths, Silversmiths and Bankers – Innovation and the Transfer of Skill, 1550–1750*. London: Alan Sutton Publishing Ltd/Centre of Metropolitan History.

Quinn, S. 1997. Goldsmith-banking – mutual acceptance and interbanker clearing in restoration London. *Explorations in Economic History*, **34**(4), 411–432.

Rebonato, R. 1998. *Interest Rate Option Models*. Chichester: John Wiley & Sons Ltd.

Redington, F.M. 1952. Review of the principles of life office valuations. *Journal of the Institute of Actuaries*, **78**(3), 286–340.

Renshaw, A.E. and Haberman, S. 2006. A cohort-based extension to the Lee–Carter model for mortality reduction factors. *Insurance: Mathematics and Economics*, **38**(3), 556–570.

Richards, S.J. 2008. Applying survivor models to pensioner mortality data. *British Actuarial Journal*, **14**(2), 257–326.

Sandström, A. 2006. *Solvency: Models, Assessment and Regulation*. Boca Raton, Florida: Chapman & Hall/CRC.

Saucier, G. 2001. *Beyond Compliance – Building a Governance Culture*. Canadian Institute of Chartered Accountants, the Canadian Venture Exchange and the Toronto Stock Exchange.

Scarsini, M. 1984. On measures of concordance. *Stochastica*, **8**, 201–218.

Sharpe, W.F. 1964. Capital asset prices – a theory of market equilibrium under conditions of risk. *Journal of Finance*, **19**(3), 425–442.

Sharpe, W.F. 1976. Corporate pension funding policy. *Journal of Financial Economics*, **3**(3), 183–193.

Sklar, A. 1959. Fonctions de répartition à *n* dimensions et leurs marges. *Publications de l'Institut Statistique de l'Université de Paris*, **8**, 229–231.

Smith, R. 2003. *Audit Committees – Combined Code Guidance*. Financial Reporting Council.

Smith, R. 2005. *Guidance on Audit Committees (The Smith Guidance)*. Financial Reporting Council.

Society of Actuaries. 2004. *Speciality Guide on Economic Capital*.

Standard and Poor's Global Fixed Income Research. 2010a. *2009 Annual Global Corporate Default Study and Rating Transitions*.

Standard and Poor's Global Fixed Income Research. 2010b. *Rating Methodology – Evaluating the Issuer*.

Standards Australia and Standards New Zealand. 2004. *Risk Management Guidelines – AS/NZS4360:2004*.

Sweeting, P.J. 2006. Correlation and the pension protection fund. *Fiscal Studies*, **27**(2), 157–182.

Sweeting, P.J. and Fotiou, F. 2011. Calculating and communicating the risk of extreme loss. Paper presented to the Institute and Faculty of Actuaries.

Sweeting, P.J., Christiansen, C., Dyer, D., Harbord, P., Joubert, P., Markou, E., Murray, C., Ng, H.I., Procter, K. and Tay, A. 2004. An analysis and critique of the methods used by rating agencies. Paper presented to the 2004 UK Actuarial Profession Finance and Investment Conference.

Talmor, E., Haugen, R. and Barnea, A. 1985. The value of the tax subsidy on risky debt. *Journal of Business*, **58**(2), 191–202.

Taylor, N. 2000. Making actuaries less human – Lessons from behavioural finance. Paper presented to the Staple Inn Actuarial Society.

Tepper, I. 1981. Taxation and corporate pension policy. *Journal of Finance*, **36**(1), 1–13.

Treasury Board of Canada. 2001. *Integrated Risk Management Framework*.

Treasury Board of Canada. 2010. *Framework for the Management of Risk*.

Treynor, J.L. 1961. Toward a theory of the market value of risky assets. Unpublished working paper.

Treynor, J.L. 1977. The principles of corporate pension finance. *Journal of Finance*, **32**(2), 627–38.

Turnbull, N. 1999. *Internal Control – Guidance for Directors on the Combined Code*. Institute of Chartered Accountants in England and Wales.

Turnbull, N. 2005. *Internal Control – Guidance for Directors on the Combined Code*. Financial Reporting Council.

US Government. 1933a. *Banking (Glass–Steagall) Act*.

US Government. 1933b. *Securities Act*.

US Government. 1934. *Securities Exchange Act*.

US Government. 1974. *Employee Retirement Income Security Act. (ERISA)*.

US Government. 1980. *Depository Institutions Deregulation and Monetary Control Act*.

US Government. 1986. *Single Employer Pension Plan Amendments Act*.

US Government. 1999. *Financial Services Modernization (Gramm–Leach–Bliley) Act*.

US Government. 2002. *Public Company Accounting Reform and Investor Protection (Sarbanes–Oxley) Act*.

VanDerhei, J.L. 1990. An empirical analysis of risk-related insurance premiums for the PBGC. *Journal of Risk and Insurance*, **57**(2), 240–259.

Vasicek, O. 1977. An equilibrium characterisation of the term structure. *Journal of Financial Economics*, **5**(2), 177–188.

Venter, G.G. 2002. Tails of copulas. *Proceedings of the Casualty Actuarial Society*, **89**(1), 68–113.

Webb, S. and Webb, B. 1920. *History of Trade Unionism*. London: Longmans & Co.

Wilmott, P. 2000. *Paul Wilmott on Quantitative Finance*. Chichester: John Wiley & Sons Ltd.

Wooldridge, J.M. 2002. *Econometric Analysis of Cross Section and Panel Data*. Cambridge, Massachusetts: MIT Press.

Würthrich, M.V. and Merz, M. 2008. *Stochastic Claims Reserving Methods in Insurance*. Chichester: John Wiley & Sons Ltd.

Index